W
T
F
WHAT'S THE FUTURE AND WHY IT'S UP TO US

TIM O'REILLY

BUSINESS BOOKS

1 3 5 7 9 10 8 6 4 2

Random House Business Books
20 Vauxhall Bridge Road
London SW1V 2SA

Random House Business Books is part of the Penguin Random House
group of companies whose addresses can be found
at global.penguinrandomhouse.com.

Tim O'R... ...or of this
Work in a... ...Act 1988.

Fir... ...17
This p... ...usiness

www.penguin.co.uk

A CIP catalogue record for this book is available from the British Library.

ISBN 9781847941862

Printed and bound by Clays Ltd, St Ives plc

Penguin Random House is committed to a sustainable future for
our business, our readers and our planet. This book is made
from Forest Stewardship Council® certified paper.

MIX
Paper from
responsible sources
FSC® C018179

For all who work to make tomorrow better than today.

CONTENTS

INTRODUCTION: THE WTF? ECONOMY

THIS MORNING, I SPOKE OUT LOUD TO A $150 DEVICE IN MY kitchen, told it to check if my flight was on time, and asked it to call a Lyft to take me to the airport. A car showed up a few minutes later, and my smartphone buzzed to let me know it had arrived. And in a few years, that car might very well be driving itself. Someone seeing this for the first time would have every excuse to say, "WTF?"

At times, "WTF?" is an expression of astonishment. But many people reading the news about technologies like artificial intelligence and self-driving cars and drones feel a profound sense of unease and even dismay. They worry about whether their children will have jobs, or whether the robots will have taken them all. They are also saying "WTF?" but in a very different tone of voice. It is an expletive.

Astonishment: phones that give advice about the best restaurant nearby or the fastest route to work today; artificial intelligences that write news stories or advise doctors; 3-D printers that make replacement parts—for humans; gene editing that can cure disease or bring extinct species back to life; new forms of corporate organization that marshal thousands of on-demand workers so that consumers can summon services at the push of a button in an app.

Dismay: the fear that robots and AIs will take away jobs, reward their owners richly, and leave formerly middle-class workers part of a new underclass; tens of millions of jobs here in the United States that don't pay people enough to live on; little-understood financial products and profit-seeking algorithms that can take down the entire world economy and drive millions of people from their homes; a surveillance society that tracks our every move and stores it in corporate and government databases.

Everything is amazing, everything is horrible, and it's all moving too fast. We are heading pell-mell toward a world shaped by technology in ways that we don't understand and have many reasons to fear.

WTF? Google AlphaGo, an artificial intelligence program, beat the world's best human Go player, an event that was widely predicted to be at least twenty years in the future—until it happened in 2016. If AlphaGo can happen twenty years early, what else might hit us even sooner than we expect? For starters: An AI running on a $35 Raspberry Pi computer beat a top US Air Force fighter pilot trainer in combat simulation. The world's largest hedge fund has announced that it wants an AI to make three-fourths of management decisions, including hiring and firing. Oxford University researchers estimate that up to 47% of human tasks, including many components of white-collar jobs, may be done by machines within as little as twenty years.

WTF? Uber has put taxi drivers out of work by replacing them with ordinary people offering rides in their own cars, creating millions of part-time jobs worldwide. Yet Uber is intent on eventually replacing those on-demand drivers with completely automated vehicles.

WTF? Without owning a single room, Airbnb has more rooms on offer than some of the largest hotel groups in the world. Airbnb has under 3,000 employees, while Hilton has 152,000. New forms of corporate organization are outcompeting businesses based on best practices that we've followed for the lifetimes of most business leaders.

WTF? Social media algorithms may have affected the outcome of the 2016 US presidential election.

WTF? While new technologies are making some people very rich, incomes have stagnated for ordinary people, and for the first time, children in developed countries are on track to earn less than their parents.

What do AI, self-driving cars, on-demand services, and income inequality have in common? They are telling us, loud and clear, that we're in for massive changes in work, business, and the economy.

But just because we can see that the future is going to be very different doesn't mean that we know exactly *how* it's going to unfold, or when. Perhaps "WTF?" really stands for "What's the Future?" Where is technology taking us? Is it going to fill us with astonishment or dismay? And most important, what is our role in deciding that future? How do we make choices today that will result in a world we want to live in?

I've spent my career as a technology evangelist, book publisher, conference producer, and investor wrestling with questions like these. My company, O'Reilly Media, works to identify important innovations, and by spreading knowledge about them, to amplify their impact and speed their adoption. And we've tried to sound a warning when a failure to understand how technology is changing the rules for business or society is leading us down the wrong path. In the process, we've watched numerous technology booms and busts, and seen companies go from seemingly unstoppable to irrelevant, while early-stage technologies that no one took seriously went on to change the world.

If all you read are the headlines, you might have the mistaken idea that how highly investors value a company is the key to understanding which technologies really matter. We hear constantly that Uber is "worth" $68 billion, more than General Motors or Ford; Airbnb is "worth" $30 billion, more than Hilton Hotels and almost as much as Marriott. Those huge numbers can make the companies seem inevitable, with their success already achieved. But it is only when a business becomes profitably self-sustaining, rather than subsidized by investors, that we can be sure that it is here to stay. After all, after eight years Uber is still losing $2 billion every year in its race to get to worldwide scale. That's an amount that dwarfs the losses of companies like Amazon (which lost $2.9 billion over its first five years before showing its first profits in 2001). Is Uber losing money like Amazon, which went on to become a hugely successful company that transformed retailing, publishing, and enterprise computing, or like a dot-com company that was destined to fail? Is the enthusiasm of its investors a sign of a fundamental restructuring of the nature of work, or a sign of an investment mania like the one leading up to the dot-com bust in 2001? How do we tell the difference?

Startups with a valuation of more than a billion dollars understandably get a lot of attention, even more so now that they have a name, *unicorn*, the term du jour in Silicon Valley. *Fortune* magazine started keeping a list of companies with that exalted status. Silicon Valley news site *TechCrunch* has a constantly updated "Unicorn Leaderboard."

But even when these companies succeed, they may not be the surest guide to the future. At O'Reilly Media, we learned to tune in to very different signals by watching the innovators who first brought us the Internet and the open source software that made it possible. They did what they did out of love and curiosity, not a desire to make a fortune. We saw that radically new industries don't start when creative entrepreneurs meet venture capitalists. They start with people who are infatuated with seemingly impossible futures.

Those who change the world are people who are chasing a very different kind of unicorn, far more important than the Silicon Valley billion-dollar valuation (though some of them will achieve that too). It is the breakthrough, once remarkable, that becomes so ubiquitous that eventually it is taken for granted.

Tom Stoppard wrote eloquently about a unicorn of this sort in his play *Rosencrantz & Guildenstern Are Dead*:

> A man breaking his journey between one place and another at a third place of no name, character, population or significance, sees a unicorn cross his path and disappear. . . . "My God," says a second man, "I must be dreaming, I thought I saw a unicorn." At which point, a dimension is added that makes the experience as alarming as it will ever be. A third witness, you understand, adds no further dimension but only spreads it thinner, and a fourth thinner still, and the more witnesses there are the thinner it gets and the more reasonable it becomes until it is as thin as reality, the name we give to the common experience.

The world today is full of things that once made us say "WTF?" but are already well on their way to being the stuff of daily life.

The Linux operating system was a unicorn. It seemed downright impossible that a decentralized community of programmers could build a world-class operating system and give it away for free. Now billions of people rely on it.

The World Wide Web was a unicorn, even though it didn't make Tim Berners-Lee a billionaire. I remember showing the World Wide Web at a technology conference in 1993, clicking on a link, and saying, "That picture just came over the Internet all the way from the University of Hawaii." People didn't believe it. They thought we were making it up. Now everyone expects that you can click on a link to find out anything at any time.

Google Maps was a unicorn. On the bus not long ago, I watched one old man show another how the little blue dot in Google Maps followed us along as the bus moved. The newcomer to the technology was amazed. Most of us now take it for granted that our phones know exactly where we are, and not only can give us turn-by-turn directions exactly to our destination—by car, by public transit, by bicycle, and on foot—but also can find restaurants or gas stations nearby or notify our friends where we are in real time.

The original iPhone was a unicorn even before the introduction of the App Store a year later utterly transformed the smartphone market. Once you experienced the simplicity of swiping and touching the screen rather than a tiny keyboard, there was no going back. The original pre-smartphone cell phone itself was a unicorn. As were its predecessors, the telephone and telegraph, radio and television. We forget. We forget quickly. And we forget ever more quickly as the pace of innovation increases.

AI-powered personal agents like Amazon's Alexa, Apple's Siri, the Google Assistant, and Microsoft Cortana are unicorns. Uber and Lyft too are unicorns, but not because of their valuation. Unicorns are the kinds of apps that make us say, "WTF?" in a good way.

Can you still remember the first time you realized that you could get the answer to virtually any question with a quick Internet search, or that your phone could route you to any destination? How cool that was, before you started taking it for granted? And how quickly did you move from taking it for granted to complaining about it when it doesn't work quite right?

We are layering on new kinds of magic that are slowly fading into the ordinary. A whole generation is growing up that thinks nothing of summoning cars or groceries with a smartphone app, or buying something from Amazon and having it show up in a couple of hours, or talking to AI-based personal assistants on their devices and expecting to get results.

It is this kind of unicorn that I've spent my career in technology pursuing.

So what makes a real unicorn of this amazing kind?

1. It seems unbelievable at first.
2. It changes the way the world works.
3. It results in an ecosystem of new services, jobs, business models, and industries.

We've talked about the "at first unbelievable" part. What about changing the world? In *Who Do You Want Your Customers to Become?* Michael Schrage writes:

> Successful innovators don't ask customers and clients to do something different; they ask them to become someone different. . . . Successful innovators ask users to embrace—or at least tolerate—new values, new skills, new behaviors, new vocabulary, new ideas, new expectations, and new aspirations. They transform their customers.

For example, Schrage points out that Apple (and now also Google and Microsoft and Amazon) asks their "customers to become the sort of people who wouldn't think twice about talking to their phone as a sentient servant." Sure enough, there is a new generation of users who think nothing of saying things like:

"Siri, make me a six p.m. reservation for two at Camino."
"Alexa, play 'Ballad of a Thin Man.'"
"Okay, Google, remind me to buy currants the next time I'm at Piedmont Grocery."

Correctly recognizing human speech alone is hard, but listening and then performing complex actions in response—for millions of simultaneous users—requires incredible computing power provided by massive data centers. Those data centers support an ever-more-sophisticated digital infrastructure.

For Google to remind me to buy currants the next time I'm at my local supermarket, it has to know where I am at all times, keep track of a particular location I've asked for, and bring up the reminder in that context. For Siri to make me a reservation at Camino, it needs to know that Camino is a restaurant in Oakland, and that it is open tonight, and it must allow conversations between machines, so that my phone can lay claim to a table from the restaurant's reservation system via a service like OpenTable. And then it may call other services, either on my devices or in the cloud, to add the reservation to my calendar or to notify friends, so that yet another agent can remind all of us when it is time to leave for our dinner date.

And then there are the alerts that I didn't ask for, like Google's warnings:

"Leave now to get to the airport on time. 25 minute delay on the Bay Bridge."

or

"There is traffic ahead. Faster route available."

All of these technologies are additive, and addictive. As they interconnect and layer on each other, they become increasingly powerful, increasingly magical. Once you become accustomed to each new superpower, life without it is like having your magic wand turn into a stick again.

These services have been created by human programmers, but they will increasingly be enabled by artificial intelligence. That's a scary word to many people. But it is the next step in the progression of the unicorn from the astonishing to the ordinary. While the term *artificial intelligence* or *AI* suggests a truly autonomous intelligence, we are far, far from that eventuality. AI is still just a tool, still subject to human direction.

The nature of that direction, and how we must exercise it, is a key subject of this book. AI and other unicorn technologies have the potential to make a better world, in the same way that the technologies of the first industrial revolution created wealth for society that was unimaginable two centuries ago. AI bears the same relationship to previous programming techniques that the internal combustion engine does to the steam engine. It is far more versatile and powerful, and over time we will find ever more uses for it.

Will we use it to make a better world? Or will we use it to amplify the worst features of today's world? So far, the "WTF?" of dismay seems to have the upper hand.

"Everything is amazing," and yet we are deeply afraid. Sixty-three percent of Americans believe jobs are less secure now than they were twenty to thirty years ago. By a two-to-one ratio, people think good jobs are difficult to find where they live. And many of them blame technology. There is a constant drumbeat of news that tells us that the future is one in which increasingly intelligent machines will take over more and more human work. The pain is already being felt. For the first time, life expectancy is actually declining in America, and what was once its rich industrial heartland has too often become a landscape of despair.

For everyone's sake, we must choose a different path.

Loss of jobs and economic disruption are not inevitable. There is a profound failure of imagination and will in much of today's economy. For every Elon Musk—who wants to reinvent the world's energy infrastructure, build revolutionary new forms of transport, and settle humans on Mars—there are far too many companies that are simply using technology to cut costs and boost their stock price, enriching those able to invest in financial markets at the expense of an ever-growing group that may never be able to do so. Policy makers seem helpless, assuming that the course of technology is inevitable, rather than something we must shape.

And that gets me to the third characteristic of true unicorns: They create value. *Not just financial value, but real-world value for society*.

Consider past marvels. Could we have moved goods as easily or as quickly without modern earthmoving equipment letting us bore tunnels through mountains or under cities? The superpower of humans + machines made it possible to build cities housing tens of millions of people, for a tiny fraction of our people to work producing the food that all the rest of us eat, and to create a host of other wonders that have made the modern world the most prosperous time in human history.

Technology is going to take our jobs! Yes. It always has, and the pain and dislocation are real. *But it is going to make new kinds of jobs possible.* History tells us technology kills professions, but does not kill jobs. We will find things to work on that we couldn't do before but now can accomplish with the help of today's amazing technologies.

Take, for example, laser eye surgery. I used to be legally blind without huge Coke-bottle glasses. Twelve years ago, my eyes were fixed by a surgeon who would never have been able to do the job without the aid of a robot, who was now able to do something that had previously been impossible.

After more than forty years of wearing glasses so strong that I was legally blind without them, I could see clearly on my own. I kept saying to myself for months afterward, "I'm seeing with my own eyes!"

But in order to remove my need for prosthetic vision, the surgeon ended up relying on prosthetics of her own, performing the surgery on my cornea with the aid of a computer-controlled laser. During the actual surgery, apart from lifting the flap she had cut by hand in the surface of my cornea and smoothing it back into place after the laser was done, her job was to clamp open my eyes, hold my head, utter reassuring words, and tell me, sometimes with urgency, to keep looking at the red light. I asked what would happen if my eyes drifted and I didn't stay focused on the light. "Oh, the laser would stop," she said. "It only fires when your eyes are tracking the dot."

Surgery this sophisticated could never be done by an unaugmented human being. The human touch of my superb doctor was paired with the superhuman accuracy of complex machines, a twenty-first-century hybrid freeing me from assistive devices first invented eight centuries earlier in Italy. The revolution in sensors, computers, and control technologies is going to make many of the daily activities of the twentieth century seem quaint as, one by one, they are reinvented in the twenty-first. This is the true opportunity of technology: It extends human capability.

In the debate about technology and the shape of the future, it's easy to forget just how much technology already suffuses our lives, how much it has already changed us. As we get past that moment of amazement, and it fades into the new normal, we must put technology to work solving new problems. We must commit to building something new, strange to our past selves, but better, if we commit to making it so.

> We must keep asking: What will new technology let us do that was previously impossible? Will it help us build the kind of society we want to live in?

This is the secret to reinventing the economy. As Google chief economist Hal Varian said to me, "My grandfather wouldn't recognize what I do as work."

What are the new jobs of the twenty-first century? Augmented reality—the overlay of computer-generated data and images on what we see—may give us a clue. It definitely meets the WTF? test. The first time a venture capitalist friend of mine saw one unreleased augmented reality platform in the lab, he said, "If LSD were a stock, I'd be shorting it." That's a unicorn.

But what is most exciting to me about this technology is not the LSD factor, but how augmented reality can change the way we work. You can imagine how augmented reality could enable workers to be "upskilled." I'm particularly fond of imagining how the model used by Partners in Health could be turbocharged by augmented reality and telepresence. The organization provides free healthcare to people in poverty using a model in which community health workers recruited from the population being served are trained and supported in providing primary care. Doctors can be brought in as needed, but the bulk of care is provided by ordinary people. Imagine a community health worker who is able to tap on Google Glass or some next-generation wearable, and say, "Doctor, you need to see this!" (Trust me. Glass will be back, when Google learns to focus on community health workers, not fashion models.)

It's easy to imagine how rethinking our entire healthcare system along these lines could reduce costs, improve both health outcomes and patient satisfaction, and create jobs. Imagine house calls coming back into fashion. Add in health monitoring by wearable sensors, health advice from an AI made as available as Siri, the Google Assistant, or Microsoft Cortana, plus an Uber-style on-demand service, and you can start to see the outlines of one small segment of the next economy being brought to us by technology.

This is only one example of how we might reinvent familiar human activities, creating new marvels that, if we are lucky, will eventually fade into the texture of everyday life, just like wonders of a previous age such as airplanes and skyscrapers, elevators, automobiles, refrigerators, and washing machines.

• • •

Despite their possible wonders, many of the futures we face are fraught with unknown risks. I am a classicist by training, and the fall of Rome is always before me. The first volume of Gibbon's *Decline and Fall of the Roman Empire* was published in 1776, the same year as the American Revolution. Despite Silicon Valley's dreams of a future singularity, an unknowable fusion of minds and machines that will mark the end of history as we know it, what history teaches us is that economies and nations, not just companies, can fail. Great civilizations do collapse. Technology can go backward. After the fall of Rome, the ability to make monumental structures out of concrete was lost for nearly a thousand years. It could happen to us.

We are increasingly facing what planners call "wicked problems"—problems that are "difficult or impossible to solve because of incomplete, contradictory, and changing requirements that are often difficult to recognize."

Even long-accepted technologies turn out to have unforeseen downsides. The automobile was a unicorn. It afforded ordinary people enormous freedom of movement, led to an infrastructure for transporting goods that spread prosperity, and enabled a consumer economy where goods could be produced far away from where they are consumed. Yet the roads we built to enable the automobile carved up and hollowed out cities, led to more sedentary lifestyles, and contributed mightily to the overpowering threat of climate change.

Ditto cheap air travel, container shipping, the universal electric grid. All of these were enormous engines of prosperity that brought with them unintended consequences that only came to light over many decades of painful experience, by which time any solution seems impossible to attempt because the disruption required to reverse course would be so massive.

We face a similar set of paradoxes today. The magical technologies of today—and choices we've already made, decades ago, about what we value as a society—are leading us down a path with complex contingencies, unseen dangers, and decisions that we don't even know we are making.

AI and robotics in particular are at the heart of a set of wicked problems that are setting off alarm bells among business and labor leaders, policy makers and academics. What happens to all those people who drive for a living when the cars start driving themselves? AIs are flying planes, advising doctors on the best treatments, writing sports and financial news, and telling us all, in real time, the fastest way to get to work. They are also telling human workers when to show up and when to go home, based on real-time measurement of demand. Computers used to work for humans; increasingly it's now humans working for computers. The algorithm is the new shift boss.

What is the future of business when technology-enabled networks and marketplaces let people choose when and how much they want to work? What is the future of education when on-demand learning outperforms traditional universities in keeping skills up to date? What is the future of media and public discourse when algorithms decide what we will watch and read, making their choice based on what will make the most profit for their owners?

What is the future of the economy when more and more work can be done by intelligent machines instead of people, or only done by people in partnership with those machines? What happens to workers and their families? And what happens to the companies that depend on consumer purchasing power to buy their products?

There are dire consequences to treating human labor simply as a cost to be eliminated. According to the McKinsey Global Institute, 540 to 580 million people—65 to 70% of households in twenty-five advanced economies—had incomes that had fallen or were flat between 2005 and 2014. Between 1993 and 2005, fewer than 10 million people—less than 2%—had the same experience.

Over the past few decades, companies have made a deliberate choice to reward their management and "superstars" incredibly well, while treating ordinary workers as a cost to be minimized or cut. Top US CEOs now earn 373x the income of the average worker, up from 42x in 1980. As a result of the choices we've made as a society about how to share the benefits of economic growth and technological productivity gains, the gulf between the top and the bottom has widened enormously, and the middle has largely disappeared. Recently published research by Stanford economist Raj Chetty shows that for children born in 1940, the chance that they'd earn more than their parents was 92%; for children born in 1990, that chance has fallen to 50%.

Businesses have delayed the effects of declining wages on the consumer economy by encouraging people to borrow—in the United States, household debt is over $12 trillion (80% of gross domestic product, or GDP, in mid-2016) and student debt alone is $1.2 trillion (with more than seven million borrowers in default). We've also used government transfers to reduce the gap between human needs and what our economy actually delivers. But of course, higher government transfers must be paid for through higher taxes or through higher government debt, either of which political gridlock has made unpalatable. This gridlock is, of course, a recipe for disaster.

Meanwhile, in hopes that "the market" will deliver jobs, central banks have pushed ever more money into the system, hoping that somehow this will unlock business investment. But instead, corporate profits have reached highs not seen since the 1920s, corporate investment has shrunk, and more than $30 trillion of cash is sitting on the sidelines. The magic of the market is not working.

We are at a very dangerous moment in history. The concentration of wealth and power in the hands of a global elite is eroding the power and sovereignty of nation-states while globe-spanning technology platforms are enabling algorithmic control of firms, institutions, and societies, shaping what billions of people see and understand and how the economic pie is divided. At the same time, income inequality and the pace of technology change are leading to a populist backlash featuring opposition to science, distrust of our governing institutions, and fear of the future, making it ever more difficult to solve the problems we have created.

That has all the hallmarks of a classic wicked problem.

Wicked problems are closely related to an idea from evolutionary biology, that there is a "fitness landscape" for any organism. Much like a physical landscape, a fitness landscape has peaks and valleys. The challenge is that you can only get from one peak—a so-called local maximum—to another by going back down. In evolutionary biology, a local maximum may mean that you become one of the long-lived stable species, unchanged for millions of years, or it may mean that you become extinct because you're unable to respond to changed conditions.

And in our economy, conditions are changing rapidly. Over the past few decades, the digital revolution has transformed media, entertainment, advertising, and retail, upending centuries-old companies and business models. Now it is restructuring every business, every job, and every sector of society. No company, no job—and ultimately, no government and no economy—is immune to disruption. Computers will manage our money, supervise our children, and have our lives in their "hands" as they drive our automated cars.

The biggest changes are still ahead, and every industry and every organization will have to transform itself in the next few years, in multiple ways, or fade away. We need to ask ourselves whether the fundamental social safety nets of the developed world will survive the transition, and more important, what we will replace them with.

Andy McAfee, coauthor of *The Second Machine Age*, put his finger on the consequence of failing to do so while talking with me over breakfast about the risks of AI taking over from humans: "The people will rise up before the machines do."

This book provides a view of one small piece of this complex puzzle, the role of technology innovation in the economy, and in particular the role of WTF? technologies such as AI and on-demand services. I lay out the difficult choices we face as technology opens new doors of possibility while closing doors that once seemed the sure path to prosperity. But more important, I try to provide tools for thinking about the future, drawn from decades on the frontiers of the technology industry, observing and predicting its changes.

The book is US-centric and technology-centric in its narrative; it is not an overview of all of the forces shaping the economy of the future, many of which are centered outside the United States or are playing out differently in other parts of the world. In *No Ordinary Disruption*, McKinsey's Richard Dobbs, James Manyika, and Jonathan Woetzel point out quite correctly that technology is only one of four major disruptive forces shaping the world to come. Demographics (in particular, changes in longevity and the birth rate that have radically shifted the mix of ages in the global population), globalization, and urbanization may play at least as large a role as technology. And even that list fails to take into account catastrophic war, plague, or environmental disruption. These omissions are not based on a conviction that Silicon Valley's part of the total technology innovation economy, or the United States, is more important than the rest; it is simply that the book is based on my personal and business experience, which is rooted in this field and in this one country.

The book is divided into four parts. In the first part, I'll share some of the techniques that my company has used to make sense of and predict innovation waves such as the commercialization of the Internet, the rise of open source software, the key drivers behind the renaissance of the web after the dot-com bust and the shift to cloud computing and big data, the Maker movement, and much more. I hope to persuade you that understanding the future requires discarding the way you think about the present, giving up ideas that seem natural and even inevitable.

In the second and third parts, I'll apply those same techniques to provide a framework for thinking about how technologies such as on-demand services, networks and platforms, and artificial intelligence are changing the nature of business, education, government, financial markets, and the economy as a whole. I'll talk about the rise of great world-spanning digital platforms ruled by algorithm, and the way that they are reshaping our society. I'll examine what we can learn about these platforms and the algorithms that rule them from Uber and Lyft, Airbnb, Amazon, Apple, Google, and Facebook. And I'll talk about the one master algorithm we so take for granted that it has become invisible to us. I'll try to demystify algorithms and AI, and show how they are not just present in the latest technology platforms but already shape business and our economy far more broadly than most of us understand. And I'll make the case that many of the algorithmic systems that we have put in place to guide our companies and our economy have been *designed* to disregard the humans and reward the machines.

In the fourth part of the book, I'll examine the choices we have to make as a society. Whether we experience the WTF? of astonishment or the WTF? of dismay is not foreordained. It is up to us.

It's easy to blame technology for the problems that occur in periods of great economic transition. But both the problems and the solutions are the result of human choices.

During the industrial revolution, the fruits of automation were first used solely to enrich the owners of the machines. Workers were often treated as cogs in the machine, to be used up and thrown away. But Victorian England figured out how to do without child labor, with reduced working hours, and their society became more prosperous.

We saw the same thing here in the United States during the twentieth century. We look back now on the good middle-class jobs of the postwar era as something of an anomaly. But they didn't just happen by chance. It took generations of struggle on the part of workers and activists, and growing wisdom on the part of capitalists, policy makers, political leaders, and the voting public. In the end we made choices as a society to share the fruits of productivity more widely.

We also made choices to invest in the future. That golden age of postwar productivity was the result of massive investments in roads and bridges, universal power, water, sanitation, and communications. After World War II, we committed enormous resources to rebuild the lands destroyed by war, but we also invested in basic research. We invested in new industries: aerospace, chemicals, computers, and telecommunications. We invested in education, so that children could be prepared for the world they were about to inherit.

The future comes in fits and starts, and it is often when times are darkest that the brightest futures are being born. Out of the ashes of World War II we forged a prosperous world. By choice and hard work, not by destiny. The Great War of a generation earlier had only amplified the cycle of dismay. What was the difference? After World War I, we punished the losers. After World War II, we invested in them and raised them up again. After World War I, the United States beggared its returning veterans. After World War II, we sent them to college. Wartime technologies such as digital computing were put into the public domain so that they could be transformed into the stuff of the future. The rich taxed themselves to finance the public good.

In the 1980s, though, the idea that "greed is good" took hold in the United States and we turned away from prosperity. We accepted the idea that what was good for financial markets was good for everyone and structured our economy to drive stock prices ever higher, convincing ourselves that "the market" of stocks, bonds, and derivatives was the same as Adam Smith's market of real goods and services exchanged by ordinary people. We hollowed out the real economy, putting people out of work and capping their wages in service to corporate profits that went to a smaller and smaller slice of society.

We made the wrong choice forty years ago. We don't need to stick with it. The rise of a billion people out of poverty in developing economies around the world at the same time that the incomes of ordinary people in most developed economies have been going backward should tell us that we took a wrong turn somewhere.

The WTF? technologies of the twenty-first century have the

potential to turbocharge the productivity of all our industries. But making what we do now more productive is just the beginning. We must share the fruits of that productivity, and use them wisely. If we let machines put us out of work, it will be because of a failure of imagination and a lack of will to make a better future.

PART I USING THE RIGHT MAPS

The map is not the territory.
—Alfred Korzybski

1

SEEING THE FUTURE IN THE PRESENT

IN THE MEDIA, I'M OFTEN PEGGED AS A FUTURIST. I DON'T think of myself that way. I think of myself as a mapmaker. I draw a map of the present that makes it easier to see the possibilities of the future. Maps aren't just representations of physical locations and routes. They are any system that helps us see where we are and where we are trying to go. One of my favorite quotes is from Edwin Schlossberg: "The skill of writing is to create a context in which other people can think." This book is a map.

We use maps—simplified abstractions of an underlying reality, which they represent—not just in trying to get from one place to another but in every aspect of our lives. When we walk through our darkened home without the need to turn on the light, that is because we have internalized a mental map of the space, the layout of the rooms, the location of every chair and table. Similarly, when an entrepreneur or venture capitalist goes to work each day, he or she has a mental map of the technology and business landscape. We dispose the world in categories: friend or acquaintance, ally or competitor, important or unimportant, urgent or trivial, future or past. For each category, we have a mental map.

But as we're reminded by the sad stories of people who religiously follow their GPS off a no-longer-existent bridge, maps can be wrong. In business and in technology, we often fail to see clearly what is ahead because we are navigating using old maps and sometimes even bad maps—maps that leave out critical details about our environment or perhaps even actively misrepresent it.

Most often, in fast-moving fields like science and technology, maps are wrong simply because so much is unknown. Each entrepreneur, each inventor, is also an explorer, trying to make sense of what's possible, what works and what doesn't, and how to move forward.

Think of the entrepreneurs working to develop the US transcontinental railroad in the mid-nineteenth century. The idea was first proposed in 1832, but it wasn't even clear that the project was feasible until the 1850s, when the US House of Representatives provided the funding for an extensive series of surveys of the American West, a precursor to any actual construction. Three years of exploration from 1853 to 1855 resulted in the *Pacific Railroad Surveys*, a twelve-volume collection of data on 400,000 square miles of the American West.

But all that data did not make the path forward entirely clear. There was fierce debate about the best route, debate that was not just about the geophysical merits of northern versus southern routes but also about the contested extension of slavery. Even when the intended route was decided on and construction began in 1863, there were unexpected problems—a grade steeper than previously reported that was too difficult for a locomotive, weather conditions that made certain routes impassable during the winter. You couldn't just draw lines on the map and expect everything to work perfectly. The map had to be refined and redrawn with more and more layers of essential data added until it was clear enough to act on. Explorers and surveyors went down many false paths before deciding on the final route.

Creating the right map is the first challenge we face in making sense of today's WTF? technologies. Before we can understand how to deal with AI, on-demand applications, and the disappearance of middle-class jobs, and how these things can come together into a future we want to live in, we have to make sure we aren't blinded by old ideas. We have to see patterns that cross old boundaries.

The map we follow into the future is like a picture puzzle with many of the pieces missing. You can see the rough outline of one pattern over here, and another there, but there are great gaps and you can't quite make the connections. And then one day someone pours out another set of pieces on the table, and suddenly the pattern pops into focus. The difference between a map of an unknown territory and a picture puzzle is that no one knows the full picture in advance. It doesn't exist until we see it—it's a puzzle whose pattern we make up together as we go, invented as much as it is discovered.

Finding our way into the future is a collaborative act, with each explorer filling in critical pieces that allow others to go forward.

LISTENING FOR THE RHYMES

Mark Twain is reputed to have said, "History doesn't repeat itself, but it often rhymes." Study history and notice its patterns. This is the first lesson I learned in how to think about the future.

The story of how the term *open source software* came to be developed, refined, and adopted in early 1998—what it helped us to understand about the changing nature of software, how that new understanding changed the course of the industry, and what it predicted about the world to come—shows how the mental maps we use limit our thinking, and how revising the map can transform the choices we make.

Before I delve into what is now ancient history, I need you to roll back your mind to 1998.

Software was distributed in shrink-wrapped boxes, with new releases coming at best annually, often every two or three years. Only 42% of US households had a personal computer, versus the 80% who own a smartphone today. Only 20% of the US population had a mobile phone of any kind. The Internet was exciting investors—but it was still tiny, with only 147 million users worldwide, versus 3.4 billion today. More than half of all US Internet users had access through AOL. Amazon and eBay had been launched three years earlier, but Google was only just founded in September of that year.

Microsoft had made Bill Gates, its founder and CEO, the richest man in the world. It was the defining company of the technology industry, with a near-monopoly position in personal computer software that it had leveraged to destroy competitor after competitor. The US Justice Department launched an antitrust investigation against the company in May of that year, just as it had done nearly thirty years earlier against IBM.

In contrast to the proprietary software that made Microsoft so successful, open source software is distributed under a license that allows anyone to freely study, modify, and build on it. Examples of open source software include the Linux and Android operating systems; web browsers like Chrome and Firefox; popular programming languages like Python, PHP, and JavaScript; modern big data tools like Hadoop and Spark; and cutting-edge artificial intelligence toolkits like Google's TensorFlow, Facebook's Torch, or Microsoft's CNTK.

In the early days of computers, most software was open source, though not by that name. Some basic operating software came with a computer, but much of the code that actually made a computer useful was custom software written to solve specific problems. The software written by scientists and researchers in particular was often shared. During the late 1970s and 1980s, though, companies had realized that controlling access to software gave them commercial advantage and had begun to close off access using restrictive licenses. In 1985, Richard Stallman, a programmer at the Massachusetts Institute of Technology, published *The GNU Manifesto*, laying out the principles of what he called "free software"—not free as in price, but free as in freedom: the freedom to study, to redistribute, and to modify software without permission.

Stallman's ambitious goal was to build a completely free version of AT&T's Unix operating system, originally developed at Bell Labs, the research arm of AT&T.

At the time Unix was first developed in the late 1970s, AT&T was a legal monopoly with enormous profits from regulated telephone services. As a result, AT&T was not allowed to compete in the computer industry, then dominated by IBM, and in accord with its 1956 consent decree with the Justice Department had licensed Unix to computer science research groups on generous terms. Computer programmers at universities and companies all over the world had responded by contributing key elements to the operating system.

But after the decisive consent decree of 1982, in which AT&T agreed to be broken up into seven smaller companies ("the Baby Bells") in exchange for being allowed to compete in the computer market, AT&T tried to make Unix proprietary. They sued the University of California, Berkeley, which had built an alternate version of Unix (the

Berkeley Software Distribution, or BSD), and effectively tried to shut down the collaborative barn raising that had helped to create the operating system in the first place.

While Berkeley Unix was stalled by AT&T's legal attacks, Stallman's GNU Project (named for the meaningless recursive acronym "Gnu's Not Unix") had duplicated all of the key elements of Unix except the kernel, the central code that acts as a kind of traffic cop for all the other software. That kernel was supplied by a Finnish computer science student named Linus Torvalds, whose master's thesis in 1990 consisted of a minimalist Unix-like operating system that would be portable to many different computer architectures. He called this operating system Linux.

Over the next few years, there was a fury of commercial activity as entrepreneurs seized on the possibilities of a completely free operating system combining Torvalds's kernel with the Free Software Foundation's re-creation of the rest of the Unix operating system. The target was no longer AT&T, but rather Microsoft.

In the early days of the PC industry, IBM and a growing number of personal computer "clone" vendors like Dell and Gateway provided the hardware, Microsoft provided the operating system, and a host of independent software companies provided the "killer apps"—word processing, spreadsheets, databases, and graphics programs—that drove adoption of the new platform. Microsoft's DOS (Disk Operating System) was a key part of the ecosystem, but it was far from in control. That changed with the introduction of Microsoft Windows. Its extensive Application Programming Interfaces (APIs) made application development much easier but locked developers into Microsoft's platform. Competing operating systems for the PC like IBM's OS/2 were unable to break the stranglehold. And soon Microsoft used its dominance of the operating system to privilege its own applications—Microsoft Word, Excel, PowerPoint, Access, and, later, Internet Explorer, their web browser (now Microsoft Edge)—by making bundling deals with large buyers.

The independent software industry for the personal computer was slowly dying, as Microsoft took over one application category after another.

> This is the rhyming pattern that I noticed: The personal computer
> industry had begun with an explosion of innovation that broke
> IBM's monopoly on the first generation of computing, but had
> ended in another "winner takes all" monopoly. Look for repeating
> patterns and ask yourself what the next iteration might be.

Now everyone was asking whether a desktop version of Linux could
change the game. Not only startups but also big companies like IBM,
trying to claw their way back to the top of the heap, placed huge bets
that they could.

But there was far more to the Linux story than just competing with
Microsoft. It was rewriting the rules of the software industry in ways
that no one expected. It had become the platform on which many of
the world's great websites—at the time, most notably Amazon and
Google—were being built. But it was also reshaping the very way that
software was being written.

In February 1997, at the Linux Kongress in Würzburg, Germany,
hacker Eric Raymond delivered a paper, called "The Cathedral and the
Bazaar," that electrified the Linux community. It laid out a theory of
software development drawn from reflections on Linux and on Eric's
own experiences with what later came to be called open source software
development. Eric wrote:

> Who would have thought even five years ago that a world-class
> operating system could coalesce as if by magic out of part-time
> hacking by several thousand developers scattered all over the
> planet, connected only by the tenuous strands of the Internet? . . .
> [T]he Linux community seemed to resemble a great babbling
> bazaar of differing agendas and approaches (aptly symbolized by
> the Linux archive sites, who'd take submissions from anyone) out
> of which a coherent and stable system could seemingly emerge
> only by a succession of miracles.

Eric laid out a series of principles that have, over the past decades,
become part of the software development gospel: that software should

be released early and often, in an unfinished state rather than waiting to be perfected; that users should be treated as "co-developers"; and that "given enough eyeballs, all bugs are shallow."

Today, whether programmers develop open source software or proprietary software, they use tools and approaches that were pioneered by the open source community. But more important, anyone who uses today's Internet software has experienced these principles at work. When you go to a site like Amazon, Facebook, or Google, you are a participant in the development process in a way that was unknown in the PC era. You are not a "co-developer" in the way that Eric Raymond imagined—you are not another hacker contributing feature suggestions and code. But you are a "beta tester"—someone who tries out continually evolving, unfinished software and gives feedback—at a scale never before imagined. Internet software developers constantly update their applications, testing new features on millions of users, measuring their impact, and learning as they go.

Eric saw that something was changing in the way software was being developed, but in 1997, when he first delivered "The Cathedral and the Bazaar," it wasn't yet clear that the principles he articulated would spread far beyond free software, beyond software development itself, shaping content sites like Wikipedia and eventually enabling a revolution in which consumers would become co-creators of services like on-demand transportation (Uber and Lyft) and lodging (Airbnb).

I was invited to give a talk at the same conference in Würzburg. My talk, titled "Hardware, Software, and Infoware," was very different. I was fascinated not just with Linux, but with Amazon. Amazon had been built on top of various kinds of free software, including Linux, but it seemed to me to be fundamentally different in character from the kinds of software we'd seen in previous eras of computing.

Today it's obvious to everyone that websites are applications and that the web has become a platform, but in 1997 most people thought of the web browser as the application. If they knew a little bit more about the architecture of the web, they might think of the web server and associated code and data as the application. The content was something managed by the browser, in the same way that Microsoft Word manages a document or that Excel lets you create a spreadsheet. By contrast, I was convinced that the content itself was an essential part of the application, and that the dynamic nature of that content was leading to an entirely new architectural design pattern for a next stage beyond software, which at the time I called "infoware."

Where Eric was focused on the success of the Linux operating system, and saw it as an alternative to Microsoft Windows, I was particularly fascinated by the success of the Perl programming language in enabling this new paradigm on the web.

Perl was originally created by Larry Wall in 1987 and distributed for free over early computer networks. I had published Larry's book, *Programming Perl*, in 1991, and was preparing to launch the Perl Conference in the summer of 1997. I had been inspired to start the Perl Conference by the chance conjunction of comments by two friends. Early in 1997, Carla Bayha, the computer book buyer at the Borders bookstore chain, had told me that the second edition of *Programming Perl*, published in 1996, was one of the top 100 books in any category at Borders that year. It struck me as curious that despite this fact, there was virtually nothing written about Perl in any of the computer trade papers. Because there was no company behind Perl, it was virtually invisible to the pundits who followed the industry.

And then Andrew Schulman, the author of a book called *Unauthorized Windows 95*, told me something I found equally curious. At that time, Microsoft was airing a series of television commercials about the way that their new technology called Active/X would "activate the Internet." The software demos in these ads were actually mostly done with Perl, according to Andrew. It was clear to me that Perl, not Active/X, was actually at the heart of the way dynamic web content was being delivered.

I was outraged. I decided that I needed to make some noise about Perl. And so, early in 1997, I had announced my first conference as a publicity stunt, to get people to pay attention. And that's also what I had come to the Linux Kongress in Würzburg to talk about.

In the essay I later based on that talk, I wrote: "Perl has been called 'the duct tape of the Internet,' and like duct tape, it is used in all kinds of unexpected ways. Like a movie set held together with duct tape, a website is often put up and torn down in a day, and needs lightweight tools and quick but effective solutions."

I saw Perl's duct tape approach as an essential enabler of the "infoware" paradigm, in which control over computers was through an information interface, not a software interface per se. A web link, as I described it at the time, was a way to embed commands to the computer into dynamic documents written in ordinary human language, rather than, say, a drop-down software menu, which embedded little bits of human language into a traditional software program.

The next part of the talk focused on a historical analogy that was to obsess me for the next few years. I was fascinated by the parallels between what open source software and the open protocols of the Internet were doing to Microsoft and the way that Microsoft and an independent software industry had previously displaced IBM.

When I had first entered the industry in 1978, it was shaking off IBM's monopoly, not dissimilar to the position that Microsoft occupied twenty years later. IBM's control over the industry was based on integrated computer systems in which software and hardware were tightly coupled. Creating a new type of computer meant inventing both new hardware and a new operating system to control it. What few independent software companies existed had to choose which hardware vendor they would be satellite to, or to "port" their software to multiple hardware architectures, much as phone developers today need to create separate versions for iPhone and Android. Except the problem was much worse. In the mid-1980s, I remember talking with one of the customers of my documentation consulting business, the author of a mainframe graphics library called DISSPLA (Display Integrated Software System and Plotting Language). He told me that he had to maintain more than 200 different versions of his software.

The IBM Personal Computer, released in August 1981, changed all that. In 1980, realizing that they were missing out on the new microcomputer market, IBM set up a skunkworks project in Boca Raton, Florida, to develop the new machine. They made a critical decision: to cut costs and accelerate development they would develop an open architecture using industry-standard parts—including software licensed from third parties.

The PC, as it was soon called, was an immediate hit when it was released in the fall of 1981. IBM's projections had called for sales of 250,000 units in the first five years. They were rumored to have sold 40,000 on the first day; within two years, more than a million were in customers' hands.

However, the executives at IBM failed to understand the full consequences of their decisions. At the time, software was a small player in the computer industry, a necessary but minor part of an integrated computer, often bundled rather than sold separately. So when it came time to provide an operating system for the new machine, IBM decided to license it from Microsoft, giving them the right to resell the software to the segment of the market that IBM did not control.

The size of that segment was about to explode. Because IBM had published the specifications for the machine, its success was followed by the development of dozens, then hundreds of PC-compatible clones. The barriers to entry to the market were so low that Michael Dell built his eponymous company while still a student at the University of Texas, assembling and selling computers from his dorm room. The IBM personal computer architecture became the standard, over time displacing not only other personal computer designs, but, over the next two decades, minicomputers and mainframes.

As cloned personal computers were built by hundreds of manufacturers large and small, however, IBM lost its leadership in the new market. Software became the new sun that the industry revolved around; Microsoft became the most important company in the computer industry.

Intel also forged a privileged role through bold decision making. In order to ensure that no one supplier became a choke point, IBM had required that every component in the PC's open hardware architecture be available from at least two suppliers. Intel had gone along with this mandate, licensing their 8086 and 80286 chips to rival AMD, but in 1985, with the release of the 80386 processor, they made the bold decision to stand up to IBM, placing the bet that the clone market was now big enough that IBM's wishes would be overridden by the market. Former Intel CTO Pat Gelsinger told me the story. "We took a vote in the five-person management committee. It was three to two against. But Andy [Grove, Intel's CEO] was one of the two, so we did it anyway."

> That's another lesson about the future. It doesn't just happen. People make it happen. Individual decisions matter.

By 1998, the story had largely repeated itself. Microsoft had used its position as the sole provider of the operating system for the PC to establish a monopoly on desktop software. Software applications had become increasingly complex, with Microsoft putting up deliberate barriers to entry against competitors. It was no longer possible for a single programmer or small company to make an impact in the PC software market.

Open source software and the open protocols of the Internet were now challenging that dominance. The barriers to entry into the software market were crashing down. History may not repeat itself, but yes, it does rhyme.

Users could try a new product for free—and even more than that, they could build their own custom version of it, also for free. Source code was available for massive independent peer review, and if someone didn't like a feature, they could add to it, subtract from it, or re-implement it. If they gave their fix back to the community, it could be adopted widely very quickly.

What's more, because developers (at least initially) weren't trying to compete on the business end, but instead focused simply on solving real problems, there was room for experimentation. As has often been said, open source software "lets you scratch your own itch." Because of the distributed development paradigm, with new features being added by users, open source programs "evolve" as much as they are designed. And as I wrote in my 1998 paper, "Hardware, Software, and Infoware," "Evolution breeds not a single winner, but diversity."

That diversity was the reason that the seeds of the future were found in free software and the Internet rather than in the now-establishment technologies offered by Microsoft.

> It is almost always the case that if you want to see the future, you have to look not at the technologies offered by the mainstream but by the innovators out at the fringes.

Most of the people who launched the personal computer software industry four decades ago weren't entrepreneurs; they were kids to whom the idea of owning their own computer was absurdly exciting. Programming was like a drug—no, it was better than a drug, or joining a rock band, and it was certainly better than any job they could imagine. So too Linux, the open source operating system now used as a PC operating system by 90 million people, and by billions as the operating system on which most large Internet sites run, and as the underlying code in every Android phone. The title of Linus Torvalds's book about how he developed Linux? *Just for Fun.*

The World Wide Web got its start the same way. At first, no one took it seriously as a place to make money. It was all about the joy of sharing our work, the rush of clicking on a link and connecting with another computer half the world away, and constructing similar destinations for our peers. We were all enthusiasts. Some of us were also entrepreneurs.

To be sure, it is those entrepreneurs—people like Bill Gates, Steve Jobs, and Michael Dell in the personal computer era; Jeff Bezos, Larry Page, Sergey Brin, and Mark Zuckerberg in the web era—who saw that this world driven by a passion for discovery and sharing could become the cradle of a new economy. They found financial backers, shaped the toy into a tool, and built the businesses that turned a movement into an industry.

> The lesson is clear: Treat curiosity and wonder as a guide to the future. That sense of wonder may just mean that those crazy enthusiasts are seeing something that you don't . . . yet.

The enormous diversity of software that had grown up around free software was reflected in the bestselling books that drove my publishing business. Perl wasn't alone. Many of the most successful technology books of the 1990s, books I published with names only a programmer could love—*Programming Perl, Learning the Vi Editor, Sed & Awk, DNS and Bind, Running Linux, Programming Python*—were all about software written by individuals and distributed freely over the Internet. The web itself had been put into the public domain.

I realized that many of the authors of these programs didn't actually know each other. The free software community that had coalesced around Linux didn't mix much with the Internet crowd. Because of my position as a technology publisher, I traveled in both circles. So I resolved to bring them together. They needed to see themselves as part of the same story.

In April 1998, I organized an event that I originally called "the Freeware Summit" to bring together the creators of many of the most important free software programs.

The timing was perfect. In January, Marc Andreessen's high-profile web company, Netscape, built to commercialize the web browser, had decided to provide the source code to its browser as a free software project using the name Mozilla. Under competitive pressure from Microsoft, which had built a browser of its own and had given it away for free (but *without* source code) in order to "cut off Netscape's air supply," Netscape had no choice but to go back to the web's free software roots.

At the meeting, which was held at the Stanford Court Hotel (now the Garden Court) in Palo Alto, I brought together Linus Torvalds, Brian Behlendorf (one of the founders of the Apache web server project), Larry Wall, Guido van Rossum (the creator of the Python programming language), Jamie Zawinski (the chief developer of the Mozilla project), Eric Raymond, Michael Tiemann (the founder and CEO of Cygnus Solutions, a company that was commercializing free software programming tools), Paul Vixie (the author and maintainer of BIND [Berkeley Internet Name Daemon], the software behind the Internet Domain Name System), and Eric Allman (the author of Sendmail, the software that routed a majority of the Internet's email).

At the meeting, one of the topics that came up was the name *free software*. Richard Stallman's free software movement had created many enemies with its seemingly radical proposition that all software source code must be given away freely—because it was immoral to do otherwise. Even worse, many people had taken *free software* to mean that its developers were hostile to commercial use. At the meeting, Linus Torvalds remarked, "I didn't realize that *free* had two meanings in English: 'libre' and 'gratis.'"

Linus wasn't the only one who had different notions about what *free* meant. In a separate meeting, Kirk McKusick, the head of the Berkeley Unix project, which had developed many of the key Unix features and utilities that had been incorporated into Linux, had told me: "Richard Stallman likes to say that copyright is evil, so we need this new construct called copyleft. Here at Berkeley we use copycentral—that is, we tell people to go down to Copy Central [the local photocopy shop] and copy it." The Berkeley Unix project, which had provided my own introduction to the operating system in 1983, was part of the long academic tradition of knowledge sharing. Source code was given away so people could build on it, and that included commercial use. The only requirement was attribution.

Bob Scheifler, director of MIT's X Window System project, followed the same philosophy. The X Window System had been started in 1984, and by the time I encountered it in 1987, it was becoming the standard window system for Unix and Linux, adopted and adapted by virtually every vendor. My company developed a series of programming manuals for X that used the MIT specifications as a base, rewriting and expanding them, and then licensed them to companies shipping new Unix and X-based systems. Bob encouraged me. "That's exactly what we want companies to do," he said. "We're laying a foundation, and we want everyone to build on it."

Larry Wall, creator of Perl, was another of my mentors in how to think about free software. When I asked him why he had made Perl free software, he explained that he had gotten so much value from the work of others that he felt an obligation to give something back. Larry also quoted to me a variation of Stewart Brand's classic observation, saying, "Information doesn't want to be free. Information wants to be valuable." Like many other free software authors, Larry had discovered that one way to make his information (that is, his software) more valuable was to give it away. He was able to increase its utility not only for himself (because others who took it up made changes and enhancements that he could use), but for everyone else who uses it, because as software becomes more ubiquitous it can be taken for granted as a foundation for further work.

Nonetheless, it was also clear to me that proprietary software creators, including those, such as Microsoft, who were regarded by most free software advocates as immoral, had found that they could make their information valuable by restricting access to it. Microsoft had created enormous value for itself and its shareholders, but it was also a key enabler of ubiquitous personal computing, a necessary precursor to the global computing networks of today. That was value for all of society.

I saw that Larry Wall and Bill Gates had a great deal in common. As the creators (albeit with a host of co-contributors) of a body of intellectual work, they had made strategic decisions about how best to maximize its value. History has proven that each of their strategies can work. The question for me became one of how to maximize *value creation* for society, rather than simply *value capture* by an individual or a company. What were the conditions under which giving software away was a better strategy than keeping it proprietary?

This question has recurred, ever more broadly, throughout my career: *How can a business create more value for society than it captures for itself?*

WHAT'S IN A NAME?

As we wrestled with the name *free software*, various alternatives were proposed. Michael Tiemann said that Cygnus had begun using the term *sourceware*. But Eric Raymond argued for *open source*, a new term that had been coined only six weeks earlier by Christine Peterson of the Foresight Institute, a nanotechnology think tank, at a meeting convened by Larry Augustin, the CEO of a Linux company called VA Linux Systems.

Eric and another software developer and free software activist, Bruce Perens, had been so excited about Christine's new term that they had formed a nonprofit organization called the Open Source Initiative to reconcile the various free software licenses that were being used into a kind of metalicense. But as of yet, the term was largely unknown.

Not everyone liked it. "Sounds too much like 'open sores,'" one participant commented. But we all agreed that there were serious problems with the name *free software* and that wide adoption of a new name could be an important step forward. So we put it to a vote. *Open source* won handily over *sourceware* and we all agreed to use the new term going forward.

It was an important moment, because at the end of the day I'd arranged a press conference with reporters from the *New York Times*, *Wall Street Journal*, *San Jose Mercury News* (at the time the daily paper of Silicon Valley), *Fortune*, *Forbes*, and many other national publications. Because I'd earlier built relationships with many of these reporters during my time in the early 1990s promoting the commercialization of the Internet, they showed up even though they didn't know what the news was going to be.

I lined up the participants in front of the assembled reporters and told a story that none of them had heard before. It went something like this:

When you hear the term *free software*, you think that it's a rebel movement that is hostile to commercial software. I'm here to tell you that every big company—including your own—already uses free software every day. If your company has an Internet domain name—say nytimes.com or wsj.com or fortune.com—that name only works because of BIND, the software written by this man— Paul Vixie. The web server you use is probably Apache, created by a team co-founded by Brian Behlendorf, sitting here. That website also makes heavy use of programming languages like Perl and Python, written by Larry Wall, here, and Guido van Rossum, here. If you send email, it was routed to its destination by Sendmail, written by Eric Allman. And that's before we even get to Linux, which you've all heard about, which was written by Linus Torvalds here.

And here's the amazing thing: All of these guys have dominant market share in important categories of Internet software without any venture capitalist giving them money, without any company behind them, just on the strength of building great software and giving it away to anyone who wants to use it or to help them build it.

Because *free software* has some negative connotations as a name, we've all gotten together here today and decided to adopt a new name: *open source software*.

Over the next couple of weeks, I gave dozens of interviews in which I explained that all of the most critical pieces of the Internet infrastructure were "open source." I still remember the disbelief and surprise in many of the initial interviews. After a few weeks, though, it was accepted wisdom, the new map. No one even remembers that the event was originally called the Freeware Summit. It was thereafter referred to as "The Open Source Summit."

This is a key lesson in how to see the future: bring people together who are already living in it. Science fiction writer William Gibson famously observed, "The future is already here. It's just not evenly distributed yet." The early developers of Linux and the Internet were already living in a future that was on its way to the wider world. Bringing them together was the first step in redrawing the map.

ARE YOU LOOKING AT THE MAP OR THE ROAD?

There's another lesson here too: Train yourself to recognize when you are looking at the map instead of at the road. Constantly compare the two and pay special attention to all the things you see that are missing from the map. That's how I was able to notice that the narrative about free software put forward by Richard Stallman and Eric Raymond had ignored the most successful free software of all, the free software that underlies the Internet.

Your map should be an aid to seeing, not a replacement for it. If you know a turn is coming up, you can be on the lookout for it. If it doesn't come when you expect, perhaps you are on the wrong road.

My own training in how to keep my eyes on the road began in 1969, when I was only fifteen years old. My brother Sean, who was seventeen, met a man named George Simon, who was to have a shaping role in my intellectual life. George was a troop leader in the Explorer Scouts, the teenage level of the Boy Scouts—no more than that and yet so much more. The focus of the troop, which Sean joined, was on nonverbal communication.

Later, George went on to teach workshops at the Esalen Institute, which was to the human potential movement of the 1970s what the Googleplex or Apple's Infinite Loop is to the Silicon Valley of today. I taught at Esalen with George when I was barely out of high school, and his ideas have deeply influenced my thinking ever since.

George had this seemingly crazy idea that language itself was a kind of map. Language shapes what we are able to see and how we see it. George had studied the work of Alfred Korzybski, whose 1933 book, *Science and Sanity*, had come back into vogue in the 1960s, largely through the work of Korzybski's student S. I. Hayakawa.

Korzybski believed that reality itself is fundamentally unknowable, since *what is* is always mediated by our nervous system. A dog perceives a very different world than a human being, and even individual humans have great variability in their experience of the world. But at least as importantly, our experience is shaped by the words we use.

I had a vivid experience of this years later when I moved to Sebastopol, a small town in Northern California, where I kept horses. Before that, I'd look out at a meadow and I'd see something that I called "grass." But over time, I learned to distinguish between oats, rye, orchard grass, and alfalfa, as well as other types of forage such as vetch.

Now, when I look at a meadow, I see them all, as well as other types whose names I don't know. Having a language for grass helps me to see more deeply.

Language can also lead us astray. Korzybski was fond of showing people how words shaped their experience of the world. In one famous anecdote, he shared a tin of biscuits wrapped in brown paper with his class. As everyone munched on the biscuits, some taking seconds, he tore off the paper, showing that he'd passed out dog biscuits. Several students ran out of the class to throw up. Korzybski's lesson: "I have just demonstrated that people don't just eat food, but also words, and that the taste of the former is often outdone by the taste of the latter."

Korzybski argued that many psychological and social aberrations can be seen as problems with language. Consider racism: It relies on terms that deny the fundamental humanity of the people it describes. Korzybski urged everyone to become viscerally aware of the process of abstraction, by which reality is transformed into a series of statements about reality—maps that can guide us but can also lead us astray.

This insight seems particularly important in the face of the fake news that bedeviled the 2016 US presidential election. It wasn't just the most outrageous examples, like the child slavery ring supposedly being run by the Clinton campaign out of a Washington, DC, pizza joint, but the systematic and increasingly algorithmic selection of news to fit and amplify people's preconceived views. Whole sectors of the population are now led by vastly divergent maps. How are we to solve the world's most pressing problems when we aren't even trying to create maps that reflect the actual road ahead, but instead drive toward political or business goals?

After working with George for a few years, I got a near-instinctive sense of when I was wrapped in the coils of the words we use about reality and when I was paying attention to what I was actually experiencing, or even more, reaching beyond what I was experiencing now to the thing itself. When faced with the unknown, a certain cultivated receptivity, an opening to that unknown, leads to better maps than simply trying to overlay prior maps on that which is new.

It is precisely this training in how to look at the world directly, not simply to reshuffle the maps, that is at the heart of original work in science—and as I try to make the case in this book, in business and technology.

As recounted in his autobiography, *Surely You're Joking, Mr. Feynman*, fabled physicist Richard Feynman was appalled by how many students in a class he visited during his sabbatical in Brazil couldn't apply what they had been taught. Immediately after a lecture about the polarization of light, with demonstrations using strips of polarizing film, he asked a question whose answer could be determined by looking through the film at the light reflected off the bay outside. Despite their ability to recite the relevant formula when asked directly (something called Brewster's Angle), it never occurred to them that the formula provided a way to answer the question. They'd learned the symbols (the maps) but just couldn't relate them back to the underlying reality sufficiently to use them in real life.

"I don't know what's the matter with people: they don't learn by understanding; they learn by some other way—by rote, or something," Feynman wrote. "Their knowledge is so fragile!"

Recognizing when you're stuck in the words, looking at the map rather than looking at the road, is something that is surprisingly hard to learn. The key is to remember that this is an experiential practice. You can't just read about it. You have to practice it. As we'll see in the next chapter, that's what I did in my continuing struggle to understand the import of open source software.

2

TOWARD A GLOBAL BRAIN

MY FOCUS ON THE INTERNET RATHER THAN ON LINUX EVEN-
tually led me in very different directions from other open source advocates. They wanted to argue about the best open source licenses. I
thought that licenses didn't matter as much as everyone else thought
they did.

I was fascinated by the massive next-generation infrastructure and
business processes Google was building. Others were interested in these
things too, but they thought very few companies would need Google's
kind of infrastructure, or to use its techniques. They were wrong.

> This is my next lesson. If the future is here, but just not evenly
> distributed yet, find seeds of that future, study them, and ask
> yourself how things will be different when they are the new
> normal. What happens if this trend keeps going?

Over the next few years, I refined my argument, eventually developing a talk called "The Open Source Paradigm Shift," which I delivered
hundreds of times to business and technical audiences. I always started
the talk with a question: "How many of you use Linux?" Sometimes
only a few hands would go up; sometimes many. But when I asked my
next question, "How many of you use Google?" almost every hand in
the room would go up. "You just told me," I said, "that you still think of
the software you use as what is running on your local computer. Google
is built on top of Linux. You're *all* using it."

The way you view the world limits what you can see.

Microsoft had come to define a paradigm in which competitive advantage and user control came through proprietary software running
on a desktop computer. Most free and open source advocates had accepted that map of the world, and were looking to Linux to compete

with Microsoft Windows as an operating system for desktop and lap-
top computers. Instead, I argued, open source was becoming the "Intel
Inside" of the next generation of computer applications. I was thinking
about what was different in how that next generation of software was
going to work, and how that would upset the power dynamics of the
computer industry far more deeply than any competition by Linux on
the desktop.

As open source developers gave away their software for free, many
could see only the devaluation of something that was once a locus of
enormous value. Thus Red Hat founder Bob Young told me, "My goal
is to shrink the size of the operating system market." (Red Hat, however, aimed to own a large part of that smaller market.) Defenders of
the status quo, such as Microsoft VP Jim Allchin, claimed that "open
source is an intellectual property destroyer," and painted a bleak picture
in which a great industry is destroyed, with nothing to take its place.

The commoditization of operating systems, databases, web servers and browsers, and related software was indeed threatening to Microsoft's core business. But that destruction created the opportunity
for the killer applications of the Internet era. It is worth remembering
this history when contemplating the effect of on-demand services like
Uber, self-driving cars, and artificial intelligence.

I found that Clayton Christensen, the author of *The Innovator's Dilemma* and *The Innovator's Solution*, had developed a framework that
explained what I was observing. In a 2004 article in *Harvard Business
Review*, he articulated "the law of conservation of attractive profits" as
follows:

> **"When attractive profits disappear at one stage in the value
> chain because a product becomes modular and commoditized,
> the opportunity to earn attractive profits with proprietary
> products will usually emerge at an adjacent stage."**

I saw Christensen's law of conservation of attractive profits at work
in the paradigm shifts required by open source software. Just as IBM's
commoditization of the basic design of the personal computer led to

opportunities for attractive profits "up the stack" in software, new fortunes were being made up the stack from the commodity open source software that underlies the Internet, in a new class of proprietary applications.

Google and Amazon provided a serious challenge to the traditional understanding of free and open source software. Here were applications built on top of Linux, but they were fiercely proprietary. What's more, even when using and modifying software distributed under the most restrictive of free software licenses, the GPL (GNU Public License), these sites were not constrained by any of its provisions, all of which were framed in terms of the old paradigm. The GPL's protections were triggered by the act of software distribution, yet web-based applications don't distribute any software: It is simply performed on the Internet's global stage, delivered as a service rather than as a packaged software application.

But even more important, even if these sites gave out their source code, users would not easily be able to create a full copy of the running application. I told free software advocates like Richard Stallman that even if they had all of the software that Amazon or Google had built on top of Linux, they wouldn't have Amazon or Google. These sites didn't just consist of a set of software programs. They consisted of massive amounts of data and the people and business processes used to gather, manage, and build ongoing services using that data.

As I had been exploring this line of argument, the tectonic processes of technology were adding new continents that had to be reflected in the map. In June 1999, Internet file-sharing site Napster turned the industry on its head by allowing users to share music files with each other free of charge across the net. What was most interesting from the technical point of view was that Napster, and soon other file-sharing networks like FreeNet and Gnutella (and a bit later, BitTorrent), didn't keep all the files in one place like existing online music sites. Instead they stored them on the hard drives of millions of users across the Internet. Andy Oram, one of the editors at my publishing company, made the point to me that the architectural implications of these programs were more important than their business implications. (This is a history that has repeated itself fifteen years later with bitcoin and the block-chain.)

This was a kind of decentralization beyond even the World Wide Web. It was becoming clear that the future demanded even more extreme rethinking of what the Internet could become as a platform for next-generation software applications and content.

Nor was this future limited to file sharing. The SETI@home project, started in mid-1999, was signing up Internet users to contribute unused computing power on their home PCs to analyze radio telescope signals for signs of extraterrestrial intelligence. Computation, not just files and data, could be split across thousands of computers. And developers were increasingly beginning to understand that the powerful applications of the web could be treated as components callable by other programs—what we now call "web services." An API was no longer just a way for an operating system vendor like Microsoft to provide developers with access to its system services, but a sort of door that an Internet site could leave open for others to come and securely retrieve data.

Jon Udell, a prescient observer of technology, had actually given a talk on this topic at the first Perl Conference in 1997. He noted that when a website called a back-end database to retrieve information, it encoded the information that it wanted into the URL (the web's Uniform Resource Locator format), and that this URL could be constructed by a program, essentially turning any website into a program-callable component.

Programmers had been using these kinds of hidden clues to remotely control websites since the early days. "Web spidering," using a program to visit and copy millions of websites in succession, was an essential part of search engines, for instance, but people were now starting to think about how to generalize the process of calling remote websites to allow more specific functions.

All of this was adding up to a completely new paradigm in computing. The Internet was replacing the personal computer as the platform for a new generation of applications. The World Wide Web was the most powerful face of this platform, but peer-to-peer file sharing, distributed computation, and Internet messaging systems like ICQ demonstrated that an even bigger story was afoot.

So in September 2000, I once again pulled together a set of people who I thought ought to be in the same room to find out what they had in common.

Early the next year, based on the insights drawn from that "peer-to-peer" summit, we launched the O'Reilly Peer-to-Peer and Web Services Conference. By 2002 we had renamed it the O'Reilly Emerging Technology Conference and reframed its theme as "Building the Internet Operating System."

I still remember the perplexity with which some people responded to my choice of keynotes for the event: One was on Napster and Internet file sharing, one was on distributed computation, and the third was on web services. "What do these things have to do with each other?" people asked. It was clear to me that they were all aspects of the evolution of the Internet into a generalized platform for new kinds of applications.

| Remember, putting the right pieces of the puzzle on the table is the first step toward assembling them into a coherent picture.

At that first Peer-to-Peer and Web Services Conference in 2001, Clay Shirky memorably summarized the shift to network computing by telling an apocryphal story about Thomas Watson Sr., the head of IBM during the birth of the mainframe computer. Watson was said to have remarked that he saw no need for more than five computers worldwide. Clay noted, "We now know that Thomas Watson was wrong." We all laughed as we thought of the hundreds of millions of personal computers that had been sold. But then Clay socked us with the punch line: "He overstated the number by four."

Clay Shirky was right: For all practical purposes, there is now only one computer. Google is now running on well over a million servers, using services distributed across those servers to deliver instant access to documents and services available from nearly a hundred million other independent web servers—to users running on billions of smartphones and PCs. It is all woven into one seamless whole. John Gage, the chief scientist at Sun Microsystems, had first uttered this prescient insight in 1985 when he coined Sun's slogan: "The Network is the Computer."

WEB 2.0

The last piece of the puzzle arrived in 2003, and much as had happened with *open source*, it was a term, *Web 2.0*, coined by someone else.

Dale Dougherty, one of my earliest employees, who had played a key role in transforming O'Reilly & Associates (later renamed O'Reilly Media) from a technical writing consulting company into a technology book publishing company in the late 1980s, and whom I'd come to consider a cofounder, had gone on to explore online publishing. He created our first ebook project in 1987, and in trying to develop a platform for ebook publishing that would be open and available to all publishers, had discovered the nascent World Wide Web.

Dale had brought the web to my attention, introducing me to Tim Berners-Lee in the summer of 1992. We quickly became convinced that the web was a truly important technology that we had to cover in our forthcoming book about the Internet, which was just then opening up for commercial use. Ed Krol, the author, didn't yet know much about the web, so Mike Loukides, his editor at O'Reilly, wrote the chapter and we added it to the book just before its publication in October 1992.

Ed's book, *The Whole Internet User's Guide & Catalog*, went on to sell over a million copies and be named by the New York Public Library as one of the most significant books of the twentieth century. It was that book that introduced the World Wide Web to the world. There were only about 200 websites worldwide when we published it in the fall of 1992. Within a few years, there were millions.

Dale had gone on to create the Global Network Navigator (GNN), O'Reilly's online magazine about the people and trends behind the web and a catalog of the most interesting websites. It was the first web portal (launched a year before Yahoo!) and the first website to carry advertising. Realizing that the web was growing faster than we could keep up with as a private company, and not wanting to lose control of O'Reilly by taking in venture capital, we instead sold GNN to AOL in 1995, in the first content transaction of what was to become the dot-com boom.

Even after that boom went bust in 2000, while venture capitalists were downcast and the markets swooned, we remained convinced that it was still just the early days of the web. In 2003, at an O'Reilly Media company management retreat, we identified our key strategic goal as "reigniting enthusiasm in the computer industry." Dale was the one who found out how to do that. While brainstorming with Craig Cline, an executive for a conference company called MediaLive International, which had long wanted to partner with O'Reilly on conferences, Dale had come up with the name Web 2.0 to describe the second coming of the World Wide Web after the bust. I gave Dale the go-ahead to partner with MediaLive on the new event, which we launched a year later as the Web 2.0 Conference, with John Battelle, the author and media entrepreneur, as the host and third partner.

As with my work on open source, we began to build a new map by trying to put projects like Google's pay-per-click ad model, Wikipedia, file-sharing systems like Napster and BitTorrent, web services, and syndicated content systems like blogging into the same frame. And as had happened with open source, the introduction of the new term caught the zeitgeist perfectly and was quickly embraced. Companies rebranded themselves as "Web 2.0 companies," distancing themselves from the old "dot-com" moniker whether they actually were doing something new or not. Consultants came out of the woodwork promising to help companies adopt the new paradigm.

By 2005, I realized that I needed to put a bit more substance behind the term, and wrote an essay that summarized everything I'd learned since "The Open Source Paradigm Shift." The essay was called "What is Web 2.0?" It was that essay, more than anything else, that gave me a reputation as a futurist, because I identified so many of the key trends that came together in this next generation of computing.

I didn't predict the future. I drew a map of the present that identified the forces shaping the technology and business landscape.

THE INTERNET AS PLATFORM

The first principle of Web 2.0 was that the Internet was replacing Windows as the dominant platform on which the next generation of applications was being built. Today this is so obvious as to make you wonder how anyone could have missed it. Yet Netscape, the great challenger to Microsoft's dominance in the late 1990s, had failed because they had accepted the rules of the game as it was being played by Microsoft. They were using the old map. The defining company of Web 2.0, Google, was using the new map.

Netscape had also used the term *the web as platform*, but they had framed it in terms of the old software paradigm: Their flagship product was the web browser, a desktop application, and their strategy was to use their dominance in the browser market to establish a market for high-priced server products. Much as the "horseless carriage" framed the automobile as an extension of the familiar, Netscape promoted a "webtop" to replace the desktop, and planned to populate that webtop with information updates and applets pushed to the webtop by information providers who would purchase Netscape servers. Control over standards for displaying content and applications in the browser would, in theory, give Netscape the kind of market power enjoyed by Microsoft in the PC market.

In the end, both web browsers and web servers turned out to be commodities, and value moved up the stack to services delivered over the true web platform.

Google, by contrast, began its life as a native web application, never sold or packaged, but delivered as a service, with customers paying, directly or indirectly, for the use of that service. None of the trappings of the old software industry were present. No scheduled software releases, just continuous improvement. No licensing or sale, just usage. No porting to different platforms so that customers can run the software on their own equipment, just a massively scalable collection of commodity PCs running open source operating systems plus homegrown applications and utilities that no one outside the company ever gets to see.

"Google's service is not a server," I wrote, "though it is delivered by a massive collection of Internet servers—nor a browser—though it is experienced by the user within the browser. Nor does its flagship search service even host the content that it enables users to find. Much like a phone call, which happens not just on the phones at either end of the call, but on the network in between, Google happens in the space between browser and search engine and destination content server, as an enabler or middleman between the user and his or her online experience."

While both Netscape and Google could be described as software companies, Netscape belonged to the same software world as Lotus, Microsoft, Oracle, SAP, and other companies that got their start in the 1980s software revolution, while Google's fellows were other Internet applications like eBay, Amazon, Napster, DoubleClick, and Akamai.

As we moved from the Web 2.0 era into the "mobile-social" era and now into the "Internet of Things," the same principle continues to hold true. Applications live on the Internet itself—in the space between the device and remote servers—not just on the device in the user's hands. This idea was expressed by another of the principles I laid out in the paper, which I called "Software Above the Level of a Single Device," using a phrase first introduced by Microsoft open source lead David Stutz in his open letter to the company when he left in 2003.

The implications of this principle continue to unfold. When I first wrote about the idea of software above the level of a single device, I wasn't just thinking about web applications like Google but also hybrid applications like iTunes, which used three tiers of software—a cloud-based music store, a personal PC-based application, and a handheld device (at the time, the iPod). Today's applications are even more complex. Consider Uber. The system (it's hard to call it an "application" anymore) simultaneously spans code running in Uber's data centers, on GPS satellites and real-time traffic feeds, and apps on the smartphones of hundreds of thousands of drivers and of millions of passengers, in a complex choreography of data and devices.

HARNESSING COLLECTIVE INTELLIGENCE

Another key to what distinguished the web applications that survived the dot-com bust from those that died was that the survivors all, in one way or another, worked to harness the collective intelligence of their users. Google is an aggregator of hundreds of millions of websites built by people all over the world, and uses hidden signals from its own users and from the people who create those websites to rank and organize them. Amazon not only aggregates products from a worldwide network of suppliers, but allows its customers to annotate its product database with reviews and ratings, using the power of the crowd to ferret out the best products.

I'd originally seen this pattern in the way that the Internet turbo-charged the global collaboration around open source projects. And as the future continued to unfold, once again, the pattern held true. The iPhone leapt to dominance in the early mobile era not just because of its touch-screen interface and sleek, innovative design but because the App Store enabled a worldwide developer community to add features in the form of apps. Social media platforms like YouTube, Facebook, Twitter, Instagram, and Snapchat all gain their power by aggregating the contributions of billions of users.

When people asked me what came after Web 2.0, I was quick to answer "collective intelligence applications driven by data from sensors rather than from people typing on keyboards." Sure enough, advances in areas like speech recognition and image recognition, real-time traffic and self-driving cars, all depend on massive amounts of data harvested from sensors on connected devices.

The current race in autonomous vehicles is a race not just to develop new algorithms, but to collect larger and larger amounts of data from human drivers about road conditions, and ever-more-detailed maps of the world created by millions of unwitting contributors. It's easy to forget that in 2005, when Stanford won the DARPA Grand Challenge for self-driving vehicles, they did so by completing a seven-mile course in seven hours. Yet by 2011, Google had managed more than a million miles on ordinary highways. One of their secret weapons: Google

Street View cars, driven by human drivers, using cameras, GPS, and LIDAR (LIght Detection And Ranging) to collect data. As Peter Norvig, a director of research at Google, once said to me, "It is a hard AI problem to pick a traffic light out of a video image. It's much easier to tell whether it's green or red when you already know it's there." (In the years since Peter said that, the first problem has gotten easier too, but you get the idea.)

Today, companies like Tesla and Uber have a shot at leadership in self-driving cars because they have large fleets of instrumented vehicles, vehicles whose sensors are used not just for the task at hand, but as input to the algorithmic systems of the future. But remember: Those vehicles are driven by humans. The data they capture is the next stage in harnessing the collective intelligence of billions of instrumented humans going about their daily lives.

DATA AS THE NEXT INTEL INSIDE

Contribution of user data for collective intelligence sounds like kumbaya, and in the first years of the new century many of the people celebrating user-contributed sites like Wikipedia or new media networks like blogging saw only the utopian possibilities. I argued that data would turn out to be the key to market dominance for companies like Google and Amazon. As I put it in one talk at the time, "'Harnessing collective intelligence' is how the Web 2.0 revolution begins; 'Data is the Intel Inside' is how it ends."

Intel, of course, was the company that, along with Microsoft, had captured a monopoly position in the personal computer market, such that every PC bore the sticker INTEL INSIDE. Intel had done this by becoming the sole source for the processor, the brain of the PC. Microsoft had done it by controlling access to its software operating system.

Open source software and the open communications protocols of the Internet had changed the game for Microsoft and Intel. But my map told me that the game didn't end there. Per Clayton Christensen's Law of Conservation of Attractive Profits, I knew that something else was going to become valuable. In a word: *data*. In particular, I thought that building a critical mass of user-contributed data led to self-reinforcing network effects.

The term *network effect* generally refers to systems that gain in utility the more people use them. A single telephone is not very useful, but once enough people have them, it is very hard *not* to join the network. So too, the competition in social networks has been to assemble massive user bases, because the lock-in is not via software but through the number of other people using the same service.

The network effects that I observed in data were more indirect, and had to do with the way that companies were learning to harvest value from the users of their systems. Barnes & Noble had all the same products as Amazon, but Amazon had vastly more user reviews and comments. People came not just for the products but for the intelligence added by other users. So too, in addition to Google's superior algorithms and commitment to constantly improving the product, Google Search kept getting better because more people were using it, which meant that Google could accumulate more data and therefore learn faster than competitors, keeping them perpetually ahead.

Returning to the question of who will win in self-driving cars, one has to ask not just who will have the best software, but who will have the most data.

In a 2016 conversation with Uber executives, they argued that their hundreds of millions of miles of data collected from driver and passenger apps will give them the edge. However, it's hard to believe that the data from smartphone apps alone will match the level of detail that Google has been collecting with its specially equipped vehicles. That's why Uber believes it is so urgent to get self-driving vehicles offered as part of their service, even if these remain crewed by drivers for many years to come. Tesla too has detailed telemetry from every vehicle, and in the case of the second-generation vehicles with self-driving features, that does include detailed camera and radar data. The big question for automakers without this edge is whether the sensors used for accident avoidance or automated parking will be sufficient for them to collect enough data to compete.

A lot depends not just on how much data you have, of course, but how able you are to make sense of it. There Google, Tesla, and Uber have a big edge on traditional auto companies.

THE END OF THE SOFTWARE RELEASE CYCLE

In the PC era, we were accustomed to thinking of software as an artifact. Companies had to start thinking of software as a service. This meant we'd see a whole new approach to software development. While I didn't develop this idea as fully as the previous three, it was clear even in 2005 that what we now call "iterative, user-centered, data-driven development" would be the new normal. Software built in what we now call "the cloud" is constantly updated.

But it's not just updated many times faster than PC-era software. Today's software is developed by watching what users do in real time—with A/B testing of features on subsets of users, measurement of what works and what doesn't work informing development on an ongoing basis. In this way, the collaborative model of open source software development—"given enough eyeballs, all bugs become shallow"—has been taken to its logical conclusion, and completely divorced from the original licensing model of free and open source software.

> In the end, I was able to see the future more clearly because my map was more useful than one based on a battle between proprietary software and free software licensing models. Having the right orientation matters. But even then, it had taken years to explore the landscape sufficiently to fill in all the blank spaces on the map.

THINKING IN VECTORS

We all know that the world is changing, but too often we take refuge in the familiar, and fail to stretch our thinking to look at current trends and ask ourselves, "What happens if this goes on?" We also fail to take into account that some trends are potentially much more powerful than others, developing at a faster rate, or taking things in a radically different direction rather than as a simple continuation of the familiar.

The path I traveled from noticing these trends to predicting the future began with the fact that the free software narrative had left out the software behind the Internet. Putting that observation together with my knowledge of the early history of the PC and the rise of Microsoft, and thinking about the long arc of Internet-enabled collaboration, is an example of what I call "thinking in vectors."

A vector is defined in mathematics as a quantity that can only be fully described by both a magnitude and a direction. You have to take both into account. Some of the most famous "laws" that have been cited in the computer industry are essentially descriptions of vectors.

Moore's Law, originally formulated by Intel cofounder Gordon Moore in 1965, noted that the number of transistors on an integrated circuit had roughly doubled every year, and looked to continue for the foreseeable future. In 1975, Moore revised his prediction to predict a doubling of the transistor count every two years. Intel executive David House proposed that the actual performance increase would be closer to a doubling every eighteen months, due to an increase in processor speed as well as the increase in chip density, and it is that version that largely held true for many decades.

One of my favorite popular definitions of Moore's Law came in a conversation I had with Reid Hoffman, the founder and chairman of LinkedIn, and Senator Sheldon Whitehouse (D-RI) over dinner in San Francisco seven or eight years ago. "We need to start seeing Moore's Law apply to healthcare," I said. "What's Moore's Law?" the senator asked. "You have to understand, Senator," Reid interjected, "that in Washington, you assume that every year things cost more and do less. In Silicon Valley, everyone expects our products to cost less every year but do more."

Whether through Moore's Law proper, or through related advances, like the speed and density of memory storage, hard disk density, networking interconnections, display pixels per dollar, and many other systematic advances, that broader "Hoffman's Law," as I now dub it, that every year technology products cost less and do more, has generally held true for a very long time.

In the case of Hoffman's Law and some of the other fundamental drivers of progress in the computer industry, the vector is clear. We don't necessarily know where the next increment will come from, but the line has been drawn through enough data points that there is a reasonable expectation that it will continue.

You must always be alert, though, for an inflection point where the old gives way to something profoundly new. For example, we know that Moore's Law proper cannot continue forever, because of the physical limits of transistor density. Without some breakthrough like quantum computing, which uses subatomic particles for computation, transistor density is limited by the size of the atom, which we will approach in only a few more generations of Moore's Law. That being said, as Moore's Law is slowing down, multi-core processors have provided the industry with a temporary workaround, so even though we hit limits in transistors and clock speed, we are still increasing throughput.

Vectors are not only a productive way to think about well-defined trends like Moore's Law, but also a way to make sense of virtually everything that changes. The future is the outcome of millions of intersecting vectors, which add up in unexpected ways. The art is to pick out important vectors and weave a net from them in which to catch a view of the future.

At O'Reilly Media, when we first take note of a new trend but don't have the quantification yet to fully characterize it as a vector, with both a magnitude and a direction, we still begin to plot a line, extending it as each new data point comes in. This needn't be entirely conscious. Instead, it requires an attitude of receptivity, in which new information is always coming in, in which multiple scenarios, multiple futures, are unfolding, all still possible, but gradually collapsing into the present. Lawrence Wilkinson, one of the cofounders of Global Business Network, the company that pioneered a technique called scenario planning, whom I met in 2005, introduced me to a wonderful phrase that captured how my mind works: "news from the future."

So, for example, consider how the "Harnessing Collective Intelligence" vector became clear to us:

1. In the late 1980s and early 1990s, we were exposed to the "barn raising" style of collaborative software development of the early Unix community—what we later came to call open source software.

2. In developing our first books, we practiced a version of this kind of crowdsourcing ourselves. In 1987, I wrote a book called *Managing UUCP and Usenet*, which described how to use a program called the Unix-to-Unix Copy Program (UUCP) to connect to Usenet, a distributed dial-up precursor to today's social web. It was on Usenet that the world's software developers conversed about their work, shared tips and advice, and, increasingly, talked about everything from sex to politics. At first the book was based on my own experience connecting systems to Usenet, but that experience was limited. Readers sent me information about how to use additional equipment that I didn't have access to and the fine points of geekery ("Here's the 'chat script' for calling in through a Develcon switch," or "Here are the pins you need to connect in an RS-232 cable" for some particular brand of modem.)

   We reprinted the book every six months or so, and every time we reprinted, it grew by another thirty or forty pages, almost entirely composed of contributions from readers. Over its first three years, it went from about 80 pages to over 200. You might say it was an early, printed-on-paper wiki.

   In 1992, trying to create a print book that emulated the link style of the World Wide Web, I designed and coauthored a book called *Unix Power Tools*, which wove together tips and tricks harvested from hundreds of Internet contributors into a hyperlinked web of short articles, each of which could be read independently because it also contained links to additional articles providing tutorial and background information that my coauthors Jerry Peek and Mike Loukides and I felt was needed to make sense of the crowdsourced lessons.

3. In 1992 and 1993, as we turned "the Whole Internet Catalog" into GNN, the Global Network Navigator, every day we

sought out the best of the new sites joining the World Wide Web, curating them into a rich catalog of experiences created, as if by magic, by a distributed network of people pursuing their own passions.

4. We watched the early search engines, starting with Webcrawler in 1994, automatically collect links not just to the best websites, but to *every* website. And in 1998, when Google launched, with far better results, it became clear that they had found hidden intelligence in web links. A link wasn't just a pointer to a page that might previously have gone unnoticed by the crawler, it was a way of discovering new content on the web. The number of links was also a vote about the value of that site. And the site making the link also had links pointing to it; the nature and quality of those links could tell the search engine something about the value of the page making the connection. How long had that site been on the net? How many people pointed to it? How valuable did people find the links that it made? Not only that, but there was further human intent signaled by the "anchor text"—the words in the source document that hyperlinked to another one. Google found a gold mine of data, and never looked back.

I still remember a blog post by Robert Scoble in which he gleefully demonstrated how human contribution was central to search engines. "I just discovered a new restaurant in Seattle. Its website isn't in Google. But it will be tomorrow, because I just linked to it!"

5. In 1995, we saw how eBay and Craigslist brought crowdsourcing to products and services, and began to realize that the magical aggregation of millions of people into new kinds of services wasn't limited to "content," but could also be used in the physical world.

6. We watched how Amazon ran rings around Barnes & Noble and Borders in online bookselling by applying the same principles that Google used to make a better search engine to more effective e-commerce. While Barnes & Noble followed its retail

store practices of letting publishers buy search placement, so that a search for, say, a computer book on JavaScript or Perl would turn up the book of whichever publisher had paid them the most to feature it, Amazon used multiple signals to choose the book that had the most "relevance," as defined by a mix of sales, positive reviews, inbound links from "Associates," and other factors based on collective intelligence. We were always pleased to find our books at the top of Amazon searches, because we knew that that meant they were seen as the best by tens of thousands of readers.

As a result of all these prior data points, in 2004, when I sought to define "Web 2.0," and thought about what distinguished the companies that had survived the dot-com bust from those that had failed, it was clear that all of the survivors were, in one way or another, harnessing the power of their users to create their product.

And in 2009, when I wrote "Web Squared: Web 2.0 Five Years On," it was straightforward to see what was coming next. "The smartphone revolution has moved the Web from our desks to our pockets," I wrote. "Collective intelligence applications are no longer being driven solely by humans typing on keyboards but, increasingly, by sensors. Our phones and cameras are being turned into eyes and ears for applications; motion and location sensors tell where we are, what we're looking at, and how fast we're moving. Data is being collected, presented, and acted upon in real time. The scale of participation has increased by orders of magnitude.

"The Web is no longer a collection of static pages of HTML that describe something in the world," I continued. "Increasingly, the Web *is* the world—everything and everyone in the world casts an 'information shadow,' an aura of data which, when captured and processed intelligently, offers extraordinary opportunity and mind bending implications."

What's important to note, though, is that even when you've spotted a vector, it doesn't mean that you understand all of its implications. Yes, I was able to identify in 2009 that sensors would be the key to the next generation of applications, but that didn't lead me to "predict" Google's breakthrough with self-driving cars, or that Uber was about to realize the potential of the sensors in the phone to revolutionize on-demand transportation.

I also often didn't take the time to act on my own insights. Technology journalist John Dvorak once reminded me that very early in the history of the web, I had confidently predicted to him that there would be a market for buying and selling domain names. They would be extremely valuable. Yet I never bothered to go out and buy any myself.

Once you've identified a trend, though, it's easier to recognize early which new developments are important, because they are the next step in the continued acceleration along the vector, as entrepreneurs and inventors continue, in Wallace Stevens's magnificent phrase, to "search a possible for its possibleness." In other words, the news from the future mindset helps you to pay attention to the right things, and learn from them.

TWITTER MAKES IT REAL FOR NONPROGRAMMERS

The notion of the Internet as an operating system for the next generation of applications had taken me a long way. By 2010, the idea had taken hold in the industry. Developers were routinely writing applications that relied on data from Internet services—about location, search results, social networks, music, products, and so much more. Startups were no longer building local applications in their own data centers but rather in what was now called the cloud. I didn't need to keep preaching that gospel.

And frankly, I was ready to move on. As T. S. Eliot so memorably put it:

. . . one has only learnt to get the better of words
For the thing one no longer has to say, or the way in which
One is no longer disposed to say it. And so each venture
Is a new beginning, a raid on the inarticulate

I was tired of talking about Web 2.0. And there was more to what was happening than just a cloud-based platform for computer applications. Social media was showing how the Internet connects people on a global scale, and I began to see the power of a different metaphor. A metaphor too is a kind of map; it may be all you have when you are first encountering a new territory shrouded in mist.

Increasingly, I'd been watching a kind of Cambrian explosion in applications for collective intelligence that were qualitatively different from those of the desktop web. Smartphones had put a camera in everyone's hand, and Twitter had created a real-time platform from which those photos and text updates could be instantly disseminated to the world. Billions of connected humans and devices were being woven into a global brain. That brain was all of us, augmented and connected.

Twitter was an especially fertile ground for reinvention. Three features we now take for granted were all created by users and only later adopted by the platform. The @ symbol to reply to another user first appeared in November 2006; it was formally adopted by the platform in May 2007, turning Twitter into a place for conversations as well as status updates. The first "retweet" of someone else's tweet happened in April 2007, though it wasn't formally adopted as a feature until 2009.

In August 2007, Chris Messina proposed the use of the # symbol as a way to label events or groups of tweets on Twitter. It became clear just how powerful an amplifier this was of collective knowledge and sentiment during the San Diego wildfires a few months later. Before long, hashtags, as they came to be called, were everywhere. Many of them didn't stick, but if enough people adopted one, it became the real world equivalent of Obi-Wan Kenobi's words in *Star Wars*: "I felt a great disturbance in the Force . . . as if millions of voices suddenly cried out."

And the voices cried out: #iranelection #haitiearthquake #occupy wallstreet.

Beginning in July 2009, Twitter responded to the outside-in innovation and began hyperlinking hashtags, so users could search on them. The app had already begun showing "trending topics" (using algorithms to detect common events even if they do not have the same hashtag), but hashtags added fuel to the fire.

When photos were added to Twitter (again by an outside developer providing features that the platform developer itself hadn't imagined), Twitter's power to reveal the real-time pulse of the world increased even further. On January 15, 2009, four minutes after Captain "Sully" Sullenberger ditched US Airways Flight 1549 in the Hudson after multiple bird strikes had disabled the engines, Jim Hanrahan posted the first tweet. Janis Krums snapped an iPhone photo of passengers standing on the wing of the downed plane a few minutes later and shared it on Twitter via a third-party app called TwitPic, and it went worldwide long before the story appeared on the television news.

Facebook also began to have an effect on global affairs. In 2010, an Egyptian Google employee named Wael Ghonim created a Facebook page called "We Are All Khaled Said," commemorating a young Egyptian who'd been tortured to death by police. The page became a focus for activism that led to antigovernment protests culminating in the revolution of January 25, 2011.

Wikipedia too had become a fulcrum for real-time collective intelligence about the world. After the 2011 Tohuku earthquake and tsunami in Japan, which led to the meltdown of the Fukushima nuclear plant, I had watched in awe as the Wikipedia page grew from a single line in fractured, misspelled English to a full-featured encyclopedia entry. The first entry appeared a mere thirty-two minutes after the earthquake, before the tsunami had struck. Over a short period, hundreds, then thousands of contributors made more than 5,000 edits, creating a comprehensive and authoritative account of the disaster. I still show animations of the transformation in some of my talks. It is a WTF? moment for anyone who sees it.

The debates behind the scenes on Wikipedia "talk" pages, about controversial elements on the published pages, are also eye-opening. In *Reinventing Discovery*, his wonderful book about lessons from the consumer Internet for the practice of science, Michael Nielsen writes, "Wikipedia is not an encyclopedia. It is a virtual city, a city whose main export to the world is its encyclopedia articles, but with an internal life of its own."

In response to the speed of blogging and social media, Google sped up its web crawl, and Google's search results too became increasingly real-time. This led to a qualitative difference in how quickly information was transmitted, and magnified its impact. Now news, ideas, and images propagate across the global brain in seconds rather than weeks or months.

In one sense, this is nothing new. There has always been a global brain. As recounted by Jeff Bezos in a 2005 talk at my Emerging Technology Conference, computer scientist Danny Hillis once said to him that "global consciousness is that thing that decided that decaffeinated coffeepots should be orange." The idea that "orange means decaffeinated" originated during World War II, when Sanka promoted its decaffeinated coffee brand by giving away orange-rimmed coffeepots to restaurants across America. The idea took hold—not universally, to be sure, but sufficiently that the pattern propagates. At some point, it no longer belonged to Sanka but to the world.

The association of "orange" with "decaffeinated" is an example of what Richard Dawkins called a "meme"—a self-replicating idea. Today people often think of memes as images and slogans shared on social media, but any great idea that takes hold is a meme. In 1880, "Darwin's Bulldog" Thomas Henry Huxley wrote, "The struggle for existence holds as much in the intellectual as in the physical world. A theory is a species of thinking, and its right to exist is coextensive with its power of resisting extinction by its rivals."

Knowledge spread from mind to mind even before the advent of writing. But the printed word made it possible for ideas and news to reach people in distant lands, first at the speed of walking, then riding, and eventually of steamships and rail. The first electronic transmissions by telephone and telegraph cut a delay of weeks or months to minutes. With radio and television, transmission became almost instantaneous, but creation and vetting of what was to be transmitted was still slow, done in offices and boardrooms, because the channels for dissemination of instantaneous media were so limited. The Internet, and in particular the combination of the Internet and the smartphone, changed all that. Anyone could share anything at any time; others could pick it up and pass it on even more quickly.

It isn't just ideas and sensations (news of current events) that spread across the network. We talk of information as "going viral," but there are malicious programs designed to do exactly that, to reproduce themselves whether we wish it or not. But perhaps more important than hostile viruses are those with which we willingly cooperate.

In his magnificent history of the origins of modern computing, *Turing's Cathedral*, George Dyson notes that some of the earliest thinkers about digital computing realized that the spread of "codes"—that is, programs—from computer to computer is akin to the spread of viruses, and perhaps of more complex living organisms, that take over a host and put its machinery to work reproducing that program.

"Numerical organisms were replicated, nourished, and rewarded according to their ability to go out and do things: they performed arithmetic, processed words, designed nuclear weapons, and accounted for money in all its forms. They made their creators fabulously wealthy," Dyson writes. "They . . . then influenced the computational atmosphere as pervasively as the oxygen released by early microbes influenced the subsequent course of life. They coalesced into operating systems amounting to millions of lines of code—allowing us to more efficiently operate computers while allowing computers to more efficiently operate us. They learned how to divide into packets, traverse the network, correct any errors suffered along the way and reassemble themselves at the other end. By representing music, imagery, voice, knowledge, friendship, status, money, and sex—the things people value most— they secured unlimited resources, forming complex metazoan organisms running on a multitude of individual processors the way a genome runs on a multitude of cells."

When people join the web, or download a new mobile app, they reproduce its code onto their local machine; they interact with the program, and it changes their behavior. This is true of all programs, but in the network age there are a set of programs whose explicit goal is to get their users to share them more widely. Thus the global brain is actively building new capacity.

The kinds of "thoughts" that a global brain has are different from those of an individual, or of a less connected society. At their best, these thoughts allow for coordinated memory on a scale never seen before, and sometimes even for unforeseen ingenuity and new forms of cooperation; at their worst, they allow for the adoption of misinformation as truth, for corrosive attacks on the fabric of society as one portion of the network seeks advantage at the expense of others (think of spam and fraud, or of the behavior of financial markets in recent decades, or of the rash of fake news sites during the 2016 US presidential election).

But perhaps the most riveting thing to realize is that, bit by bit, the global brain is getting a body. It has eyes and ears (billions of connected cameras and microphones), a sense of position and motion (GPS and motion sensors) that is far more precise and powerful than that of humans, and with specialized sensors, data-gathering capabilities that far outstrip our own.

Now it starts to move. Self-driving cars are a manifestation of the global brain; their memory is the memory of roads traveled under the tutelage of human drivers but recorded with their uncanny senses. But not unsurprisingly, the most powerful manifestation of the global brain's ability to touch the physical world relies not on robots but on the power of networked applications to direct human activity.

> There is usually a paradigmatic company or group of companies that best exemplifies the next wave of technology. "Unpacking" the lessons of that company can help you draw your map of the future.

From 1998 to 2005, I'd built my map of the future by thinking about what we could learn from Amazon and Google. Today two of the companies that teach us the most about the trends shaping the future are Uber and its rival Lyft.

Many readers may bristle at the notion that Uber is a positive model for the technology-driven economy of the future. After all, the company has been embroiled in controversy almost from the beginning. Critics question whether it truly provides economic opportunity for drivers or traps them with deceptive promises of income that it can't deliver. Cities fume at its brazen confrontations with regulators and use of technology to deflect their investigations. Rivals sue over claims of stolen technology. Former employees make accusations of a toxic workplace culture that tolerates sexual harassment.

It is easy to forget that many of the people who invent the future do so by crashing through barriers, crushing competitors, and dominating a new industry by force of will as well as intellect. Sometimes dirty tricks come into play. Thomas Edison and John D. Rockefeller, Bill Gates and Larry Ellison, were all justifiably reviled at various points in their careers. When I began my work in computing, Microsoft was routinely referred to as "the Evil Empire."

Whatever you may think of Uber, it is hard to deny its impact on the economy. If we want to understand the future, we have to understand Uber. Like it or not, it is the poster child for many of the ways that technology is changing the world of work.

Lyft, Uber's smaller rival, is a more idealistic, worker-friendly company that, in practice, has the same business model. Each of the companies has introduced key innovations that were copied by the other. In many ways, they are co-inventing the future of urban transportation. We will consider them together throughout the book.

3

LEARNING FROM LYFT AND UBER

IN THE SUMMER OF 2000, MY EXECUTIVE TEAM AND I DID SOME work with Dan and Meredith Beam, of BEAM inc., a strategy consulting firm, on a strategic planning process for our company. We no longer had one primary business; we had three: book publishing, conferences, and online publishing, which had overlapping audiences but each of which had different demands for investment, go-to-market strategies, and paths to revenue. We needed to find a way to reconcile these different lines of business into a coherent whole.

Dan and Meredith help companies build maps of their business models—one-page pictures that describe, to use their words, *"how all the elements of a business work together to build marketplace advantage and company value."*

The Beams used Southwest Airlines as one of their examples, basing their business model map of Southwest on work originally done by Michael Porter. As you can see from the diagram on the next page, various differentiating factors in Southwest's model all go together. No seat assignments, point-to-point routes, no interconnection with other airlines: all are part of strategy that allows Southwest to offer low fares with lean ground crews and rapid turnaround.

It's easy to conclude that two companies selling similar products and services are in the same business; the Beams argued otherwise.

Yes, Southwest is an airline like any other, but its business model—the way all the pieces work together to create customer value and company advantage—is very different from that of an airline company using the more traditional hub-and-spoke model. In a similar way, we were trying to understand what made us different from our competitors in the technical book and conference business.

As part of the exercise, the Beams ask their clients to develop a vision of their core strategic positioning and a vision of who they want to be-

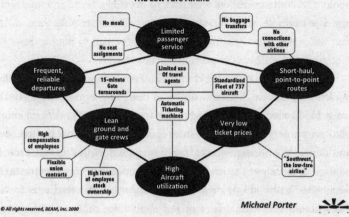

Southwest®
"THE Low Fare Airline"

Michael Porter

come. Through that process, we were able to make clear to ourselves and all our employees that we were not just a computer book publisher that also happened to do conferences and online publishing, but that our core business was something much deeper. As I came to realize, our business was really "changing the world by spreading the knowledge of innovators."

That required a set of core competencies that enabled both our publishing and events business, and could eventually enable other related businesses. As we articulated our core competencies at the time, they were:

- Knowing what's cool and important, and evangelizing it
- Recognizing influential early adopters (whom I sometimes referred to as "alpha geeks") and leveraging their expertise
- Reducing the learning curve and enhancing the depth and quality of information
- Direct connection to customers and people who impact the business
- Fostering a company and culture that make people feel their work can make the world a better place

These competencies were reflected in our book and conference business, but once we separated them in our minds from the details of books and conferences, we were able to develop a more effective strategy. For example, we understood in ways that other publishers didn't that we couldn't just sell through retailers, but had to double down on mechanisms for direct connection with customers.

We'd sold our books direct to consumers before we'd sold through bookstores, and we'd been working to develop the market for ebooks since 1987, which we also sold direct-to-consumer. As various ebook platforms emerged, most publishers ignored ebooks or treated them as a sideshow. We understood that one day the majority of our sales would be digital; if we were to maintain our direct connection to consumers, we needed to build our own digital platform. Later that year (seven years before the Kindle legitimized ebooks for other publishers) we launched Safari, a subscription-based online service for ebooks. Over the years since, we have successfully migrated Safari to a service that provides not just ebooks but also video and other learning modalities, including live online training.

We were also able to see the virtuous tie-ins between publishing and our conference business. Both required us to seek out people living on the edge of the future, people with deep expertise, and to create businesses that helped them to spread their knowledge. One of the jobs we did for that community was to help them build their status and increase their impact. All of our lines of business could be put in service to this goal. Realizing that community was a seedbed for many an entrepreneurial opportunity, we launched O'Reilly Ventures as an internal venture firm, which by 2005 had grown into O'Reilly AlphaTech Ventures (OATV), an independent early-stage venture capital firm. And in 2003 we launched "Foo Camp," our annual unconference (an event for which the program is not set in advance, but constructed by the attendees on the spot) in which we ask our community of "alpha geeks" to show us what they are working on.

In addition, we recognized that evangelizing new technologies and encouraging people to work toward a better future was profoundly motivating to the network of experts, employees, and customers that we were trying to build. We saw that the kind of activism and community building that we'd done in the early days of the commercial web and what we'd done with open source could and should be replicated as an ongoing part of our business. In 2004 we began our storytelling about Web 2.0. In 2005 we launched a magazine (*Make:* magazine) and in 2006 a "county fair for robots" (Maker Faire) that were expressions yet again of a catalytic movement (the Maker movement) using cheap, reusable components to enable a torrent of combinatorial innovation.

MEME MAPS

After working with the Beams, I realized in retrospect that I'd instinctively used a variation of their technique for building my map of open source software. Rather than mapping a single company, I'd mapped what I thought of as the key principles of the new software business model and the ecosystem of the companies that best exemplified those principles. I later did something similar in exploring what I came to call Web 2.0, trying to find the unifying principles that tied together the World Wide Web, file-sharing programs like Napster, distributed computation, and web services.

I called these meme maps. In them, I tried to represent both the canonical companies and the underlying principles defining a new wave in technology, creating a single unifying vision of a set of related technologies.

In a similar way, if we want to understand the implications of today's technologies, one good way to start is by laying out the pieces of the jigsaw that we have, pieces we are convinced have something to do with each other, but it is not quite clear where and how they fit.

What are the canonical companies and technologies that seem to be at the forefront of today's technology-driven changes to the economy? What do they teach us?

Google is still one of the key companies to understand. Its search engine is the pervasive neocortex of the information economy, a critical component of the global brain that the Internet has become, connecting billions of humans with the data and documents we collectively create. The principles that led me to make Google the poster child for Web 2.0 are still unfolding as drivers of the future: big data, algorithms, collective intelligence, software as a service, with the addition of a new focus on machine learning and AI. Understanding how algorithmic systems shape not just new services but also society is a central theme of this book.

The Android phone operating system puts Google's services into the pockets of billions of people. The company kicked off the race for self-driving cars and has been a leader in their development. And it has big ambitions in areas like healthcare, logistics, the design of cities, and robotics. And last but not least, its advertising-based business model means that almost every service it creates can be given away for free, with implications we are only beginning to grasp.

If Google was the defining company of the information age, Facebook is the defining company of the social era. The application began simply as a way for students on college campuses to find and meet each other, passed through its adolescence as a way for friends and family to keep in touch, but now, with nearly two billion members, it has challenged Google as the master of collective intelligence, uncovering an alternate routing system by which content is discovered and shared. Like Google, Facebook has invested heavily in AI, and its successes and failures have much to teach us about what it can and can't do. Contrasting the two companies teaches us something about how algorithmic systems work and how to manage them.

Amazon is also a force of nature. Jeff Bezos is arguably the greatest entrepreneur of the Internet era, reinventing industry after industry. Amazon started as an online bookseller but eventually came to dominate every aspect of online retail in the United States. Amazon also pioneered ebooks; with the Kindle it came to dominate that emerging market and gain channel control over the future of book publishing. It has become a leader in online entertainment of all kinds, rivaling Netflix as the next-generation movie and television studio. And with the Amazon Echo, it has become a force in bringing intelligent agents and AI into the consumer realm. But arguably the most important thing that Amazon did was to turn its e-commerce application into a cloud computing platform on which the bulk of Silicon Valley startups operate; as the cloud model has matured, large, established enterprises have migrated to it as well. The lessons of this business transformation alone could fill a book (and will be the subject of a later chapter in this one).

Apple led the generational shift from the personal computer to the smartphone, and from the web to mobile apps. The iPhone is the platform where most cutting-edge applications are first launched. While Apple's flood of innovations seems to have slowed since the death of Steve Jobs, it remains a dominant player in the mobile market, and its design leadership continues to challenge us to "think different" about the possibilities of the future.

There are many other companies where WTF? technologies are being birthed and brought to market. Microsoft has been reinvigorated in recent years under the leadership of Satya Nadella, and its investments in AI and "cognitive services" for developers to use in their applications are bringing it into creative conflict with Facebook, Amazon, and Google. Chinese companies like Baidu, Tencent, and Alibaba are growing at the edge of our ken here in the United States, and may well be inventing futures that will overtake ours. And there is a host of startups, large and small, not to mention technologies that have yet to make it out of the labs or the dreams of their inventors.

Over the course of the next few chapters, we will see how lessons from each of these companies, and many others, overlap and come into focus as a map of the future.

In order to look for the common patterns, it is easiest to start with a map of one of the individual companies or technologies, draw out what key principles it demonstrates, and then tease out some of the common threads that tie it together with other WTF? technologies that delight, puzzle, or alarm us today. If we've drawn the map correctly, all of its components will show up in other companies that are building twenty-first-century services.

A BUSINESS MODEL MAP OF UBER AND LYFT

One company at the center of many emerging trends is Uber, a center it shares with Lyft, its biggest competitor in the United States; Didi Chuxing in China; and other on-demand car companies around the world.

Matt Cohler, an early Facebook employee turned venture capitalist who became an early investor in Uber, noted that the smartphone is becoming "a remote control for real life." Uber and Lyft drive home the notion that *the Internet is no longer just something that provides access to media content, but instead unlocks real-world services.*

Uber began as so many startups do, not as a transformative big idea but just with an entrepreneur "scratching his own itch." In 2008, Garrett Camp began to dream of a system for summoning limousines ("black cars") on demand. He had made it big with the sale of his startup, Stumbleupon. He'd bought a nice car, but he didn't like driving, and San Francisco's notoriously deficient taxi system made it difficult for him to get around.

Over the next two years, Camp developed the idea, recruiting his friend Travis Kalanick, another successful entrepreneur, as a thought partner in the project. Camp originally planned to run his own fleet of on-demand limousines, but Kalanick argued against it. "Garrett brought the classy and I brought the efficiency," Kalanick told Brad Stone in an interview. "We don't own cars and we don't hire drivers. We work with companies and individuals who do that. . . . I want to push a button and get a ride. That's what it's all about."

When Uber was launched in the summer of 2010, it reflected the needs of its already-wealthy founders: "Everyone's private driver." It seemed to be a small niche, hardly world-changing. The service was offered in San Francisco only. Yet over the next few years, Uber developed into a force that transformed the market for on-demand transportation, and today it has more drivers providing services than the entire previous taxi and limousine industry. How did this happen?

The game changer came early in 2012 when two companies, Sidecar and Lyft, introduced a peer-to-peer model in which ordinary people, not just licensed limousine drivers, provided the service using their personal cars. It was this further innovation that reshaped the way we think about employment, with drivers who not only have no guaranteed work from the company, but also make no guarantees to the company about whether they will work when they are needed. Instead, a swarm of drivers are summoned and managed by algorithms that match drivers and passengers in a real-time online marketplace, with surge pricing to bring more drivers into the market when the algorithm determines that there are not enough of them to meet demand.

There are many historical examples of peer-to-peer public transportation. Zimride, Logan Green and John Zimmer's predecessor to Lyft, was inspired by the informal jitney systems they observed in Zimbabwe. But using the smartphone to create a two-sided, real-time market in physical space was something profoundly new.

After initial skepticism, Uber copied the peer-to-peer model a year later. Driven by an aggressive CEO, a stronger technical focus on logistics and marketplace incentives, a take-no-prisoners corporate culture, and huge amounts of capital, it has spent billions to outpace its rivals. Lyft is still a strong contender in the United States, gaining, but in distant second place.

The amount of capital raised turned out to be surprisingly important. While the transportation network companies, or TNCs, as they are sometimes called, don't have to spend money buying cars, they have spent billions on marketing, subsidized fares, and driver incentives in a race to build the biggest network of customers and drivers.

Uber's willingness to sidestep regulators was also part of its success. Sidecar and Lyft spent time working with the California Public Utilities Commission to craft new rules to legitimize their novel approach. Even earlier, companies like Taxi Magic, founded in 2008, had simply worked within the existing taxicab industry and accepted its rules. Taxi Magic, which allowed you to summon a cab and pay with your smartphone, was integrated with existing taxi dispatch systems. And there the incentives to provide better service to customers were all wrong. The next available ride was offered to the driver who had been waiting the longest, not to the one who was closest to the passenger, and even then, during busy times, a driver might prefer to pick up people on the street. Cabulous, launched in 2009, also tried to work within the confines of the highly regulated taxicab industry.

In this regard, Camp and Kalanick's start with high-end black cars was fortuitous. Limousines have fewer regulations than taxis (for example, they are able to set their own prices rather than having them be set by regulators), but they have one big regulatory limit. Unlike taxis, which can pick up passengers who hail them on the street, limousines must be scheduled in advance. With an app, though, "in advance" becomes a relative term. Drivers who previously had to wait around for a call suddenly found new opportunity with the app, and were eager to sign up. The incentives of passengers and drivers were aligned, drawing both into what would become a thriving marketplace.

The taxicab companies recognized relatively early on that the new app made limousines more competitive with taxis, and claimed that Uber was an unlicensed cab company. The company's initial name, UberCab, gave fuel to the argument. But with the small concession of dropping "Cab" from the name (something they'd wanted to do anyway), Uber was able to convince regulators that they should still be covered by the rules of the limousine market rather than by those of taxicabs.

Once Uber added peer-to-peer service, it was game over for the taxicab industry, hobbled by its existing regulatory model, which controlled both the fares that could be charged and the number of people who could provide the service. Uber had become more than a service that made black cars competitive with taxis; it represented a whole new approach to urban transportation.

The ambition expressed in the Uber "origin story" on its website hints at the possibilities:

> What started as an app to request premium black cars in a few metropolitan areas is now changing the logistical fabric of cities around the world. Whether it's a ride, a sandwich, or a package, we use technology to give people what they want, when they want it.

For the women and men who drive with Uber, our app represents a flexible new way to earn money. For cities, we help strengthen local economies, improve access to transportation, and make streets safer. When you make transportation as reliable as running water, everyone benefits.

Here's a possible business model map for Uber or Lyft like the one Dan and Meredith Beam drew for Southwest Airlines.

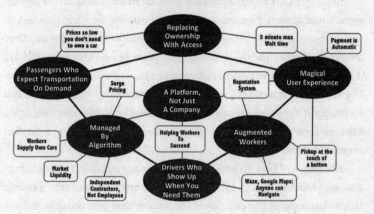

UBER
"Transportation as reliable as running water"

What are some of the core elements of this business model?

Replacing Ownership with Access. In the long run, Uber and Lyft are not competing with taxicab companies, but with car ownership. After all, if you can summon a car and driver at low cost via the touch of a button on your phone, why should you bother owning one at all, especially if you live in the city? Uber and Lyft do for car ownership what music services like Spotify did for music CDs, and Netflix and Amazon Prime did for DVDs. They are replacing ownership with access. "I tell people I live in LA like it's New York. Uber and Lyft are my public transit station," said one customer in Los Angeles.

Uber and Lyft also replace ownership with access for the companies themselves. Drivers provide their own cars, earning additional income from a resource they have already paid for that is often idle, or allowing them to help pay for a resource that they are then able to use in other parts of their lives. Meanwhile, Uber and Lyft avoid the capital expense of owning their own fleets of cars.

Passengers Who Expect Transportation On Demand. Much as Michael Schrage outlined in *Who Do You Want Your Customers to Become?*, Uber and Lyft are asking their consumers to become the kind of people who expect a car to be available as easily as they had previously come to expect access to online content. *They are asking them to redraw their map of how the world works.*

Uber and Lyft recognized early on that many young urban professionals had already given up on owning a car, but for their business to spread beyond major urban centers and wealthy demographics, they would need more people to accept this premise and make the switch. The reliability, convenience, and coverage offered by the application are not enough to achieve this ambition. That is what is behind Uber and Lyft's quest for ever-lower prices. Prices must be so low that calling an Uber or Lyft is not just vastly more convenient than owning a car, but more affordable as well.

The Magical User Experience of pulling out a phone, tapping a button, and having a car and driver find you a few minutes later, a needle in the haystack of the city, gives confidence that it is possible to have control and availability without ownership. *The WTF? moment of a brilliant new user experience is often the key to changing user behavior and turbocharging adoption.* While Lyft introduced a revolutionary part of the on-demand transportation model, Uber was the first to put it all together into a seamless experience, beautiful and easy to use.

Drivers Who Show Up When You Need Them. Transportation on demand for passengers requires a critical mass of drivers. Uber's original vision of black car drivers on demand served only a narrow slice of the potential market. As their ambitions grew, they needed a much larger supply of drivers, which the peer-to-peer model supplied.

Augmented Workers. GPS and automated dispatch technology inherently increase the supply of drivers because they make it possible for even part-time drivers to be successful at finding passengers and navigating to out-of-the-way locations. There was formerly an experience premium, whereby experienced taxi and limousine drivers knew the best way to reach a given destination or to avoid traffic. Now anyone equipped with a smartphone and the right applications has that same ability. "The Knowledge," the test required to become a London taxi driver, is famously one of the most difficult exams in the world. The Knowledge is no longer required; it has been outsourced to an app. An Uber or Lyft driver is thus an "augmented worker."

A Platform, Not Just a Company. A traditional business that wants to grow must hire people, invest in plants and equipment, and build out a management hierarchy. Instead, Uber and Lyft have created digital platforms to manage and deploy hundreds of thousands of independent drivers, trusting the marketplace itself to ensure that enough of them show up to work and bring their own equipment with them. (Imagine for a moment that Walmart or McDonald's didn't schedule their workers, but simply offered work, trusted enough people to show up, and offered higher wages when there weren't enough workers to meet demand.) This is a radically different kind of corporate organization.

There are those who argue that Uber and Lyft are simply trying to avoid paying benefits by keeping their workers as independent contractors rather than as employees. It isn't that simple. Yes, it does save them money, but independent-contractor status is also important to the scalability and flexibility of the model. Unlike taxis, which must be on the road full-time to earn enough to cover the driver's daily rental fee, the Uber and Lyft model allows many more drivers to work part-time (and to take passenger requests simultaneously from both services), leading to an ebb and flow of supply that more naturally matches demand. More drivers means better availability for customers, shorter wait times, and far better geographic coverage. These companies are able to provide a five-minute response time over a far larger geographical area than traditional taxi and limousine companies.

Management by Algorithm is central to Uber and Lyft's business. It would be impossible to marshal the workers, connect drivers and passengers in real time, automatically track and bill every ride, or provide quality control by letting the passengers rate their drivers, without the use of powerful computer algorithms. Creating and deploying these algorithms is the core of what the company does.

Every passenger is required to rate their driver after each trip; drivers also rate passengers. Drivers whose ratings fall below a certain level are dropped from the service. This can be a brutal management regime, but as political scientist Margaret Levi noted to me, from the point of view of passengers, the real-time reputation system acts as a kind of "private regulation" that outperforms traditional municipal taxi regulation in enforcing high standards of safety and customer experience.

Having enough drivers to meet demand is a marketplace management problem. Achieving "market liquidity"—enough drivers to ensure that passengers get picked up within only a few minutes, and enough passengers that drivers are willing to show up to work without being on the payroll—is a complex problem.

Unlike the taxi industry, which creates an artificial scarcity by issuing a limited number of "medallions," Uber and Lyft use market mechanisms to find the optimum number of drivers, with an algorithm that raises prices if there are not enough drivers on the road in a particular location or at a particular time. While customers initially complained, using market forces to balance the competing desires of buyers and sellers has helped Uber and Lyft to achieve an equilibrium of supply and demand in close to real time.

There are other signals in addition to surge pricing that Uber and Lyft use to tell drivers that more (or fewer) of them are needed. Incentives to drivers, especially when entering new cities, has been one reason why Uber and Lyft have had to spend so much money to enter new markets. There are those who equate this behavior to dumping— selling goods and services below cost in order to dominate the market and drive out other sellers, only to raise prices once a monopoly position is earned. They argue that raising prices or cutting the driver's share of earnings is the only way these companies will ever make money.

But from the point of view of Uber and Lyft, driving down cost is a way to grow the market to the point where the critical mass of buyers and sellers becomes self-sustaining, which will lower the cost of customer and driver acquisition. As prices get lower, new demand opens up. How many people could afford a car on demand when they were a luxury of the superrich? How many people take for granted today that they can have a car waiting whenever they need it? How many might take this for granted in a future where the cost of these services continues to come down?

The biggest strategic question in my mind is how Uber and Lyft deal with the problem of driver turnover. Are the wages and working conditions sufficient to achieve a steady-state supply of drivers or are they simply burning through a limited supply of people who try working for the service and then find other, better work? What happens when they stop providing incentives beyond the fare for drivers to sign up, to work more trips each week, or to commit to working only for one platform rather than offering their services through both? Drivers are already complaining that they are being bankrupted by fare cuts and diminished incentives.

The outcome of the contention between these platforms and labor regulators about working conditions and employment status could also play a decisive role in the success or failure of these platforms, as could disputes with traditional taxi and limousine licensing regimes, because labor regulators could, without understanding how all the pieces of the model fit together, place restrictions on these services that make it impossible for their business model to work.

• • •

One of the most important functions of a business model map is that it helps you understand how all of the pieces of a business fit together. Many taxi companies, late to the game, are now introducing apps that superficially have many of the same features as Lyft and Uber. But they are often unable to meet the expectations for price and availability that Uber and Lyft have established because they don't have a liquid marketplace. They offer a fixed number of cars and drivers, limited by city taxi licenses (medallions) and associated costs, so the supply of cars is inevitably less than is needed at times of peak demand. If they had enough cars for quick, reliable pickup at the busiest times, those cars would inevitably be idle at other times. It is no accident that a majority of Uber and Lyft drivers are part-time; it is one of the intrinsic advantages of the model that supply rises to meet demand and slacks off when demand is less.

The regulatory friction of the traditional approach makes taxi costs higher and availability worse. Uber and Lyft drivers routinely make more money per hour than taxi drivers; meanwhile, customers generally have better experiences and lower prices. Those who complain that Uber and Lyft "aren't following the rules" need to ask whether those rules are achieving their intended objective.

That doesn't mean that Lyft and Uber should get a free pass on providing employee benefits and labor protections. As we'll see in Chapter 9, the right answer is to develop a social safety net and regulatory frameworks as flexible and responsive as the on-demand business model itself. Uber and Lyft (and Airbnb) have taken the approach of asking for forgiveness rather than permission for many of their innovations, relying on swift consumer adoption to give them allies against regulators. There's no question, though, that some kind of accommodation with regulators is in the future for all of these companies. They would be wise to get ahead of the problem with regulatory proposals as innovative as their business model.

MAKING STRATEGIC CHOICES

> You can tell if a business model map is good if it helps a
> company to make sound strategic choices. That is, it frames
> the problem in such a way that a company can make conscious
> choices about what's important, rather than discovering too late
> that it broke a key part of what had made it successful.

For example, Uber and Lyft have made much of their plans to incorporate self-driving cars into their future. With a shallow understanding of their business, you might quickly conclude that the reason to do this is that eliminating the 70–80% of the fare that is paid out to drivers will make these businesses more profitable.

With the aid of the business model map I outlined above, you'd ask different questions. If the company currently depends on a liquid marketplace of drivers who bring their own cars and work only when they believe they can make a decent wage, what happens when the platforms introduce self-driving cars into the mix? They potentially destabilize their own marketplace.

There will be significant costs to achieve the kind of availability for passengers that Uber or Lyft currently have using centrally owned self-driving cars. Remember that the total number of cars in the system must be sufficient to satisfy peak demand.

If the company itself owns the self-driving cars, and uses them to compete with its human drivers for the busiest, most lucrative times, it risks making them less willing to participate. If the goal is truly to make transportation as reliable as running water or electricity, rather than simply to maximize company profit, these companies should deploy self-driving cars not to compete with their drivers but to supplement them, providing services in areas that are currently not well served, even though those cars might be utilized less often. More likely, the right answer will be to tune their mathematical models and algorithms to find the optimum mix of human and machine, in the same way that the electrical grid relies on coal, natural gas, or nuclear power for "base load" while meeting peak daytime demand with renewables.

In order to maintain the benefits of the marketplace model, rather than deploying self-driving cars itself, Uber or Lyft might instead create incentives for its drivers to purchase them and make them available to the company. In many ways, this would change their business model to one closer to that of Airbnb, in which the participants in the marketplace provide an asset they own rather than their labor. But for this plan to work, Uber or Lyft would not need to develop their own autonomous vehicles, but instead could promote interoperability between different autonomous vehicle vendors. If the plan is something like "Buy your autonomous Tesla, drive it to work, and then let us use it for the rest of the day," it would imply a mixed fleet of vehicles, requiring investments in interoperable control and dispatch. (Tesla seems to have other plans, though, forbidding their drivers from using their cars for Uber and Lyft, with the intention of rolling out its own competing service. A business model does not exist in isolation; it must adapt to the competition as well as to the needs of its customers and suppliers.)

This discussion is also important for policy makers. A world of interoperable self-driving cars would provide an opportunity for current on-demand drivers—or, in a future world of self-driving trucks, independent truckers—to participate in the marketplace as owner-operators; a world in which a company like Tesla is able to limit the ability of its car owners to drive for any competing service reduces the drivers to a real-world version of what author Nicholas Carr has called "digital sharecroppers." Ensuring interoperability of self-driving cars is as important as was the original interoperability that drove the Internet revolution. Open standards in this area will help ordinary people, not just big companies, to reap the benefits of the next wave of automation.

Betsy Masiello, who works in public policy at Uber, responded to my questions on how the peer-to-peer model might mix with autonomous vehicles by saying that right now, people think of Uber as a replacement for taxis; perhaps instead, it will end up closer to peer-to-peer fractional car rental. It is likely that the reality will be a mix of both.

Finally, if the augmented worker is indeed central to Uber and Lyft's business model, perhaps the right way to think about self-driving cars is as a further augmentation, enabling new kinds of services. Once driving itself is a commodity fulfilled cheaply by machines, what will the humans who are now augmented by that capability find to do with their new superpower? What possibilities are there for society when transportation is as cheap and reliable as running water?

A BUSINESS MODEL MAP FOR THE NEXT ECONOMY

When building a map that tries to capture the essence of multiple companies, it's important to recognize that neat categorization is a fool's errand. For example, like Uber and Lyft, Airbnb is a networked marketplace, but it's a network of apartments, homes, and rooms, and only secondarily the network of workers who clean up after a guest has come and gone. Google and Facebook are networks of people who produce, share, and consume content, and the advertisers who want to reach them. The iPhone or Google Play app store enhances a physical device with a network of apps and the ecosystem of developers who create them.

Distributed solar power generation, electric vehicles, and other signs of the transition from carbon-based energy to renewables seem like they are a bit out of the frame—or are they? After all, rooftop solar shares many of the distributed network characteristics of on-demand companies like Uber and Airbnb.

> When making sense of the future, think in terms of gravitational cores, not hard boundaries. Just as the sun's gravity well reaches out beyond the orbit of Pluto and encompasses not just the planets in the ecliptic but comets and planetoids with eccentric orbits, so too the forces shaping the future all have a gravitational core and a gradually attenuating influence. And just as the solar system has multiple gravitational subsystems, where the draw of the local giant keeps its own satellites in tow while all still partake in the larger dance, these interpenetrating trends influence each other and converge.

With that in mind, let's consider a generalized version of the Uber/Lyft map, which I've labeled below as the "Business Model Map for the Next Economy." I've intentionally left some of the boxes for contributing

Business Model Map for the Next Economy

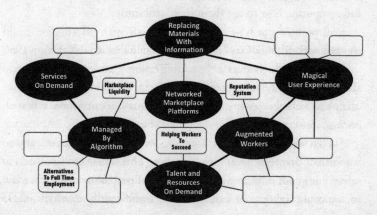

factors blank, so that you can think about how you might fill them in for your own company, or for the services you consume. I've left a few of them filled in, because they seem central to the futures that are unfolding.

Replacing Materials with Information. Giving physical assets a digital footprint allows them to be managed much more like information assets. I picked up this notion from Liam Casey, an Irish expat who runs PCH International, a design and logistics company for the electronics industry, from offices in Shenzhen, San Francisco, and Ireland. PCH provides direct-to-consumer just-in-time manufacturing for consumer electronics vendors, allowing them to take orders online that are shipped directly from China.

"We replace inventory with information," Liam said to me, showing me how his company's inventory for US customers, where he has real-time data systems, is far less than it is for Australia, where he is reliant on keeping inventory in local warehouses, even though the Australian market is orders of magnitude smaller.

I heard a related comment from my son-in-law Saul Griffith, an inventor whose company, Otherlab, develops innovative technology approaches under contracts from DARPA (Defense Advanced Research Projects Agency), NASA, the National Institutes of Health (NIH), and the US Department of Energy. "We replace materials with math," Saul told me, describing how the essence of many of his projects is computational design, understanding the nature of shapes and materials and computing how to use them more efficiently.

Saul pointed out how replacing materials with information could play out with autonomous cars. Most cars doubled in weight from 1960 to 2010. We made them safer by adding crumple zones and airbags and all sorts of clever features useful in accidents. We did not make spectacular gains in fuel economy even though engines became more efficient, because most cars got fatter—larger and heavier.

"What would happen," he asked, "if we made cars so smart, and so automated, that they never hit each other? This is biology's approach to safety. Jump out of the way or avoid the collision altogether. If we did so, we could lighten our cars up again, significantly so, which would have positive benefits on making it easier to electrify them. We could easily reduce the energy consumed in the transportation sector by two-thirds or more."

"Replacing materials with information" is a more powerful formulation than "replacing ownership with access." Yes, there is continuity between subscription models for on-demand media access and what is happening with services like Uber and Airbnb, but stating the principle more broadly allows us to understand even more of the modern world.

> When you hear a new concept like this, succinctly stated, add it to your mental toolbox. Try it on as a way of seeing the world around you. How does it help you think differently?

Might this principle even reverse the logic of labor globalization? In a recent paper, economists Laura Tyson and Michael Spence point out that for the last several decades, the logic of globalization was that manufacturing moved toward the lowest-cost sources of labor. But now, they note, as "digital-capital-intensive technologies are substituting for humans in the routine labor-intensive parts of manufacturing supply chains . . . and digital technologies make manufacturing mobile with little or no cost penalty, physical manufacturing activity will move toward market demand rather than toward labor, because there are efficiencies to be gained from proximity to the market."

Networked Marketplace Platforms. Not just Uber and Lyft, but Google, Facebook, Amazon, YouTube, Twitter, Snap, Baidu, Tencent, and Apple draw a great deal of their strength from the fact that they are networked marketplace platforms, managed by algorithm. As we'll discuss in Chapter 5, they have some fundamental differences from the twentieth-century organizations with which they compete.

Might networks and technology platforms provide a new form of organization that trumps old corporate forms, replacing them with something even more powerful?

On Demand. It's easy to put a platform like TaskRabbit, whose app allows consumers to hire occasional labor such as movers, house cleaners, or gardeners at the touch of a button, into the same map as

Uber and Lyft. Even Upwork, which lets you tap into a global marketplace of professional programmers, designers, and other skilled workers for short-term "gigs," is a clear fit. For many observers of the next economy, the map begins and ends there. But what can we call Amazon but an on-demand company, when increasingly its products are delivered same-day (sometimes even by a network of on-demand drivers in their own cars rather than by traditional package delivery companies)? What does on-demand become when companies around the world are experimenting with delivery by autonomous drone, and when Amazon's automated warehouses require only a minute of human labor per package, with most of the work being done by a complex ballet of software and machines?

On-demand delivery is an example of how WTF? services introduced by tech companies become, like Tom Stoppard's unicorn, "as thin as reality," the stuff of ordinary life. On demand is becoming a universal consumer expectation. Amazon offered fast, "free" shipping, and now it's hard for any big retailer to compete without offering the same.

Note that there are two ovals in the diagram representing on demand: *services on demand* and *talent and resources on demand*. On demand affects both sides of the networked marketplace.

Managed by Algorithm. The algorithms at the heart of a company like Uber or Lyft are computationally intensive, just like the algorithms at the heart of search engines, social networks, and financial markets. In many cases, the company with the best math wins. The cutting edge of smart algorithms is, of course, artificial intelligence, but AI is on a continuum with many other algorithmic systems, increasingly automated, that we already rely on in the modern world.

Understanding how algorithmic systems shape our society is a central theme of this book. To have a chance at making a better future for ourselves and our children, we must grasp not only how the nature of these algorithms is changing, but also why the algorithms we have most to fear may not be those of artificial intelligence but the unexamined algorithms that rule our economy. We will take up this question in Part III of this book.

Augmented Workers. The wonders of the first industrial revolution were brought about by workers partnered with new kinds of machines. Could we build skyscrapers or fly through the air or feed seven billion people without machines that make us stronger, faster, and more powerful? So too it is with today's technology. If it is being deployed correctly, it should allow us to do things that were previously impossible.

The amount of augmentation may vary. A service like TaskRabbit augments workers' ability to find customers, but not to do the job. Uber and Lyft drivers have additional augmentation in their ability to navigate and find clients. Surgeons and oncologists might be working in traditional organizations but are cognitively augmented workers, with "senses" that were not available to their forebears; so too, with the advent of augmented reality, will be building inspectors, architects, and factory workers.

> To make the future economy better than the present, find new ways to augment workers, giving them new skills and access to new opportunities. As we automate something that humans used to do, how can we augment them so that they can do something newly valuable?

The idea that Uber teaches us that augmenting workers and helping them to succeed is an essential feature of companies looking to prosper in the next economy might create some cognitive dissonance for readers who have read about Uber's abrasive, driven, former CEO, Travis Kalanick. In early 2017, Uber was rocked by a viral video that showed Kalanick berating a driver who told him that he had gone bankrupt because of Uber's falling prices. "Some people don't like to take responsibility for their own shit. They blame everything in their life on somebody else," Kalanick burst out, in clear echo of Ayn Rand's philosophy of unfettered self-interest. This is not the behavior of someone who places a high value on the humans who make the business work.

> When trying to map the future, remember that the territory is
> not an idealized landscape, but a real one, full of contradiction.
> The people who are creating the future are complex, each with a
> mix of brilliance and flaws. They see some things we don't, and
> are blind to others.

Just as I predicted in 1998 that Microsoft would one day embrace open source software (they did), I predict that Uber will one day come to realize that people are a critical component of what it has built, and make supporting them central to its competitive strategy. Lyft already knows this and uses it to its advantage.

It is also important to understand our role as customers in holding companies accountable for creating the future we want. Every time a PR crisis erupts, some of Uber's customers desert it for Lyft, but most stick around. If you want a human-centered future, support companies that demonstrate human-centered values.

Magical User Experiences. The magic goes away, we know, but its presence is a sure test of impact. That WTF? moment tells us that we are looking through a door into the future. But who opened the door, and how?

Steve Jobs, who was a master at throwing that door wide open, said, "When you grow up you tend to get told that the world is the way it is. . . . Life can be much broader once you discover one simple fact: Everything around you that you call life was made up by people that were no smarter than you. And you can change it, you can influence it. . . . Once you learn that, you'll never be the same again."

4

THERE ISN'T JUST ONE FUTURE

THE FUTURE SEEMS OBVIOUS IN RETROSPECT. HOW COULD we have missed it?

There are constant reminders in the news. Almost every time there is a terrorist attack or mass shooting, we hear that police and intelligence services received prior warning, that there was someone trying to pierce the fog of bureaucracy to report their fears about the suspect. "Were officials blind and deaf?" we ask ourselves. We forget that before an event occurs, there is a shifting complex of possibilities, any of which *might* happen. For every potential threat that is reported, hundreds never come to pass.

Once an event occurs, all those possibilities collapse into the one reality that we call the present, and then, in an instant, the past. But even the past, seemingly fixed as it appears, is an illusion constantly updated by new knowledge from the present.

This is as true in technology as it is in national security.

Back in 2000, I received an appeal from Richard Stallman. He was concerned about Amazon's 1-Click e-commerce patent, and the fact that Amazon had just sued rival Barnes & Noble for adding a similar feature to its site, barnesandnoble.com. Richard urged me, as one of Amazon's top publishers, to boycott its service. "Have you tried to talk with Jeff?" I asked.

He hadn't. So I wrote an email to Jeff Bezos (whom at that time I'd never met), asking him to reconsider:

SUBJECT: Amazon 1-Click patent
DATE: Wed, 05 Jan 2000 10:03:59–0800
FROM: Tim O'Reilly
TO: Jeff Bezos

I wanted to give you guys the heads up that I'm getting a lot of pressure from my customers (via my Ask Tim column on our website and direct customer e-mail) to comment publically [*sic*] on the Amazon 1-Click patent. I was also approached by Richard Stallman to help him publicize his Amazon boycott, and I declined, but I do want to let you know that I agree with his message although not with his methods. I will be forced to make some kind of public comment shortly, and I wanted to let you know what the substance of it will be before it goes out to the world.

First off, I think that you are reaping a harvest of ill-will with the technical community. While I know you are setting your sights on a wider consumer audience, the serious technical community represents the core of your early adopters and many of your best customers, especially in the book market. . . . And I can tell you that those customers are solidly against software patents.

Second (and this is the point most important to me), the web has grown so rapidly because it has been an open platform for experimentation and innovation. It broke us loose from the single-vendor stranglehold that Microsoft has had on much of the software industry, and created a new paradigm with opportunities for countless new players, including Amazon. The technologies that you have used to launch your amazing success would never have become widespread if the early web players, from Tim Berners-Lee on, had acted as you have acted in filing and enforcing this patent. Because, of course, you are not the only one who can play the patent game. And once the web becomes fenced in by competing patents and other attempts to make this glorious open playing field into a proprietary wasteland, the springs of further innovation will dry up. In short, I think you're pissing in the well.

Patents such as yours are the first step in vitiating the web, in raising the barriers to entry not just for your competitors, but for the technological innovators who might otherwise come up with great new ideas that you could put to use in your own business. It's a well known technology truism that all of the smart people don't work for you, and that one of the surest ways to success is to get more ideas and more work out of people outside your own fences. . . .

You've gained enormous competitive advantage by making use of technologies that were freely given to the world. If players like yourselves succeed in replacing that gift economy with a dog-eat-dog world in which everyone tries to keep their advances to themselves, and worse, tries to keep others from replicating them, you'll soon find yourself either spending a larger and larger part of your budget on developing your own technology, or, more likely, you'll find yourself hostage again to commercial software vendors whose interests may not be aligned with your own.

If you see yourselves primarily as a technology company, you might want to play the Microsoft game of trying to corner the technology market with proprietary APIs, file formats, and patents, but if you see yourself as a great customer service and marketing company, you want other people inventing technology platforms that you can build on. That's been a key part of your success so far: You've been able to take a great open platform, and build vertical applications that provide a fabulous service to your customers. Filing frivolous patents will only retard the growth of the platform.

And that's a third point: The patent is very unlikely to be upheld in the long run. It's a classic example of the kind of software patent that would never be granted if the patent office had even the slightest clue about software: A trivial application of cookies. I'd be very surprised if there isn't a fair amount of prior art even in using cookies in conjunction with saved credit card information. But even if there isn't, the basic method of saving state information about prior visitors is so fundamental that there's nothing new in what you did.

Finally, I want to say that I admire you guys tremendously. I speak and write constantly about Amazon as the paradigmatic example of "the next generation of computer applications." I think that you're a terrific competitor, delivering a terrific service, and I don't think you need to use tools like this patent to keep yourselves on top. You can win without it, and I firmly believe that in the long run, it will do you more harm than good.

I realize that having come out so strongly behind this patent, it would be very difficult for you to do an about-face and back off from it. However, I urge you to do so. . . .

As I've suggested publically [sic] on more than one occasion, I believe that the companies that have profited most from the web have an obligation to give something back. This is more than a "thank you" to the developers who made your success possible; it's also an act of self-interest, to keep the innovations coming.

After a few days, Jeff replied with a polite brush-off. At that point, I decided to go public with the issue, and published my email to Jeff along with an open letter that I asked my customers and other interested parties to sign. Within less than two days, I had 10,000 signatures, and a phone call from Jeff.

Jeff argued that the patent was valid, that Barnes & Noble—at the time, the biggest, most predatory force in the bookselling business—was copying every move that Amazon made, and that his legal counterattack was necessary for Amazon's survival. But he agreed that my arguments had merit, that open innovation was better than patent warfare. Amazon needed to file patents to protect itself, he said, but would constrain itself in the future to using them defensively—that is, in response to a patent lawsuit or threat of lawsuit from some other party.

Then, in a brilliant move of PR jujitsu, Jeff suggested that we go together to Washington, DC, to lobby for patent reform. We did that, and eventually we invested together in a startup, called BountyQuest, that promised to help surface "prior art"—that is, previously available technology that, if known to the patent office, would lead them to deny a patent application or to require that the patent applicant make clear how its application demonstrates an innovation not shown in that prior art. (BountyQuest was itself a great example of prior art for later innovations like Kickstarter, since it was one of the first examples of Internet-enabled "crowdsourcing"—even though the term itself wasn't introduced till six years later.)

With Jeff's support, BountyQuest made 1-Click the subject of its first quest for prior art. What happened next surprised me and everyone else who'd assumed that the supposed invention was completely obvious. Despite posting a bounty of $10,000, we were unable to find any previous piece of software that implemented anything as simple as Amazon's 1-Click Buy button. We did award the bounty for several potentially useful pieces of prior art, but there was no "smoking gun." The 1-Click patent was actually quite original.

What was going on here? Almost everyone in the computer industry argued that a 1-Click Buy button backed by stored credit card credentials was completely obvious. If so, why had no one done it before?

In our conversation a few days after I published my open letter, Jeff explained to me why he thought 1-Click was original enough to patent.

It had nothing to do with the implementation, which he admitted was fairly trivial to duplicate, but with the reframing of the problem. At the time he came up with 1-Click, everyone was locked in to the shopping cart metaphor, because that's what you do in the real world. You pick up an item and take it to the counter to buy it. On the web, he realized, something very different was possible: All you had to do was point to an article, and it was yours.

Phrased that way, rather than in the convoluted language of the patent, I could see that Amazon was not just cynically abusing the patent system. Jeff was claiming in good faith that he had made a seemingly small but nonetheless significant innovation.

Jeff argued further that you can't just attribute this patent, as many did, to patent office incompetence. Barnes & Noble had a chance to present prior art in court, he noted, and after a review of all the evidence B&N was able to dig up, the judge granted a preliminary injunction. Jeff felt that this was fairly strong evidence (coupled with the positive press coverage when 1-Click was introduced) that the feature was a genuine innovation.

In short, the invention was obvious only in retrospect. When Amazon launched 1-Click, we rewrote our mental map of the past to make the present state of things seem inevitable. This is the corollary to the power of redrawing the map.

> When you draw a new map successfully enough, you change the perception not only of the future but of the past. That thing that seemed unthinkable becomes the fabric of the everyday, and it's hard to remember that it once was only one of many possibilities.

We've seen other, more recent examples of this kind of creative rethinking of what is possible, which then becomes "obvious." When Garrett Camp and Travis Kalanick first conceived of Uber, the notion that you could summon a car on demand was lying latent in the field of possibilities, unexplored. All the capabilities were in place. There were already hundreds of millions of smartphones equipped with sensors able to track the location of both drivers and passengers. And there were even connected taxicabs. But all that the traditional taxi companies did with their connectivity was to put a credit card reader in the back of the taxi, and a small screen for broadcasting content and ads.

In fact, one Internet entrepreneur had thought of Camp and Kalanick's idea long before they did. Sunil Paul's patent for System and Method for Determining an Efficient Transportation Route, filed in 2000 and issued in 2002, is eerily prescient. It describes almost perfectly many of the features of modern on-demand ride-hailing systems. But all the pieces of the puzzle weren't yet on the table to build what Sunil imagined.

"I had this idea that the smartphone would replace cars," Sunil told me. "You wouldn't need a car anymore because you could coordinate your transportation through your smartphone." Early smartphones existed at that time, mostly in Europe, but they were far from ubiquitous. "In 1999, I tried starting a company around it," Sunil continued. "But after about two months, I concluded it was too early. I didn't think the technology was ready. There wasn't that much demand yet." The company was never formed, but the project name was VCar, for "Virtual Car."

It's not entirely clear what Sunil hoped to build. The patent is extraordinarily broad, covering every possible application of positional pickup and routing. It covered not only the Uber use case, but also how a phone could be used to manage fractional car rental services (a category that had already been pioneered by Mobility in Switzerland, founded in 1997, and Zipcar, founded in 1999 by Robin Chase and Antje Danielson in the United States), or to offer subscriber access to cars or even fractional use of cars owned collectively by multiple drivers.

The broad scope of Sunil's patent highlights that, as science fiction writer Frank Herbert once told me, "Ideas are a dime a dozen. It's implementation that matters." *The future isn't just imagined. It is built.* Garrett Camp and Travis Kalanick created a successful service that implemented the idea and found the market for it.

Even after Uber was launched, though, the rethinking of what was possible was far from over.

Sunil reports that it was Airbnb, which, starting in 2007, allowed people to rent out rooms in their homes, or their entire home when they went away, that inspired him to think about peer-to-peer car sharing and to return to the work that he'd begun in 1999. Getaround, founded in 2009 and incubated in a Singularity University class led by Sunil, began offering the peer-to-peer equivalent to car rental, not to taxi service. It was an updated version of Chase and Danielson's Zipcar.

In 2007 Logan Green and John Zimmer had founded a peer-to-peer service called Zimride, which was focused on matching drivers and passengers for long intercity rides. In 2012, Sunil's work inspired them to launch a new service, called Lyft, which offered the first public peer-to-peer ride-sharing service for local pickup not by professional drivers, but by "your friend with a car." Sunil, late to the party despite being way early, launched Sidecar at about the same time. (It was still in private beta when Lyft launched publicly.) But by the time Sidecar went out to raise money, Uber and Lyft had already built huge venture capital war chests, and Sidecar was unable to compete in a capital-intensive business. It went out of business at the end of 2015.

Uber responded to Lyft with UberX, and the ride-sharing landscape as we know it today was born. Lyft has continued to innovate, with Lyft Line (which Uber matched as UberPool), consistent with Zimmer and Green's original vision to create a modern version of the peer-to-peer public transportation network similar to the one they'd seen during youthful travels in Zimbabwe, and which had inspired them to create first Zimride and then Lyft.

THINKING THE UNTHINKABLE

Camp and Kalanick also realized a key payment innovation that went even beyond Amazon's 1-Click shopping: They realized that in a world infused with connected sensors, the very act of consuming a service could trigger payment. An app such as Uber knows when the ride begins and when it ends, calculates the cost in real time, and charges the stored credit card as soon as the ride is over. This innovation has still not entirely been grasped by others who could put it to use.

In 2014, more than five years after the launch of Uber, the Apple Pay announcement demonstrated that even leading-edge companies were still stuck in the old model. The Apple Pay web page gushed: "Gone are the days of searching for your wallet. The wasted moments finding the right card. The swiping and waiting. Now payments happen with a single touch."

What's wrong with this picture? It's describing the digital facsimile of a process that was already on its way to becoming obsolete.

> Truly disruptive new services don't just digitize the familiar. They do away with it.

I never search for my wallet when I take an Uber or Lyft. I never search for my wallet when I buy something from Amazon. I don't even search for my wallet when buying a song from iTunes—or, for that matter, if I'm buying an iPhone from an Apple Store. In each of these cases, my payment information is simply a stored credential that is already associated with my identity. And that identity is increasingly recognized by means other than an explicit payment process.

In the case of Uber, I summoned the car. The driver already knows my name and my face, and our phones are traveling in tandem. Uber knows what I owe based on GPS. And it charges me automatically. I "pay" simply by reaching my destination and getting out of the car. That is the future of payment, not "hold[ing] your iPhone near the contactless reader with your finger on Touch ID."

So, in a sense, Apple Pay was payments for everyone who hadn't caught up with the fact that truly disruptive services had already done away with the old payment model.

Amazon has continued to push forward with the future of payment. In late 2016, the company announced that it is developing convenience stores powered by Amazon Go and "Just Walk Out Shopping." Simply enable the Amazon Go app, and machine vision and other algorithmic systems keep track of what you take off the shelf, and automatically debit your account.

I had proposed something like this myself in 2009 or so, in a "Web

2.0" brainstorming session with Russ Daniels, chief technology officer for cloud computing strategy at Hewlett-Packard. HP was trying to figure out how to do something distinctive in the cloud computing business. I knew that HP had once owned Verifone, the point-of-sale payment equipment vendor, and that led me to suggest that the future would include smart shopping carts that could do paymentless checkout.

Foursquare had just been launched, and its magical ability to detect where you were and offer a location for "check-in" made me think that it could also be used for "checkout." A participating merchant could recognize you as a customer, pulling up your stored payment credentials. As for the products you wanted to buy, I was thinking about the possibility of bar code readers in the cart, or possibly sensors that knew the exact location of each product in the store, or identified it by weight when you put it in the cart. Computer vision wasn't yet at the point where it could reliably work the kind of magic Amazon is now practicing.

Sometimes ideas are in the air, but the technology to make them a reality hasn't yet arrived.

I've had numerous other experiences like that. One of my earliest business ideas, back in 1981, was for an interactive hotel brochure using the new RCA LaserDisc player. It would let you see the rooms in the hotel, and even the view from each room. I penned a proposal with a friend who sold video services to companies, and we pitched it to one of the hotel chains, but it never went anywhere. The idea was too far ahead of its time.

And even when the technology is there, as it was with the early World Wide Web, it's easy to apply it only in a limited way that is framed by the problem that you are familiar with. This is why the future proceeds in fits and starts, with each inventor using the idea of another as a stepping-stone to move a little further forward.

When I first had the idea that advertising could be the business model for GNN, the first web portal, which we launched in the summer of 1993, my thinking was shaped by the kind of direct-response advertising that I used extensively in my book publishing business. I still remember the copy of *Computerworld* that lay on my desk at the moment, inspiring the idea. At the time, business publications had the paper equivalent of the web hyperlink that you could follow to get more information. It was called "the bingo card." Each direct-response ad—for that attractive hotel in the Caribbean, for that new electronic device, or for O'Reilly books—had a number associated with it. In the center of the magazine was a prepaid mailing card with a matrix of numbers, like a massive multiple-choice exam. You filled in the numbers associated with the advertisements you wanted to get more information about, and the advertiser sent you a catalog in the mail, or for an expensive product, perhaps called you directly.

Web hyperlinks, I reasoned, could get rid of all the catalogs and brochures sent through the mail. Every company would one day have its own commercial website to provide information about its products. What is a company's website but an ad for its brand, its products, and its services? When we first proposed running ads on the World Wide Web, I saw ads as a specialized kind of information product, not as the kind of intrusive media bombardment they became.

When we launched GNN in the summer of 1993, the Internet was still a research network under the oversight of the National Science Foundation. Debates about whether and how to commercialize it were carried out on an online mailing list called com-priv (Commercialization and Privatization). I remember talking informally at a conference with Steve Wolff, at the time the NSF administrator in charge of the Internet, about whether our proposed idea would violate the NSF's Acceptable Use Policy (AUP), and I will always treasure his response. "Well, the Internet is supposed to be for research and education. And if anyone is about research and education, you guys are. So go for it."

Of course, we were still so early in the web that few of our potential customers had websites, so the ads we created were actually listings in our own online catalog, the commercial listings section of GNN. The closest analogue was to a Yellow Pages phone book. Our first web advertiser was our law firm at the time, Heller Ehrman White & McAuliffe. Our lawyer, Dan Appelman, whom I'd hired because he was one of the first lawyers to use email, wrote a check for $5,000 (which I considered keeping as a souvenir rather than cashing, but we needed every penny), and in return we produced a web page listing the services of the firm, how to reach them, and so on.

Customers weren't ready. Advertise? On the Internet? In 1994 we did the first ever survey of Internet users, calling 50,000 people to collect income and demographics. But even so, we didn't think about advertising as *Hotwired*, the online version of *Wired* magazine, introduced it in October 2004. Banner ads summoning people to visit other websites should have been obvious. But they weren't.

At that time, we were insanely focused on getting people to take the web seriously, to put up their own websites, and to keep up with the flood of new ones that we were listing in GNN's catalog. We were also trying to get more people to give the Internet a try. Together with Spry, a small software company, we'd just launched a product called Internet in a Box, which was designed to make it easy to get consumers onto the Internet. It included a copy of all the software you needed to get onto the net, with GNN as the easy-to-use front end and with a copy of Ed Krol's *Whole Internet User's Guide & Catalog* as the user manual.

Dale and I also reached out to all the phone companies to get them to provide Internet access with the Global Network Navigator as a front end. It seemed completely obvious to us that the Internet was a great offering for phone companies. They already had connectivity to people's homes. They already had a billing relationship with the consumer. But they wouldn't listen. People are comfortable with what they're doing, and they don't see the future coming at them.

But even we were blind. We saw the future of direct-response advertising, but not only did we not push further and reimagine display advertising, we didn't really imagine e-commerce when we launched GNN. The web was still a collection of static pages. The Common Gateway Interface (CGI), Rob McCool's hack to let the web talk to a back-end database, a key enabling technology for e-commerce, wasn't released till the end of 1993. eBay and Amazon were founded two years later.

This is not uncommon. Steve Jobs was originally opposed to the idea of third-party apps on the iPhone. Travis Kalanick was for a long time skeptical of the peer-to-peer model. After all, it was illegal for drivers to provide "livery services" without a license. It was Sunil Paul's efforts to get the California Public Utilities Commission to accept the model that made it thinkable. Lyft jumped on the opportunity. Uber eventually followed.

A more recent demonstration of how old thinking holds back even smart entrepreneurs is how long it took for the Amazon Echo to arrive, given that speech recognition has been a feature of smartphones since the 2011 launch of Apple's Siri intelligent agent. Yet it was Amazon's Alexa, not Siri or Google, that brought a seemingly minor change that made all the difference: Alexa was the first smart agent always listening to your commands without the need to first touch a button.

Tony Fadell, one of the creators of the original iPod and the founder and former CEO of Nest, the company bought by Google for $3.4 billion to be the heart of its push into the connected home, gave me a clue when I ribbed him about Amazon stealing a huge march on him. "Can you imagine," he asked, "what the backlash would have been if Google had put out a connected home device that was always listening to you?" Because of its advertising-based business model, Google's critics already paint it as a surveillance company and constantly harp on the privacy risks inherent in the amount of data the company collects on its users in the course of delivering its services. An always-listening home device was thus unthinkable. Even though the device is only listening for its command trigger phrase rather than listening to every word, there is no question that this might have posed a risk to Google's business.

Google had made the vulnerability even worse by using "Google" as the name of its intelligent agent. "Okay, Google . . ." as the trigger phrase reminds the user every time just who is listening. With the Echo, you can ask "Alexa . . ." which is far less likely to remind you that it is Amazon that is potentially listening in on every conversation in your home.

(A caveat: Google did actually go there first with its Moto X phone, which was also always listening for your commands. I found it a remarkable device and was surprised that it didn't sweep to immediate success. Google didn't continue the effort with subsequent phones and with its own Google Home system, though, until after the Echo had legitimized the market.)

Jeff Bezos is very good at thinking the unthinkable. Much as he saw in 1998 that it was time to cut through consumer fears about stored credit cards, and that you could create a far better user experience if you pushed the boundaries a little bit, he saw that the time was right for an always-listening intelligent agent in the home.

This is a key lesson for every entrepreneur. *Ask yourself: What is unthinkable?*

And if, like Tony Fadell, you aren't ready to push past that boundary of unthinkability because you believe the market isn't ready, you can still prepare.

> Keep waiting for the missing pieces of the puzzle to arrive. Even if you aren't the one to push that boundary, once someone does it successfully, there's a huge opportunity for a fast follower. Be ready!

MAGICAL USER EXPERIENCE: SEEING THE PRESENT WITH FRESH EYES

This ability to see the present with fresh eyes is central to the success of the greatest entrepreneurs. Their creativity lies in their ability to understand and apply ways that the world has changed, while everyone else is still following the old map.

Many of the key capabilities that enable the magical Uber user experience came "for free" because of the trends that I had described as Web 2.0. As the Internet has become the platform for the development of software above the level of a single device, key data subsystems are on offer from multiple providers.

Location tracking is built into every smartphone. It is trivial for applications to identify where the user is at any moment. Uber didn't have to develop anything new. What they did have to do was to realize the implications of this capability. Applications like Google Maps and Waze had long provided smartphone-based navigation, including real-time traffic detection and routing optimization; Foursquare had used real-time location capability to allow users to check in to restaurants and bars as a way to coordinate meetups with friends. Uber took the "check-in"—here's where I am right now—to the next level. While Foursquare tried to persuade users to adopt a new social behavior, Uber used the same capabilities to turbocharge an antiquated application that was just waiting to be brought into the twenty-first century.

Communication had also become a standard part of the developer's toolkit. Twilio launched in 2008, providing program-callable cloud-based communications. This is the capability that lets you reach your driver by text or phone to adjust your location or to perform last-minute coordination, without providing a phone number that could be used to reach either party directly at a later time. This is an important tool for protecting the privacy of both driver and passenger. By 2010, when Uber launched, this service was widely available.

Payment had also become a commodity. Services like Braintree, Amazon Payments, and Stripe make it routine for any developer to be able to store a credit card number and charge the card whenever a product is purchased or a service is consumed. Uber's innovation, though, was to radically simplify the purchase experience. Payment without any visible act of payment, as much as the ability to summon the car, is what makes everyone's first ride with Uber such a WTF? moment.

Understanding that what used to be hard is now free and easy due to the work of others is essential to the leapfrogging progress of technology.

Robin Chase, author of the book *Peers Inc*, describes how services ranging from Zipcar, which she founded in 1999, to Uber, Lyft, and Airbnb are all platforms for unlocking what she calls "excess capacity" and sharing it with others. They put together ordinary people ("the peers") and a platform ("the Inc") to do something neither could do alone.

In the case of Zipcar, whose cars were owned by the company, she says the "excess capacity" was the capacity for self-service: the trust that the customers themselves could be relied upon to return a car clean and full of gas for the next customer. These customers were the peers in her model. The "Inc" was, of course, her company, which provided the cars themselves, but also the reservations platform that kept track of when and where cars were available so that they could be reserved on demand for as little as an hour or two, much smaller increments than a 1990s-era rental car.

The advance of technology has made Zipcar's advances, remarkable as they were at the time, rather quaint. Where Zipcar required cars to be returned to the same location from which they were rented, newer entrants into the space, like Car2go, use modern location-tracking technology and allow customers to leave the car wherever they like. And taking the "peer" model even further, services like Getaround allow users to put up their personal cars for rental. And while the car must be returned to the original location (more or less), location-tracking technology means that users can simply find a car that is located close to them—the entire city becomes the storage lot for the excess capacity of unused vehicles for rent.

Robin's notion even extends to the idea that the smartphone revolution itself was an act of unlocking excess capacity. It's easy to forget that these devices that can now do so much once were used only for making phone calls and sending texts. One can see the progress of the car-sharing industry, for example, as an exercise in realizing just how much more is possible with the untapped capabilities of the sensors in the phone. Where Zipcar and Car2go users were originally sent a special smart card to access the car they'd reserved, Zipcar, Car2go, and Getaround users now do it with their smartphone.

And as I've outlined here, the ability of Uber to coordinate driver and passenger, communications, and payment and enable navigation relied on similar realizations of hidden capabilities just waiting to be tapped. Camp and Kalanick's brilliance was in recognizing these latent capabilities and understanding how to apply them. In an insightful 2013 tweet, box.net CEO Aaron Levie wrote: "Uber is a $3.5 billion lesson in building for how the world *should* work instead of optimizing for how the world *does* work." Uber is valued at far more than $3.5 billion today, but that only underlines Aaron's point.

> Real breakthroughs come when an entrepreneur doesn't just use new technology to duplicate what went before or to fine-tune the way the world works now, but to reimagine how it ought to work.

This is the secret power of WTF? technologies. They not only allow for, they *reward* deep rethinking of the way things work. There are many possible futures. The world as it is is not a given. We can reinvent it.

PART II
PLATFORM THINKING

When the best leader leads, the people say "We did it
ourselves."

—Lao-tzu

5

NETWORKS AND THE NATURE
OF THE FIRM

WHEN DALE AND I LAUNCHED GNN IN 1993, OUR MODEL WAS shaped by our experience as publishers. We curated a catalog that highlighted "the best of" the Web, we took over the NCSA "What's New" page to announce new sites, and we did other things that made sense in the publishing world we'd grown up in, one of whose key functions was curation.

Our eyes were opened as Yahoo! took on the far more ambitious goal of cataloging *everything* on the web. Along with the rest of the media world, we watched in awe (though many also watched in dismay) as Google (and later Facebook) became media titans by algorithmically curating what would once have been an enormous "slush pile" so that it becomes valuable to its customers and advertisers.

Today, on-demand companies like Lyft and Uber in transportation and Airbnb in hospitality bring a similar model to the physical world.

Finnish management consultant Esko Kilpi beautifully describes the power of these new technology-enabled networks in an essay on *Medium*, "The Future of Firms." Kilpi reflects on economist Ronald Coase's theory of twentieth-century business organization, which explores the question of when it makes sense to hire employees rather than simply contracting the work out to an individual or small company with specialized expertise. Coase's answer is that it makes sense to put people into one business organization because of the transaction costs of finding, vetting, bargaining with, and supervising the work of external suppliers.

But the Internet has changed that math, as Kilpi observes. "If the (transaction) costs of exchanging value in the society at large go down drastically as is happening today," he writes, "the form and logic of economic and organizational entities necessarily need to change! The core firm should now be small and agile, with a large network." He adds: "Apps can do now what managers used to do."

As far back as 2002, Hal Varian predicted that the effect might be the opposite. "Maybe the Internet's role is to provide the inexpensive communications that can support megacorporations," he wrote. In a follow-up conversation, he said to me, "If transaction costs go down, coordination within firms becomes cheaper too. It's not obvious what the outcome will be."

Of course, networks have always been a part of business. An automaker is not made up of just its industrial workers and its managers, but also of its network of parts suppliers and auto dealerships and ad agencies. Similarly, large retailers are aggregation points for a network of suppliers, logistics companies, and other suppliers. Fast-food vendors like McDonald's and Subway aggregate a network of franchisees. The entire film and TV industry consists of a small core of full-time workers and a large network of temporary on-demand workers. This is also true of publishing and other media companies. My own company, O'Reilly Media, publishes books, puts on events, and delivers online learning with a full-time staff of four hundred and an extended network of tens of thousands of contributors—authors, conference presenters, technology advisers, and other partners.

But the Internet takes the networked firm to a new level. Google, the company that ended up as the prime gateway to the World Wide Web, provides access to a universe of content that it doesn't own yet has become the largest media company in the world. In 2016, Facebook's revenues also surpassed those of the largest traditional media companies. Americans 13 to 24 years old already watch more video on YouTube, much of it user-contributed, than they watch on television. And Amazon has surpassed Walmart as the world's most valuable retailer by offering virtually unlimited selection, including marketplace items from ordinary individuals and small businesses. These are companies that, to rephrase Kilpi, "are *large* and agile, with a large network."

But perhaps most important, these companies have gone beyond being just hubs in a network. They have become platforms providing services on which other companies build, central to the operation and control of the network. And as we shall come to see in later chapters, when marketplaces become digital, they become living systems, neither human nor machine, independent of their creators and less and less under anyone's control.

THE EVOLUTION OF PLATFORMS

On-demand companies like Uber and Lyft are only the latest development in an ongoing transformation of business. Consider the evolution of the retail marketplace as exemplified first by chain stores, and then by Internet retailers like Amazon, which have largely replaced a network of small local businesses that delivered goods through retail storefronts. Cost efficiencies led to lower prices and greater selection, drawing more consumers, which in turn gave more purchasing power to larger retailers, allowing them to lower prices further and to crush rivals in a self-reinforcing cycle. National marketing of these advantages led to the rise of familiar chains. The Internet added even more leverage, reducing the need to invest in real estate, reaching customers regardless of their geographical location, and building in new habits of customer loyalty and instant gratification. With delivery now same-day in many locations, anything you need is only a few clicks away.

Internet retailers like Amazon were also able to offer even larger selections of products, not just aggregating offerings from a carefully chosen network of suppliers, but opening up self-service marketplaces in which virtually anyone can offer products. Years ago, Clay Shirky described the move from "filter, then publish" to "publish, then filter" as one of the key advantages brought by the Internet to publishing, but the lesson applies to almost every Internet marketplace. It is fundamentally an open-ended network in which filtering and curation (known in other contexts as "management") happens largely after the fact.

But that's not all. While large physical retailers cut costs by eliminating knowledgeable workers, using lower prices and greater selection to hedge against worse customer service (compare an old-time hardware store with a chain like Home Depot or Lowe's), online retailers did not make these same trade-offs. *Instead of just eliminating knowledgeable workers, they replaced and augmented them with software.*

Even though there are several orders of magnitude more products on Amazon than in physical stores, you don't need a salesperson to help you find the right product—a search engine helps you find it. You don't need a salesperson to help you understand which product is the best—Amazon has built software that lets customers rate the products and write reviews to tell you which are best, and then feeds that reputation information into their search engine so that the best products naturally come out on top. You don't need a cashier to help you check out—software lets you do that yourself.

Amazon's use of automation goes far beyond the use of robots in its warehouses (though Amazon Robotics is one of the leaders in the field). Every function the company performs is infused with software, organizing its workers, its suppliers, and its customers into an integrated workflow. Of course, every corporation is a kind of hybrid of man and machine, created and operated by humans to augment their individual efforts. But even the highest-performance traditional company has an internal combustion engine; a digital company is a Tesla, with high-torque electric engines in each wheel.

The greater labor efficiency of the online model can be seen by comparing the revenue per employee of Amazon versus Walmart. Walmart, already the most efficient offline retailer, employs 2.2 million people to achieve its $483 billion in sales, or approximately $219,000 per employee. Amazon employs 341,000 people to achieve $136 billion in sales, or approximately $399,000 per employee. Were it not for Amazon's continuing investments in expansion and R&D, that number would be far higher.

NETWORKED PLATFORMS FOR PHYSICAL WORLD SERVICES

One way to think about the new generation of on-demand companies such as Uber and Lyft is that they are networked platforms for physical world services, bringing a fragmented industry into the twenty-first century in the same way that e-commerce transformed retail. Technology is enabling a fundamental restructuring of the taxi and limousine industry from that of a network of small firms to a network of individuals, replacing many middlemen in the taxi business with software, using the freed-up resources to put more drivers on the road.

The coordination costs of the taxicab business have generally kept it local. According to the Taxicab, Limousine & Paratransit Association (TLPA), the US taxi industry consists of approximately 6,300 companies operating 171,000 taxicabs and other vehicles. More than 80% of these are small companies operating anywhere between 1 and 50 taxis. Only 6% of these companies have more than 100 taxicabs. Only in the largest of these companies do multiple drivers use the same taxicab, with regular shifts. And 88% of taxi and limousine drivers are independent contractors.

When you as a customer see a branded taxicab, you are seeing the brand not of the medallion owner (who may be a small business of as little as a single cab) but of the dispatch company. Depending on the size of the city, that brand may be sublicensed to dozens or even hundreds of smaller companies. This fragmented industry provides work not just for drivers but for managers, dispatchers, maintenance workers, and bookkeepers. The TLPA estimates that the industry employs a total of 350,000 people, which works out to approximately two jobs per taxicab. Since relatively few taxicabs are "double shifted" (these are often in the largest, densest locations, where it makes sense for the companies to own the cab and hire the driver as a full-time employee), that suggests that almost half of those employed in the industry are in secondary support roles. These are the jobs that are being replaced by the efficient new platforms. Functions like auto maintenance still have to be performed, so those jobs remain.

The fact that Uber and Lyft use algorithms and smartphone apps to coordinate driver and passenger can lead us to overlook another fact—that, at bottom, Uber and Lyft provide dispatch and branding services much like existing taxi companies, only more efficiently. And like the existing taxi industry, they essentially subcontract the job of transport—except in this case, they subcontract to individuals rather than to smaller businesses, and take a percentage of the revenue rather than charging a daily rental fee for the use of a branded taxicab.

These firms thus use technology to eliminate the jobs of what used to be an enormous hierarchy of managers (or a hierarchy of individual firms acting as suppliers), replacing them with a relatively flat network managed by algorithms, network-based reputation systems, and marketplace dynamics. These firms also rely on their network of customers to police the quality of their service. Lyft even uses its network of top-rated drivers to onboard new drivers, outsourcing what once was a crucial function of management.

But focusing on the jobs that are lost is a mistake. Jobs are not lost so much as they are displaced and transformed. Uber and Lyft now deploy more drivers (albeit a majority of them part-time) than the entire prior taxi industry. (I have been told that Uber has about 1.5 million monthly active drivers worldwide. Lyft has 700,000.) They have also provided an additional source of customers for limousine drivers at the same time that they have provided punishing competition for traditional taxi companies.

There are other on-demand employers hiding in plain sight. I have been told that at current growth rates, Flex, Amazon's network of on-demand delivery drivers, might well be larger than Lyft by 2018. Interestingly, Flex uses a model in which drivers sign up in advance for two, four, or six-hour shifts for a predetermined hourly rate. Amazon takes the risk of not having enough delivery volume to keep them busy. While drivers may earn slightly less than the most successful Uber or Lyft drivers, the greater predictability has made Flex highly desirable to drivers.

Even in a world of self-driving cars, it is possible to see how increases in the services being provided can lead to more employment, not less. If we play our cards right, jobs that are lost to automation can be equivalent to the kinds of "losses" that came to bank tellers and their managers with the introduction of the ATM. It turns out that there were fewer tellers per branch but a net increase in the total number of tellers, because automation made it cheaper to open new branches. The ATM also replaced boring, repetitive tasks with more interesting, higher-value tasks. Tellers who used to do mostly repetitive work became an important part of the "relationship banking team."

We haven't yet seen the equivalent of the "relationship banking team" in on-demand transportation (though there are signs of what that might be in Uber's early experiments in making house calls to deliver flu shots and bringing elderly patients to doctors' appointments). Uber and Lyft are on their way to becoming a generalized urban logistics system. It's important to realize that we are still exploring the possibilities inherent in the new model.

This is not a zero-sum game. The number of things that people can do for each other once transportation is cheap and universally accessible also goes up. This is the same pattern that we've seen in the world of media, where network business models have vastly increased the number of content providers despite centralizing power at firms like Google and Facebook. It is also the opposite of what happens in old-style firms, where concentration of power often led to a smaller set of goods and services at higher prices.

Similarly, robots seem to have accelerated Amazon's human hiring. From 2014 through 2016, the company went from having 1,400 robots in its warehouses to 45,000. During the same time frame, it added nearly 200,000 full-time employees. It added 110,000 employees in 2016 alone, most of them in its highly automated fulfillment centers. I have been told that, including temps and subcontractors, 480,000 people work in Amazon distribution and delivery services, with 250,000 more added at peak holiday times. They can't hire fast enough. Robots allow Amazon to pack more products into the same warehouse footprint, and make human workers more productive. They aren't replacing people; they are augmenting them.

BLINDED BY THE FAMILIAR

When the past is everything you know, it is hard to see the future. Often what keeps us from recognizing what lies before us is a kind of afterimage, superimposed on our vision even after the stimulus is gone. Afterimages occur when photoreceptors are overstimulated because you look too long at an object without the small movements (saccades) that refresh the vision, leading to a decrease in the signal to the brain. Or they may occur because your eyes are compensating for bright light, and then you suddenly move into darkness.

So too, if we wrap ourselves in the familiar without exposing our minds to fresh ideas, images are burned onto our brains, leaving shadows of the past overlaid on the present. Familiar companies, technologies, ideas, and social structures hide others with a vastly dissimilar structure, and we see only ghostly images until the new comes into focus. Once your eyes have adjusted to the new light, you see what was previously invisible to you.

Science fiction writer Kim Stanley Robinson captures this moment perfectly in his novel *Green Mars*, when one of the original settlers of Mars has a shock of insight: "He realized then that history is a wave that moves through time slightly faster than we do." If we are honest with ourselves, each of us has many such moments, when we realize that the world has moved on and we are stuck in the past.

It is this mental hiccup that leads to many a failure of insight. Famously, Jaron Lanier (and many others) have made the comparison between Kodak, which at its height had 140,000 employees, and Instagram, which had only 13 when it was sold to Facebook for $1 billion in 2012. It's easy to overlay the afterimage of Kodak, and say, as Lanier did, that the jobs have gone away. Yet for Instagram to exist and thrive, every phone had to include a digital camera and to be connected to a communications network, and that network had to be pervasive and data centers had to provide hosting services that allow tiny startups to serve tens of millions of users. (Instagram had perhaps 40 million users when it was bought; it has 500 million today.) Add up the employees of Apple and Samsung, Cisco and Huawei, Verizon and AT&T, Amazon Web Services (where Instagram was originally hosted) and Facebook's own data centers, and you see the size of the mountain range of employment of which Instagram itself is a boulder on one small peak.

But that's not all. These digital communications and content creation technologies have made it possible for a new class of media company—Facebook, Instagram, YouTube, Twitter, Snap, WeChat,

Tencent, and a host of others around the world—to turn ordinary people into "workers" producing content for their advertising business. We don't see these people as workers, because they start out unpaid, but over time, an increasing number of them see economic opportunity on the platform for which they originally volunteered, and before long, the platform supports an ecosystem of small businesses.

Of course, there were networks of people who didn't work for Kodak either—camera manufacturers, film processors, chemical suppliers, retailers. Not to mention news, portrait, and fashion photographers. But the number of people whose jobs and lives were impacted by film photography was tiny by comparison with digital. The Internet sector now represents more than 5% of GDP in developed countries. For consumers at least, digital photography is a major driver of online activity, central to how people communicate, share, buy, sell, and learn about the world. More than 1.5 trillion digital photographs are shared online each year, up from 80 billion in the days of Kodak.

The cascade of combinatorial effects continues. Without digital photography, would there be Amazon, eBay, Etsy, or Airbnb?

Digital photography certainly played a role in the success of e-commerce, not to mention a host of hotel, restaurant, and travel sites. Being able to see a picture of a product is the next-best thing to seeing it for yourself. But for Airbnb, we have a definitive answer. Photography played a key role in its success.

The company was founded in 2008 by two designers, Brian Chesky and Joe Gebbia, and an engineer, Nathan Blecharczyk. The original idea came to them in 2007, when, as Joe described it, "our rent went up for our San Francisco apartment and we had to figure out a way to bring in some extra income. There was a design conference coming to the city, but hotels were sold out. The size of our apartment could easily fit airbeds on the floor, so we decided to rent them out."

They built a simple website of their own rather than listing the space on Craigslist, the venerable online classified site founded in 1995 by Craig Newmark. The experiment was so successful that they decided to build out a short-term room, apartment, and home rental service for the upcoming SXSW technology conference in Austin, Texas, because they knew that every hotel room in the city would be sold out. They followed that up by doing the same thing for the 2008 Democratic National Convention, held in Denver, Colorado.

In 2009, they were accepted into Y Combinator, the prestigious Silicon Valley startup incubator, and then received funding from one of Silicon Valley's top venture firms, Sequoia Capital. But despite a promising start, they were still struggling with acquiring users fast enough. The breakthrough came when they realized that hosts were taking lousy photographs of their properties, leading to lower trust and thus lower interest by possible renters. So in the spring of 2009, Brian and Joe rented a high-end digital camera, went to New York, Airbnb's top city at the time, and took as many professional photos as they could. Listings on the website doubled, even tripled. So they invested in a program to hire professional photographers in top cities around the world and never looked back. The company now has more rooms available every night than the largest hotel chains in the world.

BUILDING A THICK MARKETPLACE

What made Airbnb's achievement possible, of course, was not just digital photography, making it easy for hosts to show off their property, but the World Wide Web, online credit card payments, and the experience of other sites that had built reputation systems and ratings to help users build trust with strangers. Airbnb had to wrap these services into a new platform, which you can define as the set of digital services that enables its hosts to find and serve guests.

The primary platform service provided by Airbnb, though, is not to build a pretty web page showing off a property, to schedule rentals, or to take payments. Anyone with a modicum of web experience can do all those things in an afternoon. The essential job of an Internet service like Airbnb is to build what Alvin E. Roth, the economist whose work on labor marketplaces earned him the Nobel Prize, calls a "thick marketplace," a critical mass of consumers and producers, readers and writers, or buyers and sellers. There is many a brilliant and beautiful site that for no obvious reason never attracts users, while others, seemingly inferior in design or features, flourish.

If you're lucky, and your timing is just right, a thick marketplace can happen organically, seemingly without deliberate effort. The first website went live on August 6, 1991. It contained a simple description of Tim Berners-Lee's hypertext project, complete with source code for a web server and a web browser. The site could be accessed by Telnet, a remote log-in program, and using that, you could download the source code for a web server and set up your own site. By the time Dale Dougherty and I had lunch with Tim in Boston a year later, there were perhaps a hundred websites. Yet by the time Google launched in September 1998, there were millions.

Because the World Wide Web had been put into the public domain, Tim Berners-Lee didn't have to do all the work himself. The National Center for Supercomputing Applications (NCSA), located at the University of Illinois, built an improved web server and browser. Marc Andreessen, who wrote the browser while a student there, left to found Mosaic Communications Corporation (later renamed Netscape Communications). A group of users, abandoned by the original developers, took over the server project, pooling all their patches (shared improvements to the source code) to create the Apache server, which eventually became the world's most widely used web server. (A pun: It was "a patchy server.")

The web became a rich marketplace of writers and readers. And from there, entrepreneurs layered on marketplaces for buying and selling everything from books and music to travel, homes, and automobiles. And for advertising them.

There were other online hypertext systems competing with the web. Microsoft had launched a series of successful CD-ROM-based information products, starting in 1992 with Cinemania, an interactive movie guide, and Encarta, a full encyclopedia released the following year. Their multimedia hypertext experience was far ahead of the nascent World Wide Web.

Dale Dougherty had gone up to Microsoft in the fall of 1993 to show them GNN, and as he recalls, they were brutally dismissive. As

Dale remembers it, he had been invited to present GNN and the web to a team at Microsoft. A man to whom he was never introduced "arrived late, never sat down, but interrupted me as he paced the room, dismissing the web and saying it wasn't important to Microsoft. I recall that the others in the room knew very little about the web and they seemed curious, but upon this fellow's abrupt dismissal, they grew quiet and the conversation ended."

Microsoft realized, though, that there was an online hypertext opportunity after all. The Microsoft Network (MSN), a proprietary network similar to AOL, was launched in the fall of 1995. In the spring of 1996, Nathan Myhrvold, then Microsoft's chief technology officer, gave a talk about the Microsoft Network at Esther Dyson's influential PC Forum. I remember him showing a graph with the number of documents on one axis and the number of readers on the other, and saying, "There are a few documents that are read by millions of people, and millions of documents that are read by one or two people. But there's this huge space in the middle, and that's what we're serving with MSN."

I stood up during the Q&A and said to Nathan, "I totally agree with your insight about the huge opportunity, but you're talking about the World Wide Web." I had been asked by Microsoft to publish content on their new network. "Pay us $50,000, and we'll make you rich and famous" was essentially their pitch. But the alternative was far easier: Get on the Internet if you weren't already on it, download and set up Apache, format your content with HTML, and you're off to the races. No contract needed. The web was a *permissionless network*.

Microsoft had begun to experiment with the web as early as 1994, but their big bet was on MSN. "Microsoft developed MSN to compete with AOL, something they would control in terms of content and access," Dale recalled. "The web as an open system undermined that control, and they did not want to imagine a world without them at the center, both from a technological and business point of view."

Permissionless networks, like open source software projects or the World Wide Web, often grow faster and more organically than those that require approval, and the web soon left MSN and AOL far behind.

The web grew to hundreds of millions of websites, hosting trillions of web pages.

> **This is a central pattern of the Internet age: More freedom leads to more growth.**

Of course, on a permissionless network like the Web, anyone could bring content. That was a boon to anyone who had content to post online (including bad actors peddling porn, scams, and pirated content), who could now reach millions of people virtually for nothing. It was also a boon to users, who had access to vast amounts of free content.

Not all successful network platforms are permissionless and decentralized like the Web. Facebook owns and controls its centralized user network, but allows anyone to post on it, as long as they follow certain rules. You can be kicked off the platform, but content is not vetted beforehand. The iPhone App Store is both centralized and tightly controlled. Apps must be registered and approved before they are allowed into Apple's App Store. The Android app store is far more open. But in either case—iPhone or Android—the underlying open and decentralized network of cell phone users is what first brought one side of the market to scale. With hundreds of millions of smartphone users and a clear economic opportunity for paid apps, there was plenty of incentive for app developers to join the marketplace.

Sometimes, once the network itself is at scale, a particular node on the network takes off and spawns a new network of its own. In 2007, Craig Newmark recalled the process by which Craigslist grew from a simple listing of arts and technology events in San Francisco to the world's largest online classified network: "We built something, we get feedback, we try to figure out what make sense out of the suggestions, and then we do something about it and then we listen some more." That is a great description of how Internet software is typically built today, with what is now called a "build-measure-learn" cycle, in which the users of a minimally useful service teach its creators what they want from them. But even that is not really the secret to Craigslist's success.

Classified advertising in newspapers was expensive and most Craigslist ads were free. If Craigslist hadn't been a labor of love, Craig's service to his community, it might not have come out as the winner. Would-be competitors, being venture funded, had a fatal flaw: They needed to charge money to pay back their investors. So they had fewer ads, and because they had fewer ads, they had fewer visitors. Despite being bare-bones, with a minimalist design, and having only nineteen employees, Craigslist was at one time the seventh-most-trafficked site on the web. (Today it is still No. 49.)

Later startups turned growth into a religion, seeking revenue only after massive user scale has been achieved. This is an incomplete map, which leads companies to get lots of users and then have to sell out to someone else. Networks often turn out to be two-sided marketplaces, in which one party pays for access to the other, trading money for attention. If you are unable to develop the matching side of the market, in the form of a network of advertisers, you are in trouble. This is why, for example, YouTube was sold to Google despite beating Google's own video product in attracting viewers, and why Instagram and WhatsApp were sold to Facebook. It is why Twitter is still struggling. Ultimately, network businesses need to develop both sides of the market.

Uber, Lyft, and Airbnb didn't have the luxury of user growth without revenue. Unlike advertising-based startups that could sell out to an existing giant in a well-developed industry segment, they have had to build both sides of a new marketplace. Uber and Lyft started out with organic growth, but later accelerated it by deploying huge amounts of capital to acquire new drivers and new customers.

Once a marketplace reaches critical mass, it tends to become self-sustaining, at least as long as the marketplace provider remembers that its primary job is to provide value for marketplace participants, not just for itself. Once marketplaces achieve scale, they often forget this essential point, and this is where decline begins to set in. I'd first noticed this with Microsoft's abuse of its monopoly position in the personal computer industry. In the early years, there was a thriving ecosystem of application vendors built on top of Microsoft Windows; by the time Microsoft reached its zenith, it had taken over many of the most lucrative application categories, using its platform dominance to drive the former leaders out of business. Entrepreneurs naturally went elsewhere, finding opportunity in the green fields of the as-yet noncommercial Internet.

I've watched this same dynamic unfolding on the web. Google began its life as a kind of switchboard, solely directing people to content produced by others. But over time, more of the most frequently sought information is offered directly via Google. There's a fine balance here. Google is trying to serve its users; embedding information directly into search results may be the right answer. But marketplace providers must tread carefully, because ultimately, the health of the entire ecosystem must be their concern.

A robust ecosystem is good not just for the participants but also for the marketplace platform owner. Internet entrepreneur and investor John Borthwick made a prescient comment to me when Twitter ended access to its data "firehose" for many of its third-party app providers in 2012. "It's a big mistake," he said, "for Twitter to shut down its ecosystem before someone in it invents their real business model."

Amazon needs to be especially responsible because of its dominance in so many e-commerce markets. More than 63 million Americans (roughly half of all households) are now enrolled in Amazon Prime, the company's free shipping service. Amazon has more than 200 million active credit card accounts; 55% of online shoppers now begin their search at Amazon, and 46% of all online shopping happens on the platform.

Yet Amazon too often competes with its own marketplace participants, creating private-label versions of bestselling products from its vendors, and using its control of the platform to remove the "Buy" button from the products of vendors who don't go along with its demands. This is their privilege, just as it is the privilege of any store to stock or fail to stock any product. And Amazon is far from the first large retailer to create its own private-label products. But once a company reaches monopoly status, it is no longer a marketplace participant. It *is* the market. As Olivia LaVecchia and Stacy Mitchell write in their report "Amazon's Stranglehold": "In effect, Amazon is turning an open, public marketplace into a privately controlled one."

> Over time, as networks reach monopoly or near-monopoly
> status, they must wrestle with the issue of how to create more
> value than they capture—how much value to take out of the
> ecosystem, versus how much they must leave for other players in
> order for the marketplace to continue to thrive.

Google and Amazon are both fiercely committed to creating value for
one side of the marketplace—users—and justify their actions to themselves on that basis. But as they replace more and more of the supplier
side of the network with their own services, they risk weakening the
marketplace as a whole. After all, someone else invented and invested
in those products or services that they are copying. This is why antitrust
law can't just use lower costs for consumers as its primary benchmark,
rather than the overall level of competition in the market. Lower costs
are only one outcome of competition. Innovation withers when only one
party can afford to innovate, or when there's only one place to bring new
products to market. The mental map used by regulators shapes their
decisions, and thus the future.

There is also systemic risk to the economy when a pervasive marketplace begins to compete with its participants. In 2008, not long before the financial crash, I organized a conference called Money:Tech
to explore what we could learn about the future of the Internet from
the larger and older networked economy of finance. What I learned
alarmed me.

In my 2007 research leading up to the event, Bill Janeway, the
former vice chairman of private equity firm Warburg Pincus and the
author of *Doing Capitalism in the Innovation Economy*, who began his
career on Wall Street, pointed out that Wall Street firms had moved
from being brokers to being active players who "began to trade against
their clients for their own account, such that now, the direct investment
activities of a firm like Goldman Sachs dwarf their activities on behalf
of outside customers." The events that came to a head later that year
told us just how far Wall Street firms had gone in trading against their
clients, and even more alarmingly, that their trading involved the creation of complex instruments that far outstripped their creators' ability

to understand or control them. Our economy and our politics have not yet recovered from the damage.

CENTRALIZATION VS. DECENTRALIZATION

The tension between centralized and decentralized networks, and between closed and open platforms, first became clear to me when I was exploring the difference between the personal computer industry as dominated by Microsoft in the 1980s and 1990s and the emerging world of open source software and the Internet. At the heart of these two worlds were two competing architectures, two competing platforms. One of them, like Tolkien's "one ring to bind them," was a tool of control. The other had what I call "an architecture of participation," open and inclusive.

I was deeply influenced by the design of the Unix operating system, the system on which I'd cut my teeth early in my career and which had sparked in me an enduring love of computing. Instead of a tightly integrated operating system providing every possible feature in one big package, Unix had a small kernel (the core operating system code) surrounded by a large set of single-purpose tools that all followed the same rules and could be creatively recombined to perform complex functions. Perhaps because AT&T Bell Labs, the creator of Unix, was a communications company, the rules for interoperability between the programs were well established.

As described in the book *The Unix Programming Environment*, by Brian Kernighan and Rob Pike, two of the computer scientists who were key members of the early community that had built Unix: "Even though the UNIX system introduces a number of innovative programs and techniques, no single program or idea makes it work well. Instead, what makes it effective is the approach to programming, a philosophy of using the computer. Although that philosophy can't be written down in a single sentence, at its heart is the idea that the power of a system comes more from the relationships among programs than from the programs themselves." The Internet also had a communications-oriented architecture, in which "small pieces loosely joined" (to use

David Weinberger's wonderful phrase) cooperate to become something much bigger.

In one of the early classics of systems engineering, *Systemantics*, John Gall wrote, "A complex system that works is invariably found to have evolved from a simple system that worked. The inverse proposition also appears to be true. A complex system designed from scratch never works and cannot be made to work. You have to start over beginning with a working simple system."

> Simple, decentralized systems work better at generating new possibilities than centralized, complex systems because they are able to evolve more quickly. Each decentralized component within the overall framework of simple rules is able to seek out its own fitness function. Those components that work better reproduce and spread; those that don't die off.

"Fitness function" is a term from genetic programming, an artificial intelligence technique that tries to model the development of computer programs on evolutionary biology. An algorithm is designed to produce small programs optimized to perform a specific task. In a series of iterations, those programs that perform poorly are killed off, while new variations are "bred" from those that are most successful.

Writing in 1975, John Gall wasn't thinking in terms of fitness functions. Genetic programming wasn't introduced until 1988. But add the idea of fitness functions and a fitness landscape to his insight that simple systems are able to evolve in ways that surprise their creators and you have a powerful tool for seeing and understanding how computer networks and marketplaces work.

The Internet itself proves the point.

In the 1960s, Paul Baran, Donald Davies, Leonard Kleinrock, and others had developed a theoretical alternative called packet switching to the circuit-switched networks that had characterized the telephone and telegraph. Rather than creating a physical circuit between the two endpoints for the duration of a communication, messages are broken up into small, standardized chunks, shipped by whatever route is most convenient for each packet, and reassembled at their destination.

Networks such as NPL in the United Kingdom and ARPANET in the United States were the first packet-switched networks, but by the early 1970s there were dozens, if not hundreds, of incompatible networks, and it had become obvious that some method of interoperability was needed. (To be fair, J. C. R. Licklider, the legendary DARPA program manager, had called for interoperable networks a full decade earlier.)

In 1973, Bob Kahn and Vint Cerf realized that the right way to solve the interoperability problem was to take the intelligence out of the network and to make the network endpoints responsible for reassembling the packets and requesting retransmission if any packets had been lost. Seemingly paradoxically, they had figured out that the best way to make the network more reliable was to have it do less. Over the next five years, with the help of many others, they developed two protocols, TCP (Transmission Control Protocol) and IP (Internet Protocol), generally spoken of together as TCP/IP, which effectively bridged the differences between underlying networks. It wasn't until 1983, though, that TCP/IP became the official protocol of the ARPANET, and from there became the basis of what was sometimes called "the network of networks," and eventually the Internet we know today.

Part of the genius of TCP/IP was how little it did. Rather than making the protocols more complex to handle additional needs, the Internet community simply defined additional protocols that sat on top of TCP/IP. The design was remarkably ad hoc. Any group that wanted to propose a new protocol or data format published a "Request for Comment" (RFC) describing the proposed technology. It would be examined and voted on by a community of peers who, starting in January 1986, gathered under the name of the Internet Engineering Task Force (IETF). There were no formal membership requirements. In 1992, MIT computer science professor Dave Clark described the IETF's guiding philosophy: "We reject: kings, presidents, and voting. We believe in: rough consensus and running code."

And there was this naive, glorious statement by Jon Postel in RFC

761: "TCP implementation should follow a general principle of robustness. Be conservative in what you do. Be liberal in what you accept from others." It sounds like something out of the Bible, the Golden Rule as applied to computers.

In the 1980s, a separate, more traditionally constituted international standards committee had also gotten together to define the future of computer networking. The resulting Open Systems Interconnect (OSI) model was comprehensive and complete, and one of the industry pundits of the day wrote, in 1986: "Over the long haul, most vendors are going to migrate from TCP/IP to support Layer 4, the transport layer of the OSI model. For the short term, however, TCP/IP provides organizations with enough functionality to protect their existing equipment investment and over the long term, TCP/IP promises to allow for easy migration to OSI."

It didn't work out that way. It was the profoundly simple protocols of the Internet that grew richer and more complex, while the OSI protocol stack was relegated to the status of an academic reference model used to describe network architecture. The architecture of the World Wide Web, which echoed the radical design of the underlying Internet protocols, became the basis of the next generation of computer applications, and brought what was once an obscure networking technology to billions of people.

There's a key lesson here for networks that wish to reach maximum scale. Open source software projects like Linux and open systems like the Internet and the World Wide Web work not because there's a central board of approval giving permission for each new addition but because the original designers of the system laid down clear rules for cooperation and interoperability.

| The coordination is all in the design of the system itself.

This principle is the key to understanding not only today's Internet technology giants, but also what's wrong with today's WTF? economy.

6

THINKING IN PROMISES

IT'S EASY TO RECOGNIZE THE TRANSFORMATIVE IMPACT OF network-based businesses on society without understanding just how differently they are organized inside their walls.

During the years following my 1998 Open Source Summit, I'd developed a "stump speech" about the driving principles of open source software, hacker culture, and the Internet. One slide highlighted my vision for what made the bazaar of open source development, or a permissionless network like the Internet, so powerful:

- An architecture of participation means that your users help to extend your platform.
- Low barriers to experimentation mean that the system is "hacker friendly" for maximum innovation.
- Interoperability means that one component or service can be swapped out if a better one comes along.
- "Lock-in" comes because others depend on the benefit from your services, not because you're completely in control.

I also talked about how these platforms are born and how they evolve. First, hackers and enthusiasts explore the potential of a new technology. Entrepreneurs are attracted to it and in their quest to build a business, make things easier for ordinary users. Dominant players develop a platform, raising barriers to entry. Progress stagnates; hackers and entrepreneurs move on, looking for new frontiers. But sometimes (just sometimes) the industry builds a healthy ecosystem, in which hackers, entrepreneurs, and platforms play a creative game of "leapfrog." No one gets complete lock-in, and everyone has to improve in order to stay competitive.

That was followed up by a slide titled "History Lesson," which ended on the talk's punch line: "A platform strategy beats an application strategy every time!"

A PLATFORM BEATS AN APPLICATION EVERY TIME

Jeff Bezos heard me give this talk at my Emerging Technology Conference (ETech) and in 2003 asked me to deliver a version of it to a small group of developers at Amazon.

I had previously gone to Seattle in March 2001 to pitch Jeff on the idea that Amazon ought to be offering web services access to its data. For market research purposes, O'Reilly was "spidering" Amazon every three hours in order to download data on prices, ranks, page counts, and reviews of our books and those of our competitors. Web spidering seemed wasteful to me, since we had to download far more data than we needed and then extract just the bits we wanted. I was convinced that Amazon's vast product catalog was a perfect example of the kind of rich data that ought to be programmatically accessible via a web services API in the next-generation "Internet operating system" I was evangelizing.

Jeff was intrigued by the idea, and soon discovered that a skunkworks web services project, initiated by Amazon engineer Rob Frederick, was already under way. He discovered also that there were many other small companies like us that were spidering Amazon and building unauthorized interfaces to their data. Rather than trying to shut us down, he brought us all in to learn from each other and to help inform Amazon's strategy.

I vividly remember Jeff's disappointment in my talk at this internal Amazon developer's conference. He jumped up in the back of the room when I was done and said, "You didn't say the bit about a platform beating an application every time!" I didn't make that mistake when I gave another version of the talk at an Amazon all-hands meeting in May 2003.

The first-generation web services that the e-commerce giant rolled out in 2003 were all about access to their in-house product catalog and its underlying data, and had little to do with the infrastructure services that were launched in 2006 under the name Amazon Web Services (or AWS) and that sparked the great industry transformation that is now called "cloud computing." Those services came about for entirely different reasons, but I like to think that I planted the seeds of the idea with Jeff that if Amazon was to prosper in the years ahead, it had to become far more than just an e-commerce application. It had to become a platform.

In that marvelous way he has of taking any idea and thinking it all the way through, Jeff took the idea of platform much further than I had imagined. As Jeff described in a short 2008 interview with Om Malik: "Four years ago is when it started, and we had enough complexity inside Amazon that we were finding we were spending too much time on fine-grained coordination between our network engineering groups and our applications programming groups. Basically what we decided to do is build a [set of APIs] between those two layers so that you could just do coarse-grained coordination between those two groups." (That is, "small pieces loosely joined.")

This is important: Amazon Web Services was the answer to a problem in organizational design. Jeff understood, as every network-enabled business needs to understand in the twenty-first century, that, as HR consultant Josh Bersin once said to me, "Doing digital isn't the same as *being* digital."

> In the digital era, an online service and the organization that produces and manages it must become inseparable.

How Jeff took the idea of Amazon as a platform out of the realm of software and into organizational design ought to be taught in every business school.

The story was told by former Amazon engineer Steve Yegge in a post that he wrote for colleagues at Google, but which ended up being accidentally made public and went viral among Internet developers. It is known as "Stevey's Platform Rant." In it, Yegge describes a memo that he claimed Jeff Bezos wrote "back around 2002 I think, plus or minus a year." As Yegge described it:

His Big Mandate went something along these lines:

1. All teams will henceforth expose their data and functionality through service interfaces.
2. Teams must communicate with each other through these interfaces.
3. There will be no other form of interprocess communication allowed: no direct linking, no direct reads of another team's data store, no shared-memory model, no back-doors whatsoever. The only communication allowed is via service interface calls over the network.
4. It doesn't matter what technology they use. HTTP, Corba, Pubsub, custom protocols—doesn't matter. Bezos doesn't care.
5. All service interfaces, without exception, must be designed from the ground up to be externalizable. That is to say, the team must plan and design to be able to expose the interface to developers in the outside world. No exceptions.
6. Anyone who doesn't do this will be fired.

Jeff's first key insight was that Amazon could never turn itself into a platform unless it was itself built from the ground up using the same APIs that it would offer to external developers.

And sure enough, over the next few years, Amazon redesigned its application to rely on a comprehensive set of fundamental services—storage, computation, queuing, and eventually many more—that its own internal developers accessed via standardized application programming interfaces. By 2006, these services were robust and scalable enough, and the interfaces clearly enough defined, that they could be offered to Amazon's customers.

Uptake was swift. Amazon's low pricing and high capacity swept the market, radically lowering the barriers to entry for startups to experiment with new ideas, providing the stability and performance of top-notch Internet infrastructure for a fraction of the cost of building it yourself. The long Internet boom of the past decade can be traced back to Amazon's strategic decision to rebuild its own infrastructure and then open that infrastructure to the world. It isn't just startups either. Huge companies like Netflix host their services on top of AWS. It is now a $12 billion a year business.

Microsoft, Google, and many others have been playing catch-up in cloud computing, but they were late to the game. Amazon had one big advantage that Jeff explained to me not long after Amazon's cloud computing offerings were formally introduced in 2006: "I started out as a retailer. That's a really low-margin business. There's no way this can be worse for me. Microsoft and Google have very high margins. This is always going to be a worse business for them." By the time Microsoft and Google realized just how big a business cloud computing would be, they were far behind.

SOFTWARE AS ORGANIZATIONAL STRUCTURE

But perhaps the deepest insight about the nature of networked organizations comes from the way that Amazon structured itself internally to match the service-oriented design of its platform. Amazon Chief Technology Officer Werner Vogels described it in a 2006 blog post: "Services do not only represent a software structure but also the organizational structure. The services have a strong ownership model, which combined with the small team size is intended to make it very easy to innovate. In some sense you can see these services as small startups within the walls of a bigger company. Each of these services require a strong focus on who their customers are, regardless whether they are externally or internally."

Work is done by small teams. (Amazon famously describes these as "two-pizza teams," that is, teams small enough to be fed by two pizzas.) These teams work independently, starting with a high-level description of what they are trying to accomplish. Any project at Amazon is designed via a "working backwards" process. That is, the company, famous for its focus on the customer, starts with a press release that describes what the finished product does and why. (If it's an internal-only service or product, the "customer" might be another internal team.)

Then they write a "Frequently Asked Questions" document. They create mock-ups and other ways of defining the customer experience. They go so far as to write an actual user manual, describing how to use the product. Only then is the actual product green-lighted. Development is still iterative, informed by additional data from actual users as the product is built and tested, but the promise of the final product is where everything starts.

This is an example of what computer science and management theorist Mark Burgess calls "Thinking in Promises." He wrote: "If you imagine a cookbook, each page usually starts with a promise of what the outcome will look like (in the form of a seductive picture), and then it includes a suggested recipe for making it. It does not merely throw a recipe at you, forcing you through the steps to discover the outcome on trust. It sets your expectations first. In computer programming, and in management, we are not always so helpful."

Of course, writing the press release, or providing the picture that shows you the result of following the recipe, is only one part of what it takes to build a promise-centered organization. You have to work backward from the promise to the customer to the promises that each part of the organization needs to make to each other in order to fulfill it. The small teams are also a part of this approach. As is the design of a single, clearly-defined "fitness function" for each team (the one thing it promises to deliver, that can be measured and continuously improved).

At an Amazon management off-site, Jeff Bezos once famously responded to a suggestion that the company needed to improve communication between teams. "No, communication is terrible!" he said. The reason can be explained by the old joke: "One person sits and drinks. Two people clink and drink. The more people you add, the higher the ratio of clinking to drinking." What you want is a situation where people "clink" only with the people doing shared work, not with everyone they touch. This is simple math. Communication gets worse as team size grows.

There's a bit of a paradox here. Jeff was really asking for more effective, close communication *within* teams, coupled with highly structured communication *between* teams, mirroring the highly structured communication that makes it possible for modern Internet applications to work so well. He was arguing against the kind of backdoor communication that leads to messy workarounds that, over time, break under their own weight.

It is in this context that you can also understand why Jeff banned PowerPoint and insisted that all proposals and related presentations be made by written memos setting out the argument and evidence without relying on the artificial, misleading simplification of nested hierarchies. As Bill Janeway said to me, it seems that Jeff "wanted rich discussion up front leading to Decision Time and then highly structured communication during Execution Time."

Promise theory, as Burgess outlines it, is a framework for understanding how independent actors make promises to each other—the essence of that highly structured communication. Those actors can be software modules promising to respond in a certain way to an API call, or small teams promising to deliver a particular result. Burgess writes: "Imagine a set of principles that could help you to understand how parts combine to become a whole, and how each part sees the whole from its own perspective. If such principles were any good, it shouldn't matter whether we're talking about humans in a team, birds in a flock, computers in a datacenter, or cogs in a Swiss watch. A theory of cooperation ought to be pretty universal, so we could apply it both to technology and to the workplace."

That may sound terribly inhumane to some readers—to design an organization in such a way that people can be compared to cogs in a machine. But in fact, it's entirely the opposite. It is the traditional command-and-control organization, in which people are told what to do and expected to comply without necessarily understanding why or what the desired outcome is that ends up being inhumane. Kim Rachmeler, for many years the head of Amazon customer service, said to me that when a team defines the interface that allows others to access the services it builds and provides, "the satisfaction of the people accessing the services is entirely in their hands." Because this creates a tight feedback loop between the group and its customers, you can leave the implementation up to the creativity and the skill of the team building each function.

Kim explained to me that "writing the press release first is a mechanism to make customer obsession concrete." As are two-pizza teams producing services with hardened APIs. "Amazon does a better job of creating these kinds of mechanisms for its corporate values than any other company I've seen," Kim added. "And it starts from first principles (values) more than other companies as well."

Music streaming service Spotify is another company exploring the intersection of online service design and organizational design. Its organizational culture has also been quite influential. In its animated explainer videos, Spotify plots organizational culture along two axes: alignment and autonomy. A traditional organization has high alignment but low autonomy, because managers tell people what to do and how to do it. In the kind of organization parodied by the comic strip *Dilbert*, neither the manager nor the workers know why they are doing what they are doing. This is a low-alignment/low-autonomy organization. A modern technology engineering organization (or an entire organization like Amazon or Spotify) seeks to have high alignment *and* high autonomy. Everyone knows what the goal is, but they are empowered to find their own way to do it.

This approach was also part of the revolution in warfare developed by General Stanley McChrystal in Afghanistan in response to the rapidly changing conditions on the ground there. In a presentation I heard General McChrystal give at the New York Times New Work Summit in the summer of 2016, he said, "I tell people, 'Don't follow my orders. Follow the orders I would have given you if I were there and knew what you know.'" That is, understand our shared objective, and use your best judgment about how to achieve it.

My nephew-in-law, Peter Kromhout, who served as a US Army Infantry captain in Afghanistan, confirmed McChrystal's approach. "Before McChrystal, we were given a mission. We'd land and discover new intelligence that had to be acted on quickly, and we'd radio back to base for new instructions," he said. "By the time we got an answer, things might have changed again. After the McChrystal Doctrine was introduced, we'd land, see that the mission had changed, and radio back to tell them what we were doing about it."

This outcome-focused, outside-in approach means that, effectively, *a team is promising a result, not how they will achieve it*. As in Afghanistan, high autonomy is required by the rapidly changing conditions of a fast-growing Internet service.

High autonomy also provides a way of resolving unforeseen conflicts between separate teams. In his book *The Amazon Way*, former Amazon VP John Rossman describes how the company adopted an idea from Japanese lean manufacturing, "the Andon Cord." Any employee who encounters a problem on the assembly line in a Toyota factory can pull a cord that stops production and lights up a large sign ("Andon") to call for management attention. Once the Andon Cord was in place at Amazon, Rossman writes, "When customers began complaining about a problem with a product, customer care simply took that product down from the website and sent a message to the retail group that said, in effect, 'Fix the defect or you can't sell this product.'"

"Amazon's version of the Andon Cord started with an experience Jeff had with Customer Connection, one of the mechanisms Amazon uses to make customer obsession concrete," Kim Rachmeler told me. At the time, every level 7 or above manager had to spend some time working in customer service every two years, including Jeff. As part of this program, Amazon would pair the managers with a CS rep and have them answer a couple of phone calls.

Jeff took a call, Kim remembers. "'Hello, this is Jeff Bezos, how may I help you?' The customer didn't pick up on whom they were talking to and launched into their problem. It seems that they had received a table and the top was damaged. Jeff (with the rep's help) sent a replacement. When they hung up the call, the rep said something important: 'That table always arrives damaged.' It seems that the packaging was insufficient and so shipping usually caused problems with it. Jeff saw instantly that CS reps had knowledge that would be useful to retail but it was siloed. So he suggested the use of a mechanism like the Andon Cord—which was eventually implemented."

The Andon Cord illustrates a key principle of promise-oriented systems. It sends a simple, unambiguous signal to other groups, and stands in stark contrast to a traditional management process. While each group has its own fitness function, which it is expected to relentlessly optimize, these functions can conflict; the fitness function of any group may be checked by that of another. The art of management is to shape these functions so that they drive the entire company in the direction it wants to go, which represents an overall fitness function for the organization.

High-autonomy technical cultures have developed a technique—the stand-up meeting—by which people and groups must work together toward a common goal and review the status of their promises to each other.

In a dysfunctional organization, introducing stand-ups is a great way of understanding what has gone wrong, and of introducing new, targeted communications protocols.

Mikey Dickerson, one of the Google engineers recruited by the White House to rescue the failed healthcare.gov website in the fall of 2013 and who later became the director of the new United States Digital Service, told me the story of the hundred days' worth of stand-up meetings he held to knit the government contractors who'd built the failed site into a functioning organization that could actually make it work. It went something like this:

"Joe, you promised to get three new servers up by this morning. What's the status?" "I haven't gotten the security clearance yet from Mike." "Mike, what's the holdup?" "I never got a request for any security clearance from Joe." "What do you mean, Mike? I have it right here." "Listen, Joe. I've got the list of all my tickets [job requests] right here, and there's nothing from you!"

It was only then that "Joe" and "Mike," who worked for different contractors (there were more than thirty-three companies involved in the original healthcare.gov effort, working under sixty different contracts), discovered that they weren't using the same issue-tracking system. Teams were literally sending into the void requests for work to be done by other teams. Because everybody had hidden dependencies on everyone else, work was at a standstill, with everyone waiting on results from everyone else before they could proceed.

Whether it's through web services and APIs, or through tools like issue-tracking systems, a promise-oriented model works to increase autonomy because each autonomous agent defines and is made accountable for keeping well-documented promises.

YOU ARE ALL INSIDE THE APPLICATION

The change in software development from a model in which the goal was to produce an artifact (say, the "gold master" of the next release of Microsoft Windows, which was the target of years of development and would be duplicated onto millions of CD-ROMs and distributed to tens of thousands of retailers and corporate customers on the same day) to one in which software development was a process of continuous improvement was also a process of organizational discovery.

I still remember the wonder with which Mark Lucovsky, who'd been a senior engineering leader at Microsoft, described how different his process became when he moved to Google. "I make a change and roll it out live to millions of people at once." Mark was describing a profound transformation in how software development works in the age of the cloud. There are no more gold masters. Today software has become a process of constant, more or less incremental improvements. *From the point of view of the company offering an online service, software has gone from being a thing to a process, and ultimately, a series of business workflows.* The design of those workflows has to be optimized not just for the creators of the software but for the people who will keep them running day-to-day.

The key idea is that a company is now a hybrid organism, made up of people and machines. I had made this point too in my 2003 Amazon all-hands talk. I'd told the story of von Kempelen's Mechanical Turk, the chess-playing automaton that toured Europe in the late eighteenth and early nineteenth centuries, astonishing (and defeating) such luminaries as Napoleon and Benjamin Franklin. The supposed automaton actually had a chess master hidden inside, with a set of lenses to see the board and a set of levers to move the hands of the automaton. I thought this was a marvelous metaphor for the new generation of web applications.

As I spoke to the staff at Amazon, I reminded them that the application wasn't just software, but contained an ever-changing river of content produced by their network of suppliers, enhanced by reviews, ratings, and other contributions from their vast network of customers. Those inputs were then formatted, curated, and extended by their own staff in the form of editorial reviews, designs, and programming. And that dynamic river of content was managed, day in and day out, by all of the people who worked for Amazon. I remember saying "All of you—programmers, designers, writers, product managers, product buyers, customer service reps—are inside the application."

(For a long time, I wondered whether my telling of this story might have inspired Amazon to create the Amazon Mechanical Turk service, which uses a crowdsourced network of workers to perform small tasks that are hard for computers to do. However, while the service was launched in 2005, the patent for it was filed in 2001 though not issued until 2007, so at best I might have inspired the name. The name given to it in the patent diagrams is "Junta.")

My insight that, on the Internet, programmers were "inside the application" had unfolded gradually over time. It first came to me when I was trying to understand why the Perl programming language had become so important in the early days of the web.

One conversation particularly sticks in my mind. I'd asked Jeffrey Friedl, author of the book *Mastering Regular Expressions*, which I'd published in 1997, just what it was that he did with Perl in his day job at Yahoo! "I spend my days writing regular expressions to match up news stories with ticker symbols so that we can display them on the appropriate pages of finance.yahoo.com," he told me. (Regular expressions are like wildcards on steroids—a programming language feature that makes it possible to match any string of text using what appears to the uninitiated to be a magical incantation.) It immediately became clear to me that Jeffrey himself was as much a part of finance.yahoo.com as the Perl scripts he wrote, because he couldn't just write them once and walk away. Due to the dynamic nature of the content the website was trying to reflect, he needed to keep changing his programs every day.

By the time I spoke at Amazon in 2003, I'd extended this insight to understand that all of the employees inside the company, and all of the participants in the extended network, from suppliers to customers rating and reviewing products, were part of the application.

But it wasn't until 2006, when companies like Amazon and Microsoft were beginning to understand the possibilities of cloud computing, that another key element came into focus. I'd had a conversation with Debra Chrapaty, then VP of operations for the Microsoft Network. Her insightful comment perfectly encapsulated the change: "In the future, being a developer on someone's platform will mean being hosted on their infrastructure." For example, she talked about the competitive advantage she was developing by locating her data centers where power was cheap.

The post I wrote following our conversation was titled "Operations: The New Secret Sauce." It connected deeply with Jesse Robbins, then Amazon's "Master of Disaster," whose job was to disrupt the operations of other groups in order to force them to become more resilient. He told me that he and many of his colleagues had printed out the post and hung it on the walls of their cubes. "It's the first time anyone ever said we were important."

The next year, Jesse, together with Steve Souders from Yahoo!, Andy Oram of O'Reilly Media, and Artur Bergman, the CTO of Wikia, asked for a meeting with me. "We need a gathering place for our tribe," Jesse told me. I happily complied. We organized a summit to host the leaders of the emerging field of web operations, and soon thereafter launched our Velocity Conference to host the growing number of professionals who worked behind the scenes to make Internet sites run faster and more effectively. The Velocity Conference brought together a community working on a new discipline that came to be called DevOps, a portmanteau word combining software development and operations. (The term was coined a few months after the first Velocity Conference by Patrick Debois and Andrew "Clay"

Shafer, who ran a series of what they called "DevOps Days" in Belgium.)

The primary insight of DevOps is that there were traditionally two separate groups responsible for the technical infrastructure of modern web applications: the developers who build the software, and the IT operations staff who manage the servers and network infrastructure on which it runs. And those two groups typically didn't talk to each other, leading to unforeseen problems once the software was actually deployed at scale.

DevOps is a way of seeing the entire software life cycle as analogous to the lean manufacturing processes that Toyota had identified for manufacturing. DevOps takes the software life cycle and workflow of an Internet application and turns it into the workflow of the organization, building in measurement, identifying key choke points, and clarifying the network of essential communication.

In an appendix to *The Phoenix Project*, a novelized tutorial on DevOps created by Gene Kim, Kevin Behr, and George Spafford as homage to *The Goal*, the famous novel about the principles of lean manufacturing, Gene Kim notes that speed is one of the key competitive advantages that DevOps brings to an organization. A typical enterprise might deploy new software once every nine months, with a lead time of months or quarters. At companies like Amazon and Google, there are thousands of tiny deployments a day, with a lead time of minutes. Many of these deployments are of experimental features that may be rolled back or further modified. The capability to roll something back easily makes failure cheap and pushes decision making further down into the organization.

Much of this work is completely automated. Hal Varian calls this "computer kaizen," referring to the Japanese term for continuous improvement. "Just as mass production changed the way products were assembled and continuous improvement changed how manufacturing was done," he writes, "continuous experimentation . . . improve[s] the way we optimize business processes in our organizations."

But DevOps also brings higher reliability and better responsiveness to customers. Gene Kim characterizes what happens in a high-performance DevOps organization: "Instead of upstream Development groups causing chaos for those in the downstream work centers (e.g., QA, IT operations, and Infosec), Development is spending twenty percent of its time helping ensure that work flows smoothly through the entire value stream, speeding up automated tests, improving deployment infrastructure, and ensuring that all applications create useful production telemetry." He echoes the theme that it isn't just a technical but an organizational practice: "Everyone in the value stream shares a culture that not only values each other's time and contributions but also relentlessly injects pressure into the system of work to enable organizational learning and improvement."

The practices of DevOps have continued to evolve. Google calls its version of the discipline "Site Reliability Engineering" (SRE). As described by Benjamin Treynor Sloss, who coined the term, "SRE is fundamentally doing work that has historically been done by an operations team, but using engineers with software expertise, and banking on the fact that these engineers are inherently both predisposed to, and have the ability to, design and implement automation with software to replace human labor."

He makes the case that a traditional operations group has to scale linearly along with traffic to the service it supports. "Without constant engineering," he says, "operations load increases and teams will need more people just to keep pace with the workload." In the SRE approach, by contrast, the humans inside the machine who keep it going augment themselves by constantly teaching the machine how to duplicate what they do, at ever-increasing scale.

What we see, then, in the modern networked organization is not just a radical change in the external relationships between a company, its suppliers, and its customers, but a radical change in how the workers inside the company are organized, and how they are partnered with the software and machines they are building and operating.

As the principles of Internet-scale applications and services interpenetrate the real world, every company needs to transform itself to take advantage of techniques that were pioneered in the digital realm. This is not the work of a moment, but an ongoing exploration. At that

2003 all-day Amazon all-hands meeting, Jeff Bezos gave the opening talk. It was called "It's Still Day 1." He described the history of electricity, with vivid historical photographs of the nests of wires coming down from light sockets in the ceiling to power new kinds of electric devices. The standardized power plug hadn't yet been invented. He showed how factories still drove the machines on their assembly lines with huge centralized motors, with belts and pulleys carrying the power, just as they had in the age of steam, not yet having realized that they could bring electricity directly to small motors where the work was actually happening.

> Often, when new technology is first deployed, it amplifies the worst features of the old way of doing business. Only gradually do individuals and organizations realize, through a cascading network of innovations, how to put new technology properly to work.

Jeff was right. It's still day one, even now. But searching out the possibilities of the future isn't a job just for inventors of software or machines. It is the job of every business to ask itself what technology makes possible today not just for its customers, but for how the business itself is organized to serve them. That is also the job for other institutions, such as government.

7

GOVERNMENT AS A PLATFORM

MY FASCINATION WITH THE INTERSECTION OF GOVERNMENT and technology began with my friend Carl Malamud, a longtime advocate for technology in the public interest. In 1993, early in the history of the World Wide Web, Carl was helping Sun Microsystems give a demonstration of the capabilities of the Internet to the House Subcommittee on Telecommunications and Finance. After the demonstration, Subcommittee Chairman Representative Edward J. Markey (now a US senator from Massachusetts) told Carl that his subcommittee also had oversight of the Securities and Exchange Commission. Jamie Love, who worked with Ralph Nader on Internet matters, had been sending petitions to the subcommittee asking why SEC filings weren't available online.

The initial reaction from the SEC, Representative Markey told Carl, was that the data wasn't on the Internet because making it available was technically impossible, and, as Carl wrote in his colorful history of the event, "even if the data were available the only people interested in SEC filings were Wall Street Fatcats and they didn't really need subsidized access to data they were willing to pay for."

"If something is technically impossible, I get interested," Carl wrote. So he met with the SEC and with Chairman Markey's staff. The SEC wanted to know why in the world people would want to see the data in EDGAR (the database of the quarterly and annual filings of US public corporations). "I maintained that the Internet was full of lots of people—students, journalists, senior citizen investors—who were dying for access to this data. The SEC felt that only a few people would want to see EDGAR documents, and besides, the Internet 'didn't have the right kind of people.'"

Carl continued, "Now, this was a cheap shot, and I understood that what they meant was 'there weren't a lot of people, just a few researchers,' but I couldn't resist. 'The right kind of people?' I said, rising up in my chair. 'I think the American people are the right kind of people.'"

And so the government open data movement was born.

Carl secured a small National Science Foundation grant, which he largely used to pay the licensing fee that the SEC's "value-added resellers" charged to provide its data to Wall Street banks. Eric Schmidt, then CTO at Sun Microsystems, donated a couple of servers. Carl and his co-conspirator Brad Burdick formatted the data, put up the website, and in January 1994 launched a free version of the SEC's EDGAR system on the Internet.

Carl was an activist, not an entrepreneur. "Our goal wasn't to be in the database business," he wrote. "Our goal was to have the SEC serve their own data on the Internet." So, after operating the system for eighteen months, Carl announced that the service would be shutting down in sixty days unless the SEC took it over. Fifteen thousand people wrote to the SEC, proving Carl's point. Minds were changed, and the SEC agreed to take over the site.

Over time, with public demand for company financial statements no longer subject to debate, entrepreneurs started to build improved versions. Services like Yahoo! Finance and Google Finance, which provide public access to the data from SEC filings, are direct descendants of the work that Carl did in 1993. He has continued his activism ever since. His current challenge is to make the full text of all laws, regulations, and standards incorporated into law by reference freely available on the Internet.

My interest in lessons from Silicon Valley for government came back to the fore in 2005. Amazon hadn't yet revolutionized the industry by launching its cloud services platform, but the value of the Internet as a platform, and the nature of that platform, was becoming clearer to me. I had become convinced that the next generation Internet platform was a data platform, and I had noticed that government was the source of much of that data. The work that Carl Malamud had kicked off a decade earlier was just the tip of the iceberg. Google Maps, whose interactive JavaScript ("Ajax") interface was one of the WTF? technologies of 2005, was, like all online mapping services, built using base maps licensed from the government. And when hackers realized that they could build "mashups" placing other data on Google Maps, government data was one of the first places they turned. Adrian Holovaty's chicagocrime.org (now EveryBlock), which put crime data from the city of Chicago onto a Google map, was the second mashup ever created.

Google Maps fit perfectly with my thinking about Web 2.0. Unlike operating systems like Windows and Mac OS X or Linux, whose subsystems manage access to the hardware subsystems of a computer or a network, I was convinced that the Internet operating system would manage access to data subsystems providing services like confirming someone's identity or determining their location. If these subsystems could be made easily accessible to developers, I was convinced that an explosion of innovation would happen. I was so convinced of this thesis that I had launched a new event on location-based services, called Where 2.0. In a perfect demonstration of "catching a wave" at just the right time, Google contacted me two or three weeks before the event to ask if I could fit them into the program. Google Maps hadn't yet been announced, and it arrived just in time for its public launch to become a centerpiece of the event.

While applications could technically be built on top of other online mapping services like MapQuest, Yahoo Maps, and Microsoft Maps, developers had to apply for permission and pay an up-front licensing fee. My experience with open source software had taught me that there would be far more innovation and usage in the emerging mapping services landscape if those barriers to entry were taken down so the creativity of developers could be unleashed. So I called Microsoft and MapQuest (by then owned by AOL) to try to persuade them to make their APIs freely available, but without success. Instead they shut down hackers as "pirates."

But Google got it. When an independent developer named Paul Rademacher deciphered the Google Maps data format, he realized that he could build new custom maps by combining data from multiple sources. He built a site called housingmaps.com that showed apartment listings from Craigslist on a Google map—and Google saw the opportunity. Instead of shutting down Paul's hack, they celebrated it. They hired him, and opened up an API to make mashups easier. This was a transformative breakthrough that led to Google's dominance in online mapping. As more and more developers built applications for Google Maps, the platform got more powerful and drew more users. It became the classic thick marketplace, where users came for the apps, and apps came for the users.

The same design pattern, by which hackers had taught a company about the power of platforms, happened when Apple introduced the iPhone in June 2007. The App Store is so central to our experience of the smartphone today that it's easy to forget that the first iPhone didn't have an app store. It had a revolutionary, beautiful multi-touch interface and included iTunes, the music player that powered the iPod, but like most other phones, it had a limited number of apps. Within days, though, hackers had found their way around Apple's restrictions and had added apps of their own, a process that became known as "jail-breaking" (getting your phone out of application jail). In July 2008, in response to the spread of jailbreaking, Apple introduced the App Store as a formal mechanism for developers to add applications to the phone, and the smartphone world as we know it today took off. By the latest estimates, there are more than 2 million apps for the iPhone and they have been downloaded 130 billion times. App developers have earned nearly $50 billion in revenue.

A NEW MAP FOR GOV 2.0

The iPhone App Store was launched in July 2008; in November of that year Barack Obama was elected president and widely celebrated as "the first Internet president" because of the way he'd successfully used the Internet during his campaign. I was brainstorming with Eric Faurot, whose company, TechWeb, coproduced the Web 2.0 Summit with O'Reilly Media and John Battelle. I thought we should try to attract government innovators to our events to explore how the new administration could live up to the expectations of the first Internet presidency; Eric suggested instead that we bring a special version of the event to them. So that's what we did, coproducing the Gov 2.0 Summit and Gov 2.0 Expo in Washington, DC, in 2009 and 2010. Jennifer Pahlka, who is now my wife, became TechWeb's general manager for the project, and a crucial thought partner.

As I began developing the content for the new Gov 2.0 Summit, one of my first visits was with Eric Schmidt, then Google's CEO, whom I'd known since the days that we both worked with Carl Malamud back in 1993. I knew Eric had spent a lot of time in Washington, and I thought he'd have good advice. He did, but it wasn't the specific set of recommendations I expected. "Go to DC," he said. "Talk to a lot of people, and tell us what you make of it. You're good at it. That's what you do."

The idea that we should make "government as a platform" the centerpiece of our new event came to me in a conversation with Frank DiGiammarino, then a vice president at the National Academy of Public Administration and later a special assistant to Vice President Joe Biden involved in the administration of the 2009 Recovery Act. Frank explained to me that he believed one of the key roles of government was as a convener; once the government identified a problem, it shouldn't try to solve it directly, but should instead bring together the parties it wanted to engage with this problem. Frank contrasted this idea with the old model of government, which NAPA fellow Donald Kettl had named "vending machine government." While I didn't use the metaphor in quite the same way as Kettl, I ran with it: We pay our taxes; we get back services. In the vending-machine model, the full menu of available services is determined beforehand. A small number of providers have the ability to get their products into the machine, and as a result, the choices are limited and the prices are high. And when we don't get what we expect, our "participation" is limited to protest— essentially, shaking the vending machine.

This image of traditional government as a vending machine was the missing piece that helped make sense of everything I was exploring. A "Gov 2.0" meme had started to take hold in DC circles, but it was largely associated with getting federal agencies on social media, and social media was understood mainly as a way for politicians to get their messages out, and for citizens, another way to shake the vending machine. But to me, there was a far more profound opportunity, for government to run the Google Maps and iPhone App Store play.

We set out to redefine Gov 2.0 and draw a new map of how technology could reinvent government to be closer to the vision of our nation's founders, a model in which, as Thomas Jefferson wrote in a letter to Joseph Cabell, "every man feels that he is a participator in the government of affairs, not merely at an election one day in the year, but every day." In this model, government is ultimately a vehicle for coordinating the collective action of citizens. Jefferson was talking about governance—the creation of the rules by which we guide our society—but his participatory principle resonates with the ideas of open source software, and for that matter, of any successful platform.

To be clear, government as a platform most emphatically doesn't mean outsourcing government programs to the private sector. It does mean strategically identifying what building blocks are essential for government to provide, and providing those services, but not so many that they crowd out opportunity for the marketplace participants.

I had read a remarkable paper called "Government Data and the Invisible Hand," published in the January 2009 issue of the *Yale Journal of Law & Technology* by David Robinson, Harlan Yu, William P. Zeller, and Edward W. Felten. The paper argues that the government should get out of the business of building websites for citizens. If that call sounds familiar, you've probably heard it in the context of critics charging that government is not competent at building technology and would be much better off outsourcing everything to government contractors. But that's not what these authors meant. Robinson and company instead wanted the government to provide free access to bulk data so that *anyone who wanted to* could use it to build multiple competing services, possibly supported by a variety of business models. That's the difference between the vending machine and the platform.

This idea also echoes one of the "Eight Principles of Open Government Data" that Carl Malamud, Harvard law professor Larry Lessig, and I, together with a group of about thirty other open data activists, had published after a working group meeting in December 2007. One of those principles is that data should be published in formats that are not just machine readable but machine processable, so that the data could be reused for purposes not envisioned by its original producers.

Open data had become a key talking point of the new administration, but most people only thought of it as a tool of government transparency and accountability. A handful saw that there was a real opportunity to make data much more useful to citizens and society. They were drawing a new map, one I thought could navigate us to a better government. I saw a chance to reframe the dialog between liberals and conservatives that has so dominated political discourse in recent decades. Big government versus small government is in many ways beside the point. If government is successful as a platform, you could have small government and big services, just like Apple does with the iPhone. Apple didn't build thousands of its own apps. Apple built a platform and a marketplace, and hundreds of thousands of developers piled on.

I brought Craig Mundie, then CTO at Microsoft, to the Gov 2.0 Summit, where he pushed forcefully for the idea that killer apps drive platform adoption, using the example of how Microsoft Office had been key to the success of Windows. As it turned out, the federal government already had some killer apps built on the platform of government data; we just weren't calling them that. By 2008, GPS devices were in cars providing turn-by-turn directions, phone applications were telling us when the next bus was about to arrive, and services like Foursquare and Yelp were letting us know what restaurants were nearby. Few of the users of those services realize (even today) that GPS started out as a service built by the government. The US Air Force had originally launched GPS satellites for military purposes, but after a crucial policy decision made by President Reagan, agreed to open up the system for commercial use, much as Google decided to open up its Maps platform. No longer just an application, GPS became a platform, resulting in a wave of innovation from the private and public sector and a market now worth more than $26 billion.

Gov 2.0 started to mean something much more profound than getting federal agencies on social media. Washington insiders started talking about what we could achieve as a country with government functioning as a platform on which anyone could build.

CENTRAL PARK AND THE APP STORE

It's easy to forget just how generative government interventions can be. Larry Page and Sergey Brin's research project at Stanford, which led to Google, was funded by the National Science Foundation's Digital Library program. Were the NSF an investor rather than a grant maker for the public good, that investment alone would have repaid more than the entire NSF budget for the years the grant was made. In fact, the market value of Google is greater than the entire amount of taxpayer dollars spent on the NSF since it was first founded in 1952.

The Internet itself was originally a government-funded project. So was the Interstate Highway System. Not to mention that the government funded the original computer and memory chip development that gave us Silicon Valley, the research behind Siri and self-driving cars, and actually provided much of the capital for building out Elon Musk's bold ventures in electric vehicles, rooftop solar, and commercial space travel.

But government as a platform means far more than R&D funding. Would our cities thrive without transportation, water, power, garbage collection, and all the other services we take for granted? Like an operating system providing services for applications, government provides functions that enable private sector activity. We see this particularly at the local level, where government interfaces most directly with citizens.

On a visit to New York City in the fall of 2016, I went for a morning run in Central Park. It was beautiful in the early light, and equally beautiful to see the ways that New Yorkers use their park. Runners and bikers thronged the roads and paths, but there were also people just sitting quietly absorbing the view, taking in the dawn. And of course, there were the dog walkers.

The park is pretty clean—and on that Monday's run, I came across a crew of maintenance workers who reminded me just why it is so clean. It's not that New Yorkers look after the park. It's that it is looked after for them. This oasis of natural beauty in the center of a great city is set aside and looked after for the benefit of its people. Forty-two million visitors enjoy the park each year.

As I ran through the park, I couldn't help but think of the park as a metaphor for all that government does for its citizens. Our roads, our trains, our water and sewers, our universal access to electricity, heat, and telecommunications. Our schools. Our protection from fire and flood, from crime and from foreign enemies. Our rule of law. I know that many of these services cost more than they should, and accomplish less than they could. Some are tragically at odds with core American values—I grieve for police violence against people of color, unnecessary foreign wars, a rule of law that too often seems to favor the rich and powerful over the rights of all. Yet I also think of all the ways that government is the platform upon which our economy and our society are built, with many analogies to the way that iOS and the App Store are the platform for the Apple smartphone economy.

In the same way that it puzzled me that advocates for Linux were ignoring the Internet in building their narrative, it has puzzled me that those who celebrate the success of the great Silicon Valley platforms are critical of government for doing things that are understood to be essential when coming from Google, Facebook, Amazon, or Apple. The question shouldn't be whether or not government ought to be doing these things; it should be how to help government do a better job of fulfilling its responsibilities as a platform.

As we've discussed, creating a thick marketplace is the first requirement of any platform. This is not a given. A thick marketplace requires both producers (in Apple's case, app developers) and consumers. In the smartphone space, Apple and Google were able to build thick marketplaces, but Microsoft, for all its past success, was unable to do so. Not enough people bought their phones, which were late to market, and so app developers were unwilling to build new apps for Windows Mobile, which confirmed customers in their decision to avoid the phone.

What is the equivalent for government? For "thick marketplace," read "flourishing economy." We like to think that "the market" is a natural phenomenon, but the fact that there are poor countries with abundant natural resources and large populations and rich countries with neither abundant resources nor large populations teaches us that there is an art to creating a flourishing economy.

Where a technology platform must acquire users, a nation already has a captive set of "users": its resident population. If the population or resources are small, the country must reach outside its borders for both, but in many cases, this local population is sufficient to bootstrap a robust marketplace, with plenty of consumers and plenty of providers of goods and services.

But there's an important lesson from the wealth of nations: If the population doesn't have enough money to buy goods and services on offer either from its own sellers or from those trading from other countries, the country remains poor. The marketplace is out of balance. This is the situation in much of the world economy today, where growth is slow, because wealth has become concentrated in too few hands and there aren't enough buyers for all the goods and services that might otherwise be on offer. Let this go on long enough, and the entire marketplace withers. Producers of goods and services move on to other marketplaces. Nations rise and fall in wealth, just like technology platforms.

Getting a robust marketplace off the ground and keeping it often requires strong government intervention. In his book *Bad Samaritans*, Korean economist Ha-Joon Chang describes how South Korea used central planning and targeted investment in specific industries to build a highly successful economy. "Korea, one of the poorest places in the world, was the sorry country I was born into on October 7, 1963," he writes. "Today, I am a citizen of one of the wealthier, if not wealthiest, countries in the world. . . . The material progress I have seen in my 40-odd years is as though I had started life as a British pensioner born when George III was on the throne." This transformation was due in large part to forceful government management of the Korean economy, protecting its young industries and designing a deliberate ladder by which the country would focus on successively higher-value products. Recent work makes the same point about the early United States. In *Concrete Economics*, Stephen Cohen and Brad DeLong review the lessons of history, going back to Alexander Hamilton and identifying the role of government intervention in each great step forward in the American economy.

The rules of a technology platform can be loose, as they are with Google's Android app ecosystem, or tight, like the iPhone's more tightly managed platform. This is as true for nations as it is for smartphones. *There's more than one way to succeed in creating a successful platform.*

GOVERNING PLATFORMS, PLATFORMS FOR GOVERNING

Technology platforms and government have much to learn from each other.

Government and tech platforms must each provide core services that the "apps" or other services rely on. Despite the prevailing belief that the United States economy is largely a "free market," none of it works without fundamental infrastructure. In the 1930s, the Tennessee Valley Authority and the Rural Electrification Administration built dams and power distribution systems and established the idea that access to electricity was a fundamental right of every citizen. Telecommunications followed the same pattern, with a commitment to universal service enforced by the Federal Communications Commission. And of course, our national highway system, created in the 1950s, enabled interstate commerce and accelerated the growth of our economy. These are among the fundamental platform services of our country, just as the functions of access to the underlying processor, memory, sensors, and capabilities of the phone are platform services of the iPhone, and payment, distribution, security, discovery, and so forth are fundamental platform services of the App Store.

Each must also create and enforce a rule of law as part of those core services. If Google had let companies providing low-quality information dominate search results, people would have moved to Microsoft Bing or some other search engine, so Google has invested enormous resources in clarifying acceptable behavior and punishing bad actors. If the App Store lets you download an app that steals your personal information or your money, you will think twice before downloading another app, so Apple too has a robust security, quality assurance, and monitoring infrastructure to keep that from happening. The rule of law in platforms and in government is not just about justice and peace, it enables commerce; people don't do business where they can't rely on the rules being enforced.

Each must also invest in innovation to drive opportunity. The multitouch interface of the iPhone was an innovation that paid off not only for Apple, but for many people who chose to build on or use the platform. In the same way, government investments in breakthrough innovation pay off in unexpected ways. The fundamental technology of digital computing was developed by the military during World War II, then placed into the public domain. IBM then used it to transform itself from a manufacturer of mechanical tabulating machines into the dominating, monopolistic giant of a new era. Similar wartime investments turbocharged the aerospace industry, plastics, and chemicals. During the Cold War, the military developed technologies such as the Internet and GPS satellites that, when opened up to the private sector, led to the digital world we know today.

More recently, initiatives such as the Human Genome Project and the White House BRAIN Initiative are pushing the boundaries of basic research in areas that may well be central to the next technology boom, the next platform, and the next economy once the digital realm we are so fixated on today fades into the background of the everyday, just like previous unicorn technologies.

Each charges for its services. On a private platform like the App Store, developers have accepted that 30% is the tax they have to pay to Apple for all the services it provides to the economy it supports. People also take for granted that platforms like Uber and Lyft take a cut from their drivers, and Amazon a cut from its resellers. So too, in a democratic society, people tax themselves to pursue common goals, to finance the platform upon which society builds. In a closed society, those in power extract rents from those who use the platform. But one way or another, we must pay. The question is how much, and whether we think what we get for what we pay is worth it.

And that's why, for each, performance matters. If an app, or your phone itself, is slow, unreliable, or hard to use, you look for a better alternative. Over the recent history of the United States, we've seen a growing disdain for government and its role in our society. It is characterized as bloated, inefficient, and out of touch. Like any system that has grown by accretion for hundreds of years, our government processes, structure, and regulations are all in serious need of an overhaul. And in the 2016 election, frustration with that bloat contributed to an unprecedented change in direction whose consequences are just beginning to play out.

What the most frustrated citizens of our country likely don't realize is that the mechanisms for reinventing the platform of our government for the twenty-first century have been quietly emerging.

CODE FOR AMERICA

After the first year of the Gov 2.0 events, Jennifer Pahlka, who was TechWeb's general manager for our events partnership, became obsessed with an idea. She was spending half her time on our Web 2.0 event, with a front-row seat to the emergence of Facebook, Twitter, and the iPhone, and half her time on our Gov 2.0 event, inspired by the nexus of interest in government as a platform, but also learning for the first time how government builds and buys its software.

As Eric Schmidt had encouraged me to do, we were making the rounds in Washington and talking to a lot of people, and a lot of what we heard were stories of technology project failures, of systems many years (sometimes decades) in the making that either flat-out didn't work or worked so poorly that users preferred the previous system, even if that meant everything still had to be done on paper. The contrast between the day-to-day practices of these two worlds could not have been more different. The conference was coming together as a meaningful place for dialogue between the two, but Jen wanted to *do* something about the discrepancy she saw.

We both saw the opportunity for government to improve by applying the basic practices of the consumer tech industry, but Jen also saw the human consequences. Before Jen worked in the tech media business, she had taken a job out of college working in a child welfare agency. She could draw a straight line from the failures of these technology projects to the children in the care of the state and the social workers and administrators charged with keeping them safe, whose software and systems worked so poorly. Too often, the systems made it harder, not easier, to take care of these vulnerable children.

Jen resigned from TechWeb in late 2009 and, funded by credit card debt and small planning grants from the Sunlight Foundation and the Abrons Foundation, started Code for America, a nonprofit that aimed to bring government's technology competence up to par with that of the consumer tech world. She decided to start with cities, inspired by her friend Andrew Greenhill, the chief of staff for the mayor of Tucson, who pointed out that local government presented not only less red tape, but also more opportunity to touch the public. The cities that Code for America selected through a competitive application process would each get a team of programmers, designers, and others recruited from the consumer tech industry to do a year of service building apps.

Jen asked me to join the board of directors, and I enthusiastically agreed. Others turned out to be enthusiastic as well: 525 tech industry professionals applied to the program; we selected just twenty of them to work with four cities: Boston, Philadelphia, Seattle, and Washington, DC. The fellowship formally launched in January 2011 with a month of training, with the fellows going to their cities in February for onboarding.

Over the next few years, we helped local governments create a series of apps that, as first-year fellow and former Apple designer Scott Silverman said, were "simple, beautiful, and easy to use." A school choice website in Boston; a system for tracking blighted properties in New Orleans; a crowdsourcing app for clearing snow from fire hydrants that, being open source, spread to numerous other cities and was used for other forms of citizen participation, such as clearing storm drains and, in Honolulu, reporting back on whether the tsunami sirens were operational. In Santa Cruz, the fellows built a portal for easier small business permitting; another group of fellows, working in their spare time, built an easy way to model new public transit routes for any city.

The speed with which the fellows could build and stand up new applications shocked city staff. The first version of the Boston school choice site was built in about six weeks. City IT staff later marveled that if they had gone through a normal procurement process, the site would have cost them $2 million and taken two years. Allen Square, the CIO of New Orleans, made similar comments about the blight tracking tool.

More important, the fellows' work showed that it is possible to improve the performance of the platform of government. It's not just that these apps cost less and are developed faster, it's that they work for their users. In the case of the Boston school assignment app, the status quo was a twenty-eight-page brochure printed in eight-point type. It contained a lot of information, but like many government publications, it could not address any individual situation because each required calculating the distance from a child's address to the possible schools, so it effectively couldn't do the job its users needed it to do. The frustration parents felt trying to navigate the school selection process without these tools had led to a yearlong series of articles in the *Boston Globe* that followed the struggles of families trying to navigate the maze. The school choice app was a win not just for Boston families but for embattled politicians.

Building an app using consumer technology talent and user-centered practices (and without going through government procurement channels) was a powerful way to show that our government doesn't have to be bloated, inefficient, and out of touch. Instead of bemoaning the inevitable state of government, Code for America promised everyone (not just our government partners and the programmers and designers who raise their hands) that government could work as the public expected it to. Our theory of change was that the apps would be taken over by the local governments we built them for, and that, being open source, they could be spread by volunteers, organized in a chapter organization called the Code for America Brigade. Unlike Teach For America, which scaled by recruiting tens of thousands of volunteer teachers, our goal was primarily to scale through code, like other open source and Internet applications.

The 2012 fellowship opened up new possibilities for impact. That year, four of the fellowship teams decided that they wanted to base a startup on the project that they'd developed. Following their year of service, the teams continued to develop their projects and sell them to other cities.

Ron Bouganim, a successful venture investor, volunteered to run an incubator and an accelerator for civic startups. After two years of that, he raised a venture fund, the Govtech Fund, specifically to invest in companies that bring twenty-first-century best practices to government technology. A number of the startups spun up by Code for America fellows have been acquired; others have received significant venture funding. Remix, the app that was started as a way for citizens to reimagine transit routes in their city, developed into a powerful tool for urban planners and was funded by top VCs who gave it a valuation of $40 million.

The vision of how government could become a platform by emulating the Apple App Store was off to a good start.

APPS TO OPS

In 2013, though, the Code for America project with the city and county of San Francisco opened our eyes to an even more transformative opportunity. San Francisco asked us to work on the Supplemental Nutrition Assistance Program, or SNAP, more commonly referred to as food stamps. This is a federal program that, like many entitlement programs, is administered locally by states and counties. The problem that the Human Services Agency of San Francisco brought to Code for America was this: People were signing up for SNAP benefits, but then, a few months into the program, they were falling off the rolls, and then having to reapply.

This wasn't a problem that could be fixed with an app. The fellows were being asked to "debug" the operations of a government program. They asked if they could apply for food stamps (but not use the benefits), and began to work their way through the process as its normal applicants did, experiencing the process from the outside in.

The fellows tumbled into a world familiar to readers of Joseph Heller's *Catch-22*, or viewers of Terry Gilliam's *Brazil*. Letters arrived in the mail written in incomprehensible legal language, language that even agency employees couldn't understand, let alone the intended recipients, for some of whom English might be a second language.

Some of them might not even receive the letter, having no fixed address. Sometimes the letters were in another language—one English speaker received his letter in Chinese. Some of the letters setting an appointment for a follow-up interview, they discovered, were sent out after the date of the interview. Files requested during the application process were thrown away while the applicant was told they had been successfully filed, but at the agency it appeared that they had never been submitted.

The fellows were so energized by the project that one of them, Jake Solomon, didn't want to quit when the year was up. He was joined by two other fellows who'd worked on projects in other cities, Alan Williams and Dave Guarino. They continued to work unpaid until the organization was able to raise funding to formally continue the project. Alan slept on Jake's couch and went on food stamps himself, this time for real.

They discovered that the online application to the program was itself a stumbling block. Fifty screens long and taking forty-five minutes to complete even with the aid of a social worker, asking many questions that were irrelevant to a specific applicant, it was impossible to use on a mobile phone, even though about half of all online searches for the program come from mobile devices. Rather than realizing the possibility that a digital application could create branching sets of questions, the web application simply duplicated all of the questions on comprehensive paper forms.

Mainly as a way to collect data, the team created a user-friendly mobile application, GetCalFresh, that allows applicants to start the application, attach documents, and request an interview in less than eight minutes. This app was the key to their user research, because it allowed them to follow users through the process, keeping in touch with them by text message and, with their permission, tracking some of their data. But the app was also adopted by six other California counties, who found it superior to their existing online benefits application. It is currently being expanded to cover all fifty-eight California counties.

We had three important realizations as a result of this project.

The first was that twenty-first-century apps could only take us so far if they were built on top of a broken twentieth-century government platform. Simply putting a digital front end on a broken bureaucratic system often only makes the problem worse, because the digital system replicates existing processes without rethinking them from the ground up. Before we can build apps that really transform the experience of government for citizens, especially the neediest, we have to improve the underlying operations of government services. Just as the point of Uber ultimately isn't the experience of the app you use on your phone, but the entire service that gets you seamlessly from point A to point B, the point of food stamps isn't the online application process, but rather the ability to buy healthy food for your family. Our work on SNAP taught us that in too many government services, so much happens to users after the application that degrades or even prevents the delivery of the actual service.

The second realization was that understanding service delivery is the key to good policy making. In the course of their work, the Code for America teams encountered policies and regulations that seemed relatively innocuous, but that hindered the delivery of services and complicated matters for both the government office and the user. For example, the well-intentioned policy of adding questions to help register food stamps applicants to vote during the application process unintentionally creates confusion (and risk) for applicants who aren't actually eligible to register to vote. Systems designed to help people in difficult circumstances instead end up imposing enormous burdens on them. Because policy makers so seldom get to see what users experience, they have limited insight into the real-world effects of their policies. But when users' experiences can be made visible, the same iterative, data-driven practices that the Code for America teams used to create great apps can be used to develop or change policies and regulations to get closer to the intended outcomes.

It turned out that the apps we were building provided another way to improve government performance, by giving us insight into the processes behind them. Every Silicon Valley firm builds two intertwined systems: the application that serves users, and a hidden set of applications that they use to understand what is happening so that they can continuously improve their service. The Code for America team realized that their SNAP application app was a way to acquire users and then follow up with them to track and document their journey through the operations of the service. Once you can see what's going wrong, you can work with government to fix it. Jen calls this strategy "apps to ops."

Government can do better; it needs to reinvent itself more deeply, much like Amazon reconceived and rebuilt its e-commerce application into a cloud computing platform on which their own website now runs, along with thousands of other applications, or as Travis Kalanick and Garrett Camp rethought how taxi service could be delivered in the age of ubiquitous smartphones. In some corners, as we will see, government is doing just that.

Our third realization was brilliantly expressed in Jake's account of the project, a *Medium* essay titled "People, Not Data"—that empathy, not just technology, is the key to successfully reinventing government services. Design and user experience, not just big data and programming, are essential skills. Most of all, government decision makers have to put themselves in the shoes of those they mean to serve.

This is particularly true of services for those who need the help of government the most. These are the services that well-intentioned lawmakers and their well-heeled donors are least likely to interact with. Noting that much of the reporting about the failure of healthcare.gov mistakenly characterized it as a one-time catastrophe rather than something that is endemic, Ezra Klein writes: "One privilege the insured and well-off have is to excuse the terrible quality of services the government routinely delivers to the poor. Too often, the press ignores—or simply never knows—the pain and trouble of interfacing with government bureaucracies that the poor struggle with daily."

This realization led Code for America to change its focus to building better services for those who need them most. Code for America teams are now working to facilitate access to job training, to simplify communication for people trying to comply with the terms of their probation, and to make it easier for people with certain low-level, mostly drug-related felonies to remove them from their records so that they aren't barred from jobs, housing, and other necessities. As of this writing, the state of California, together with philanthropic donors including Reid Hoffman and the Omidyar Network, is funding an ambitious project to help Code for America build scalable digital services that we can take nationwide.

Meanwhile, back in Washington, DC, where our government-as-a-platform story started, the same realizations and transformations have been playing out in the federal government.

THE UNITED STATES DIGITAL SERVICE

While the Code for America fellows team in San Francisco was first exploring the problems with SNAP, Jen and I took a trip to London to visit the United Kingdom's Government Digital Service. While we were there, Jen got a call from Todd Park, at the time the CTO of the United States and a special assistant to the president. Todd had started a new program called the Presidential Innovation Fellows, loosely modeled on Code for America. He called to ask Jen to help him run the program, as Deputy CTO for Government Innovation.

She at first demurred, citing her commitment to Code for America. But Todd persisted. "Like water on stone," was how Nick Sinai, her soon-to-be colleague, described it. Eventually she agreed. "I'll come," she told Todd, "but only if you'll let me work on setting up a new unit like the UK GDS."

The GDS is a special unit, at the time reporting directly to the UK Cabinet Office, the group responsible for the operations of government. It had been set up in 2011 under the leadership of Mike Bracken, the former head of digital for the *Guardian*. Mike had soon attracted top talent from Britain's technology and digital media circles, and the GDS had been described by one prominent VC as "the best startup in Europe we can't invest in." Their complete redesign of the UK government's web strategy, replacing thousands of conflicting websites with one simple, user-centered hub, had won design awards that normally went to cutting-edge tech companies and had saved the UK government 60 million pounds. That turned out to be just a start.

One of the first things that struck Jen and me as we entered the

GDS office on an upper floor of an old office building high above a busy London street was a large sheet of butcher paper covering the picture window in the lobby. In the paper was a small cutout through which you could see the people on the street below. The cutout had a large arrow pointing to it, labeled "Users," reminding everyone when they walked in just whom the unit was meant to serve.

The GDS had captured this reminder in the first of its ten GDS Design Principles, which have since become a kind of bible of digital government, spelling out the key rules for designing great digital services. (These rules apply equally well to commercial services.) The first principle reads as follows:

> *Start with needs—user needs not government needs.* The design process must start with identifying and thinking about real user needs. We should design around those—not around the way the "official process" is at the moment. We must understand those needs thoroughly—interrogating data, not just making assumptions—and we should remember that what users ask for is not always what they need.

The second principle is also music to my ears, influenced by my own writing and speaking on Gov 2.0. It reads:

> *Do less.* Government should only do what only government can do. If we've found a way of doing something that works, we should make it reusable and shareable instead of reinventing the wheel every time. This means building platforms and registers others can build upon, providing resources (like APIs) that others can use, and linking to the work of others. We should concentrate on the irreducible core.

The other principles also echo so much that we had learned from technology: Design with data; Do the hard work to make it simple; Iterate. Then iterate again; Build digital services, not websites; Make things open.

The Code for America board granted Jen a one-year leave of absence, and she joined Todd at the White House in June 2013. I traveled with Jen to DC and watched as she worked through the obstacles to setting up the new service, including where it would be housed, and wrote a guiding document called *The Digital Services Playbook* with Haley Van Dyck, Charles Worthington, Nick Sinai, Ryan Panchadsaram, Casey Burns, and others. She and her colleagues and allies decided to call it the United States Digital Service, in homage to the GDS, and lobbied tirelessly for its creation.

The vision that Jen and others were championing was gaining some traction, but still far from a sure thing, when in October 2013, healthcare.gov launched . . . and fizzled. Suddenly improving government technology was not a theoretical exercise but a national emergency. The Obama administration's signature policy initiative was about to go down in flames because of the inability of the government to build a working website to process the applications.

A year and a half earlier, Tom Steinberg of UK nonprofit mySociety had offered a stark warning that now seemed eerily prescient: "[Y]ou [can] no longer run a country properly if the elites don't understand technology in the same way they grasp economics or ideology or propaganda. . . . What good governance and the good society look like is now inextricably linked to an understanding of the digital."

Certainly, the healthcare.gov crisis provided the urgency and justification for the USDS. It also provided its initial staffing and leadership. Todd Park had recruited two teams of talented technologists, primarily from Silicon Valley—one to patch together the dysfunctional website that had earned the administration such a black eye, and the second to build a much simpler version of the site using the best practices of a startup team rather than the antiquated technology procurement processes that had led to the disastrous first site. When the USDS was finally constituted in August 2014, Mikey Dickerson, the former Google Site Reliability Engineer (SRE) who had played a key role in the healthcare.gov rescue, became its first administrator.

It's significant that the first leader of the USDS was an SRE. The unit already had in its DNA a deep commitment to user-centered service design, through its initial inspiration by the UK's Government Digital Service and the Code for America team working on food stamps in California. But Site Reliability Engineering is at its core the practice of "debugging" the disconnect between software development and operations and building new connective tissue, and that is exactly what the federal government needed.

In its first two years, the USDS engaged hands-on with high-priority projects at federal agencies, including streamlining disability claim processing at the Department of Veterans Affairs, improving the visa processing system at the State Department, working with the Department of Education to help students make more informed college choices, and identifying security vulnerabilities in Department of Defense websites. In addition, it is working on modernizing procurement processes for digital services, and expanding the use of common platforms and tools. The USDS also now has branches at seven cabinet-level agencies—teams that operate using the USDS playbook, but who work directly within and for the agencies.

With all that the USDS has done to prove that government can be competent—even great—at technology, in the end the most valuable lesson we learn from this experiment with the US federal government is the same one we learn with local government and with national government in the United Kingdom. To succeed, platforms can't just offer apps, or services. They have to effectively set and adjust the rules that govern the behavior of the platform participants.

We also learned that the practices that make good apps turn out to be very relevant for making good rules as well.

Take, for example, the regulations derived from a law called MACRA (the Medicare Access and CHIP Reauthorization Act of 2015). After the near-death experience of healthcare.gov, it was not surprising that the MACRA team wanted the Department of Health and Human Services' Digital Service team to build the website that would implement this law, designed to allow Medicare to pay more for better care. But by now, leaders in the White House and beyond had learned an important lesson: As Cecilia Muñoz, head of the Domestic Policy Council under President Obama, said at a White House event on December 16, 2016, "Don't wait till you're building your website to invite the tech people to the table." When the MACRA team approached Mina Hsiang, the head of the HHS Digital Service, about the project, Mina proposed something a little different.

What usually happens is that before regulators engage a tech team around a website for users (in this case, doctors and other providers of medical care), they spend many months of study and research, producing a specification describing in great detail the rules the web application will encode. Mina proposed that the team writing the regulations give them an early draft in about a fifth of the time it would normally take them, and let her team do an early version of the website based on that draft.

It's normal practice for a tech team to test a site with users early in the development process; what was different this time was that the regulators could also see how users experienced and interpreted the rules they'd written, and change their language based on user behavior. They could then test the new language in a subsequent (still draft) version of the site, as the tech team put out new versions of the website to their test users. They did this four more times before the regulators called the rules final.

The MACRA regulators gushed that they'd just written the best rules of their career, having benefited for the first time from real-world feedback during the process.

At both Code for America and the USDS, we have learned one last lesson: As everyone in Silicon Valley's fierce contest for the best people knows, talent is essential. Mikey Dickerson put it bluntly in a call to technologists to consider public service at the 2016 South by Southwest (SXSW) conference in Austin. "Some of you, not all of you, are working right now on another app for people to share pictures of food or a social network for dogs. I am here to tell you that your country has a better use for your talents." He listed a set of urgent problems the government needs help with, and concluded, "All of these are design and information-processing problems, and all of these are matters of life or death to millions of citizens and all of them are things you can fix if you choose to."

> *Choose.* We hear that word again. The future depends on what we choose.

There are indications as I write that the work of the USDS will continue under the Trump administration. Better government is a non-partisan issue. That being said, there are troubling signs of a different choice. The Trump administration has reversed many of the open data policies of the Obama administration, and has generally favored ideology over evidence. In its quest to "deconstruct the administrative state," it is taking a wrecking ball to bureaucracy, but it isn't clear what it will be replaced with. Without proactive leadership, it is likely to be replaced with more of the same.

We have an opportunity to reinvent government. We must not let the opportunity slip by.

Clay Johnson, one of the cofounders of election technology firm Blue State Digital, later the head of Sunlight Labs at the Sunlight Foundation, a government transparency organization, and after that a White House Presidential Innovation Fellow, likes to point out that Moore's Law has an alarming consequence for government. If government's slow, change-resistant technology procurement processes mean that it is five or six years behind the private sector, the three or four exponential generations of Moore's Law that have passed will make its capabilities ten times worse.

And in classic "news from the future" style, that's exactly what we see. Amazon can deliver packages within hours of your order; Google can tell you in near-real time that there's an accident up ahead and to take a different route. Yet the VA takes eighteen months just to determine whether discharged soldiers are eligible for benefits.

Governments around the world, and in the United States at the federal, state, and local level, nonprofits like Code for America, venture funds such as the Govtech Fund and Ekistic Ventures, and an increasing number of commercial companies are working to bring the best practices of technology to government, closing the gap between what is and what could be. The flip side of every problem is an opportunity.

Abraham Lincoln said it well: "The legitimate object of government is to do for the people what needs to be done, but which they cannot, by individual effort, do at all, or do so well, for themselves."

While it's popular in Silicon Valley, which often leans libertarian, to deride the intrusive role of government, reinventing government to bring it up to date with the rest of society is one of the grand challenges of the twenty-first century.

PART III
A WORLD RULED BY ALGORITHMS

The hope is that, in not too many years, human brains and computing machines will be coupled together very tightly, and that the resulting partnership will think as no human brain has ever thought and process data in a way not approached by the information-handling machines we know today.

—J. C. R. Licklider, 1960

PART III

A WORLD RULED BY ALGORITHMS

8

MANAGING A WORKFORCE OF DJINNS

IN 2016, MIT'S *SLOAN MANAGEMENT REVIEW* ASKED ME TO contribute a short essay on the future of management. At first, I told them I had nothing much to say, or at least nothing that hadn't long ago been said. But then I realized that I was responding to the question using an old map.

If you think with a twentieth-century factory mindset, you might believe that the tens of thousands of software engineers at companies like Google, Amazon, and Facebook spend their days grinding out products just like their industrial forebears, only today they are producing software rather than physical goods. If, instead, you step back and view these companies with a twenty-first-century mindset, you realize that a large part of what they do—delivering search results, news and information, social network status updates, relevant products for purchase, and drivers on demand—is done by software programs and algorithms. *These programs are workers, and the programmers who create them are their managers.* Each day, these "managers" take in feedback about their workers' performance, as measured in real-time data from the marketplace, and if necessary, they give feedback to the workers in the form of minor tweaks and updates to the program or the algorithm.

The tasks performed by these software workers reflect the operational workflow of the digital organization. At an e-commerce site, you can imagine how one electronic worker helps the user find possible products that might match his or her search. Another shows information about the products. Yet another suggests alternative choices. Once the customer has chosen to buy a product, a digital worker presents a web form requesting payment and validates the input (for example, checking whether the credit card number provided is valid or whether the password presented matches the one that is stored). Another worker creates an order and associates it with the customer's record. Yet another constructs a warehouse pick list to be executed by a human or a robot. One more stores data about that transaction in the company's accounting system, and another sends out an email acknowledgment to the customer.

In an earlier generation of computing, these actions might be taken by a single monolithic application responding to the requests of a single user. But modern web applications may well be servicing millions of simultaneous users, and their functions have been decomposed into what are now called "microservices"—collections of individual functional building blocks that each do one thing, and do it very well. If a traditional monolithic application like Microsoft Word were reimplemented as a set of microservices, you could easily swap out the spell-checker for a better one, or add a new service that would turn web links into footnotes, or the reverse.

Microservices are an evolution of the communications-oriented design pattern that we saw in the design of Unix and the Internet, and in Jeff Bezos's platform memo. Microservices are defined by their inputs and outputs—how they communicate with other services—not by their internal implementation. They can be written in different languages, and run cooperatively on multiple machines; if designed correctly, any one of them can be swapped out for an improved component that performs the same function without requiring the rest of the application to be updated. This is what allows for continuous deployment, in which new features can be rolled out on a constant basis rather than in one big splash, and for A/B testing, in which alternate versions of the same feature can be tested on subsets of the user population.

THE UNREASONABLE EFFECTIVENESS OF DATA

As the scale and speed of Internet applications have grown, the nature of many of the software workers has also changed. It's a bit like the shift in aeronautics from propellers to jet engines. You can only go so fast with a motor that relies on mechanical pistons and rotating parts. A radically different approach was required, one that burns the fuel more directly. For a large class of applications, that jet engine has come in the form first of applied statistics and probability theory, then of machine learning and increasingly sophisticated AI algorithms.

In 2006, Roger Magoulas, O'Reilly Media's VP of research, first used the term *big data* to describe the new tools for managing data at the scale that enables the services of companies like Google. Former Bell Labs researcher John Mashey had used the term as early as 1998, but to describe the increasing scale of data that was being collected and stored, not the kind of data-driven services based on statistics, nor the software engineering breakthroughs and business processes that make these services possible.

Big data doesn't just mean a larger-scale version of a relational database like Oracle. It is something profoundly different. In their 2009 paper, "The Unreasonable Effectiveness of Data," (a homage in its title to Eugene Wigner's classic 1960 talk, "The Unreasonable Effectiveness of Mathematics in the Natural Sciences"), Google machine learning researchers Alon Halevy, Peter Norvig, and Fernando Pereira explained the growing effectiveness of statistical methods in solving previously difficult problems such as speech recognition and machine translation.

Much of the previous work had been grammar based. Could you construct what was in effect a vast piston engine that used its knowledge of grammar rules to understand human speech? Success had been limited. But that changed as more and more documents came online. A few decades ago, researchers relied on carefully curated corpora of human speech and writings that, at most, contained a few million words. But eventually, there was so much content available online that the game changed profoundly. In 2006, Google assembled a trillion-word corpus for use by language researchers, and developed a jet engine to process it. Progress since then has been swift and decisive.

Halevy, Norvig, and Pereira noted that in many ways, this corpus, taken from the web, was far inferior to the curated versions used by previous researchers. It was full of incomplete sentences, grammatical and spelling errors, and was not annotated and tagged with grammatical constructs. But the fact that it was a million times larger outweighed all those drawbacks. "A trillion-word corpus—along with other Web-derived corpora of millions, billions, or trillions of links, videos, images, tables, and user interactions—captures even very rare aspects of human behavior," they wrote. Instead of building ever-more-complex language models, researchers began to "make use of the best ally we have: the unreasonable effectiveness of data." Complex rule-based models were not the path to language understanding; they should just use statistical analysis and let the data itself tell them what the model should be.

While this paper was focused on language translation, it summed up the approach that has been essential to the success of Google's core search service. Its insight, that *simple models and a lot of data trump more elaborate models based on less data*," has been fundamental to progress in field after field, and is at the heart of many Silicon Valley companies. It is even more central to the latest breakthroughs in artificial intelligence.

In 2008, D. J. Patil at LinkedIn and Jeff Hammerbacher at Facebook coined the term *data science* to describe their jobs, naming a field that a few years later was dubbed by *Harvard Business Review* as "the sexiest job of the 21st century." Understanding the data science mindset and approach and how it differs from older methods of programming is critical for anyone who is grappling with the challenges of the twenty-first century.

How Google deals with search quality provides important lessons. Early on, Google made a commitment to build search results with statistical methods, with a strong bias against manual overrides to correct problems. A search for "Peter Norvig" should have things like his Wikipedia page and official company bio near the top. If some inferior page comes out on top, one way to fix it would be to add a rule "for the search 'Peter Norvig,' don't allow this inferior URL in the top 10." Google decided not to do that, but instead to always look for the underlying cause. In a case like this, the fix might be something like "On a search for any well-known person, give a lot of credit to high-quality encyclopedic sources (such as Wikipedia)."

The fitness function of Google's Search Quality team has always been relevance: Does the user appear to find what he or she was looking for? One of the signals Google now uses, which makes the concept very clear, is that of "the long click" versus "the short click." If a user clicks on the first search result and doesn't come back, she was presumably satisfied with the result. If the user clicks on the first search result, spends a modest amount of time away, and then comes back to click on the second result, he was likely not completely satisfied. If users come back immediately, that's a signal that what they found was not at all what they were looking for, and so on. If the long click happens on the second or third or fifth result more often than it does on the first, perhaps that result is the most relevant. When one person does this, it might be an accident. When millions of people make the same choice, it surely tells you something important.

Statistical methods are not only increasingly powerful; they are swifter and more subtle. If our software workers were once clanking robotic mechanisms, they are now becoming more like djinns, the powerful, independent spirits from Arabian mythology who can be coerced into fulfilling our wishes, but who so often artfully reinterpret the wish to their master's maximum disadvantage. Like the broom in Disney's version of *The Sorcerer's Apprentice*, algorithmic djinns do whatever it is that we ask them to do, but they are likely to be very single-minded and obtuse in interpreting it, with unintended and sometimes frightening results. How do we ensure that they do what we ask of them?

Managing them is a process of comparing the result of the programs and algorithms to some ideal target and testing to see what changes get you closer to that target. In the case of some work, such as Google's web crawl, the key functions to evaluate might be speed, completeness, and freshness. In 1998, when Google started, the crawl and the computed index of web pages was updated every few weeks. Today it happens nearly instantaneously. In the case of determining relevance, it is a matter of comparing the results of the program to what an informed user might expect. In the first implementation of Google, this practice was fairly primitive. In their original paper on Google Search, published while they were still at Stanford, Larry and Sergey wrote: "The ranking function has many parameters. . . . Figuring out the right values for these parameters is something of a black art."

Google says that the number of signals used to calculate relevance has grown to over 200, and search engine marketing guru Danny Sullivan estimates that there may be as many as 50,000 subsignals. Each of these signals is measured and calculated by a complex of programs and algorithms, each with its own fitness function it is trying to optimize. The output of these functions is a score that you can think of as the target of a master fitness function designed to optimize relevance.

Some of these functions, like PageRank, have names, and even research papers explaining them. Others are trade secrets known only to the engineering teams that create and manage them. Many of them represent fundamental improvements in the art of search. For example, Google's addition of what it called "the Knowledge Graph" allowed it to build on known associations between various kinds of entities, such as dates, people, places, and organizations, understanding for instance that a person might be "born on," an "employee of," a "daughter of" or "mother of," "living in," and so on. This work was based on a database created by a company called Metaweb, which Google acquired in 2010. When Metaweb unveiled its project in March 2007, I wrote enthusiastically, "They are building new synapses for the global brain."

Other components of the overall search algorithm were created in response to changing conditions in that global brain, the collective expression of billions of connected humans. For example, Google at first struggled to adapt to the real-time stream of consciousness coming from Twitter; the algorithms also had to be adjusted as smartphones made video and images as common on the Internet as text; as more and more searches were being made from mobile phones, devices whose precise location is known, local results became far more important; with the advent of speech interfaces, search queries became more conversational.

Google constantly tests new ideas that might give better results. In a 2009 interview, Google's then VP of search, Udi Manber, noted that they'd run more than 5,000 experiments in the previous year, with "probably 10 experiments for every successful launch." Google would launch a tweak to the algorithms or a new ranking factor on the order of 100 to 120 times a quarter, or an average of once a day. Since then, that speed has only accelerated. There were even more experiments on the advertising side.

How do they know that a change improves relevance? One way to evaluate a change is short-term user response: What are users clicking on? Another is long-term user response: Do they come back to Google for more? Another is talking to actual users one-on-one and asking them what they think.

Google also has a team of human evaluators check the results of a standardized list of common queries that are run automatically on a continuous basis. In the earliest days of Google, both the list of queries and the evaluation were done by the engineers themselves. By 2003 or 2004, Google had built a separate Search Quality team devoted to this effort. This team includes not just the search engineers but a statistically significant panel of external users who work Mechanical Turk–style, to give a thumbs-up or thumbs-down to a broad range of search results. In 2015, Google actually published the manual that they provide to their Search Quality raters.

It's important to remember, though, that when the raters find a problem, Google doesn't manually intervene to push the rank of a site up or down. When they find an anomaly—a case where the result the algorithm produces doesn't match what the human testers expect— they ask themselves, "What additional factors or different weighting can we apply in the algorithm that will produce the result we believe users are looking for?"

It's not always immediately obvious how to solve some search problems with pure ranking. At one point, the best algorithmically determined result for "Glacier Bay" turned up the Glacier Bay brand of faucets and sinks rather than the US national park of the same name. The algorithm was correct that more people were linking to and searching for Glacier Bay plumbing products, but users would be very surprised if the park didn't show up at the top of search results.

My own company, O'Reilly Media, was the subject of a similar problem. O'Reilly Media (at the time still called O'Reilly & Associates) was one of the earliest sites on the web and we published a lot of content—rich, high-quality pages that were especially relevant to the web's early adopters—so we had many, many inbound links. This gave us a very high page rank. At one point early in Google's history, someone published "the Google alphabet"—the top result for searching on a single letter. My company owned the letter *o*. But what about O'Reilly Auto Parts, a Fortune 500 company? They didn't even show up on the first page of search results.

For a brief time, until they came up with a proper algorithmic fix, Google divided pages like these into two parts. In the case of Glacier Bay, the national park occupied the top half of the search results page, with the bottom half given over to sinks, toilets, and faucets. In the case of O'Reilly, Bill O'Reilly and I came to share the top half while O'Reilly Auto Parts got the lower half. Eventually, Google improved the ranking algorithms sufficiently to interleave the results on the page.

One factor requiring constant adjustment to the algorithms is the efforts of the publishers of web pages to adapt to the system. Larry and Sergey foresaw this problem in their original search paper:

> Another big difference between the web and traditional well controlled collections is that there is virtually no control over what people can put on the web. Couple this flexibility to publish anything with the enormous influence of search engines to route traffic and companies which deliberately manipulate search engines for profit become a serious problem.

That was an understatement. Entire companies were created to game the system. Many of Google's search algorithm changes were responses to what came to be called "web spam." Even when web publishers weren't using underhanded tactics, they were increasingly struggling to improve their ranking. "Search engine optimization," or SEO, became a new field. Consultants with knowledge of best practices advised clients on how to structure their web pages, how to make sure that keywords relevant to the search were present in the document and properly emphasized, why it was important to get existing high-quality sites to link to them, and much more.

There was also "black hat SEO"—creating websites that intentionally deceive, and that violate the search engine's terms of service. Black hat SEO techniques included stuffing a web page with invisible text readable by a search engine but not by a human, and creating vast web "content farms" containing algorithmically generated low-quality content, including all the right search terms but little useful information that the user actually might want, pages cross-linked to each other to provide the appearance of human activity and interest. Google introduced numerous search algorithm updates specifically to deal with this kind of spam. The battle against bad actors is unrelenting for any widely used online service.

Google had one enormous advantage in this battle, though: its focus on the interests of the user, as expressed through measurable relevance. In his 2005 book, *The Search*, John Battelle called Google "the database of intentions." Web pages might use underhanded techniques to try to improve their standing—and many did—but Google was constantly working toward a simple gold standard: Is this what the searcher wants to find?

When Google introduced its pay-per-click ad auction in 2002, what had started out as an idealistic quest for better search results became the basis of a hugely successful business. Fortunately, unlike other advertising business models, which can pit the interests of advertisers against the interests of users, pay-per-click aligns the interests of both.

In the pay-per-impression model that previously dominated online advertising, and continues to dominate print, radio, and television, advertisers pay for the number of times viewers see or hear an ad (or in the case of less measurable media, how often they *might* see or hear it, based on estimates of readership or viewership), usually expressed as CPM (cost per thousand). But in the pay-per-click model, introduced by a small company called GoTo (later renamed Overture) in 1998, the same year Google was founded, advertisers pay only when a viewer actually clicks on an ad and visits the advertised website.

A click on an ad thus becomes similar to a click on a search result: a sign of user intention. In Overture's pay-per-click model, ads were sold to the highest bidder, with the company willing to pay the most to have their ad appear on a popular page of relevant search results getting the coveted spots. The company had achieved modest success with the model, but it didn't really take off till Google took the idea further. Google's insight was that the actual revenue from a pay-per-click ad was the combination of its price and the probability that the ad would actually be clicked on. An ad costing only $3 but twice as likely to be clicked on as a $5 ad would generate an additional dollar in expected revenue. Measuring the probability of an ad click and using it to rank the placement of an ad is obvious in retrospect, but like Amazon's 1-Click shopping or Uber's automatic payment, it was unthinkable to people wrapped in the coils of the prevailing paradigm for how advertising was sold.

This is a vast oversimplification of how Google's ad auction actually works, but it highlights the alignment of Google's search business model with its promise to users to help them find the most relevant results.

Facebook was not so lucky in finding alignment between the goals of its users and those of its advertisers.

Why? People don't just turn to social media for facts. They turn to it for connection with their friends, breaking news, entertainment, and the latest memes. In an attempt to capture these user goals, Facebook chose for its fitness functions measures of what they believe users find "meaningful." Like Google, Facebook uses many signals to determine what their users find most meaningful in their feed, but one of the strongest is what we might call "engagement." The omnipresent "Like" button on every post is one measure of engagement; users look for the endorphin rush that comes when their friends pay attention and give approval to the content they share. Facebook also measures clicks, just like Google, but the clicks they value most are not the ones that send people away, but the ones that keep them on the site, and searching for more like what they just saw.

The Facebook News Feed was originally a strict timeline of updates from the friends you'd chosen to follow. It was a neutral platform. But once Facebook realized that it could get higher engagement by promoting the most liked pages and the most clicked-on links to the top of the News Feed, sometimes showing them again and again, it became something like the television shopping channels of old.

In the early days of Internet commercialization, I had the opportunity to visit QVC, the granddaddy of television shopping, which was looking to build an online equivalent. Three rotating soundstages held products and the hosts who sold them to viewers by describing them in glowing terms. Immediately facing the stage was an analyst with a giant computer workstation, monitoring call volume and sales from each of the company's call centers in real time, giving the signal to switch to the next product only when attention and sales fell off. I was told that hosts were hired for their ability to talk nonstop about the virtues of a pencil for at least fifteen minutes.

That's the face of social media with engagement as its fitness function. Millions of nonstop hosts. Billions of personalized shopping channels for content.

And as was the case with Google, both legitimate players and bad actors soon were playing to the strengths and weaknesses of the algorithm. As Father John Culkin so aptly summarized the ideas of Marshall McLuhan, "We shape our tools, and thereafter our tools shape us." You choose the fitness function of your algorithms, and in turn, they shape your company, its business model, its customers, and ultimately our entire society. We'll explore some of the downsides of Facebook's fitness function in Chapter 10, and of financial markets in Chapter 11.

FROM JET ENGINES TO ROCKETS

If the introduction of probabilistic big data was like replacing a piston engine with a jet, the introduction of machine learning is like moving to a rocket. A rocket can go where a jet cannot, since it carries with it not only its own combustible fuel, but its own oxygen. This is a poor analogy, but it hints at the profound change that machine learning is bringing to the practices of even a company like Google.

Sebastian Thrun, the self-driving-car pioneer who led Google's early efforts in that area and is now the CEO of Udacity, an online learning platform, described how much the practice of software engineering is changing. "I used to create programs that did exactly what I told them to do, which forced me to think of every possible contingency and make a rule for every contingency. Now I build programs, feed them data, and *teach* them how to do what I want."

Using the old approach, a software engineer working on Google's search engine might have a hypothesis about a signal that would improve search results. She'd code up the algorithm, test it on some subset of search queries, and if it improved the results, it might go into deployment. If it didn't, the developer might modify her code and rerun the experiment. Using machine learning, the developer starts out with a hypothesis, just like before, but instead of producing a handcrafted algorithm to process the data, she collects a set of training data reflecting that hypothesis, then feeds the data into a program that outputs a model—a mathematical representation of features to be looked for in the data. This cycle is repeated again and again, with the program making minute adjustments to the model, gradually modifying the hypothesis using a technique such as gradient descent until it more perfectly matches the data. In short, the refined model is *learned* from the data. That model can then be turned loose on real-world data similar to that in the training data set.

Yann LeCun, a pioneer in a breakthrough machine learning technique called deep learning and now the head of the Facebook AI Research lab, uses the following analogy to explain how a model is trained to recognize images:

A pattern recognition system is like a black box with a camera at one end, a green light and a red light on top, and a whole bunch of knobs on the front. The learning algorithm tries to adjust the knobs so that when, say, a dog is in front of the camera, the red light turns on, and when a car is put in front of the camera, the green light turns on. You show a dog to the machine. If the red light is bright, don't do anything. If it's dim, tweak the knobs so that the light gets brighter. If the green light turns on, tweak the knobs so that it gets dimmer. Then show a car, and tweak the knobs so that the red light gets dimmer and the green light gets brighter. If you show many examples of the cars and dogs, and you keep adjusting the knobs just a little bit each time, eventually the machine will get the right answer every time. . . . The trick is to figure out in which direction to tweak each knob and by how much without actually fiddling with them. This involves computing a "gradient," which for each knob indicates how the light changes when the knob is tweaked. Now, imagine a box with 500 million knobs, 1,000 light bulbs, and 10 million images to train it with. That's what a typical Deep Learning system is.

Deep learning uses layers of recognizers. Before you can recognize a dog, you have to be able to recognize shapes. Before you can recognize shapes, you have to be able to recognize edges, so that you can distinguish a shape from its background. These successive stages of recognition each produce a compressed mathematical representation that is passed up to the next layer. Getting the compression right is key. If you try to compress too much, you can't represent the richness of what is going on, and you get errors. If you try to compress too little, the network will memorize the training examples perfectly, but will not generalize well to novel inputs.

Machine learning takes advantage of the ability of computers to do the same thing, or slight variations of the same thing, over and over again very fast. Yann once waggishly remarked, "The main problem with the real world is that you can't run it faster than real time." But computers do this all the time. AlphaGo, the AI-based Go player created by UK company DeepMind that defeated one of the world's best human Go players in 2016, was first trained on a database of 30 million Go positions from historical games played by human experts. It then played millions of games against itself in order to refine its model of the game even further.

Machine learning has become a bigger part of Google Search. In 2016, Google announced RankBrain, a machine learning model that helps to identify pages that are about the subject of a user's query but that might not actually contain the words in the query. This can be especially helpful for queries that have never been seen before. According to Google, RankBrain's opinion has become the third most important among the more than two hundred factors that it uses to rank pages.

Google has also applied deep learning to language translation. The results were so startlingly better that after a few months of testing, the team stopped all work on the old Google Translate system discussed earlier in this chapter and replaced it entirely with the new one based on deep learning. It isn't yet quite as good as human translators, but it's close, at least for everyday functional use, though perhaps not for literary purposes.

Deep learning is also used in Google Photos. If you have tried Google Photos, you've seen how it can recognize objects in your photos. Type "horse" and you will turn up pictures of horses, even if they are completely unlabeled. Type castle or fence, and you will turn up pictures of castles or fences. It's magical.

Remember that Google Photos is doing this on demand for the photos of more than 200 million users, photos that it's never seen before, hundreds of billions of them.

This is called supervised learning, because, while Google Photos hasn't seen your photos before, it has seen a lot of other photos. In particular, it's seen what's called a training set. In the training set, the data *is* labeled. Amazon's Mechanical Turk, or services like it, are used to send out pictures, one at a time, to thousands of workers who are asked to say what each contains, or to answer a question about some aspect of it (such as its color), or, as in the case of the Google Photos training set, simply to write a caption for it.

Amazon calls these microtasks HITs (Human Intelligence Tasks). Each one asks a single question, perhaps even using multiple choice: "What color is the car in this picture?" "What animal is this?" The same HIT is sent to multiple workers; when many workers give the same answer, it is presumably correct. Each HIT may pay as little as a penny, using a distributed "gig economy" labor force that makes driving for Uber look like a good middle-class job.

The role of Amazon's Mechanical Turk in machine learning is a reminder of just how deeply humans and machines are intertwined in the development of next-generation applications. Mary Gray, a researcher at Microsoft who has studied the use of Mechanical Turk, noted to me that you can trace the history of AI research by looking at how the HITs used to build training data sets have changed over time. (An interesting example is the update to Google's Site Rater Guidelines early in 2017, which was made, according to Paul Haahr, a Google search ranking engineer, in order to produce training data sets for the algorithmic detection of fake news.)

The holy grail in AI is unsupervised learning, in which an AI learns on its own, without being carefully trained. Popular excitement was inflamed by DeepMind's creators' claim that their algorithms "are capable of learning for themselves directly from raw experience or data." Google purchased DeepMind in 2014 for $500 million, after it demonstrated an AI that had learned to play various older Atari computer games simply by watching them being played.

The highly publicized victory of AlphaGo over Lee Sedol, one of the top-ranked human Go players, represented a milestone for AI, because of the difficulty of the game and the impossibility of using brute-force analysis of every possible move. But DeepMind cofounder Demis Hassabis wrote, "We're still a long way from a machine that can learn to flexibly perform the full range of intellectual tasks a human can—the hallmark of true artificial general intelligence."

Yann LeCun also blasted those who oversold the significance of AlphaGo's victory, writing, "most of human and animal learning is unsupervised learning. If intelligence was a cake, unsupervised learning would be the cake, supervised learning would be the icing on the cake, and reinforcement learning would be the cherry on the cake. We know how to make the icing and the cherry, but we don't know how to make the cake. We need to solve the unsupervised learning problem before we can even think of getting to true AI."

At this point, humans are always involved, not only in the design of the model but also in the data that is fed to the model in order to train it. This can result in unintended bias. Possibly the most important questions in AI are not the design of new algorithms, but how to make sure that the data sets with which we train them are not inherently biased. Cathy O'Neil's book *Weapons of Math Destruction* is essential reading on this topic. For example, if you were to train a machine learning model for predictive policing on a data set of arrest records without considering whether police arrest blacks but tell whites "don't let me catch you doing that again," your results are going to be badly skewed. The characteristics of the training data are much more important to the result than the algorithm. Failure to grasp that is itself a bias that those who have studied a lot of pre–machine learning computer science will have trouble overcoming.

This unfortunate example also provides insight into how machine learning models work. There are many feature vectors in any given model, creating an n-dimensional space into which the classifier or recognizer places each new item it is asked to process. While there is fundamental research going on to develop entirely new machine learning algorithms, most of the hard work in applied machine learning involves identifying the features that might be most predictive of the desired result.

I once asked Jeremy Howard, formerly the CTO of Kaggle, a company that carries out crowdsourced data science competitions, what distinguished the winners from the losers. (Jeremy himself was a five-time winner before joining Kaggle.) "Creativity," he told me. "Everyone is using the same algorithms. The difference is in what features you choose to add to the model. You're looking for unexpected insights about what might be predictive." (Peter Norvig noted to me, though, that the frontier where creativity must be exercised has already moved on: "This was certainly true back when random forests and support vector machines were the winning technologies on Kaggle. With deep networks, it is more common to use every available feature, so the creativity comes in picking a model architecture and tuning hyperparameters, not so much in feature selection.")

Perhaps the most important question for machine learning, as for every new technology, though, is which problems we should choose to tackle in the first place. Jeremy Howard went on to cofound Enlitic, a company that is using machine learning to review diagnostic radiology images, as well as scanning many other kinds of clinical data to determine the likelihood and urgency of a problem that should be looked at more closely by a human doctor. Given that more than 300 million radiology images are taken each year in the United States alone, you can guess at the power of machine learning to bring down the cost and improve the quality of healthcare.

Google's DeepMind too is working in healthcare, helping the UK

National Health Service to improve its operations and its ability to diagnose various conditions. Switzerland-based Sophia Genetics is matching 6,000 patients to the best cancer treatment each month, with that number growing monthly by double digits.

Tellingly, Jeff Hammerbacher, who worked on Wall Street before leading the data team at Facebook, once said, "The best minds of my generation are thinking about how to make people click ads. That sucks." Jeff left Facebook and now plays a dual role as chief scientist and cofounder at big data company Cloudera and faculty member of the Icahn School of Medicine at Mount Sinai, in New York, where he runs the Hammer Lab, a team of software developers and data scientists trying to understand how the immune system battles cancer.

The choice of the problems to which we apply the superpowers of our new digital workforce is ultimately up to us. We are creating a race of djinns, eager to do our bidding. What shall we ask them to do?

9

"A HOT TEMPER LEAPS O'ER A COLD DECREE"

I SPOKE IN EARLY 2017 AT A GATHERING OF MINISTERS FROM the Organisation for Economic Co-operation and Development (OECD) and G20 nations to discuss the digital future. One of the German ministers confidently asserted over lunch, "The only reason that Uber is successful is because it doesn't have to follow the rules." Fortunately, I was not the one who had to ask the obvious question. One of the OECD officials asked, "Have you ever ridden in an Uber?" "No," he admitted, "I have my own car and driver."

Of course, if you've ever used a service like Uber or Lyft, you know that the experience is far better than it is with taxis in most jurisdictions. The drivers are polite and friendly; they all use Google Maps or Waze to find the most efficient way to their destination; while there is no meter, you can get an estimate of the fare in advance and a detailed electronic receipt within seconds after you finish the trip, and you never have to fumble for cash or a credit card when you want to pay; but most important, you have a car on call to pick you up wherever you are, just like that German minister, except at a fraction of the price he pays.

Over the years, I've had similarly frustrating conversations with others charged with regulating or litigating a new technology. For example, during the controversy about Google Book Search back in 2005, I was asked to debate a lawyer for the Authors Guild, which had sued Google for scanning books in order to create a searchable index of their content. Only snippets of the content were shown in the book search index, just like the snippets of text from websites that show up in the normal Google index. The actual content could be viewed only with the permission of the publisher, with the exception of books that were known to be in the public domain.

"Scanning the books means they are making an unauthorized copy," she said. "They are stealing our content!" When I tried to explain that making a copy was an essential step in creating a search engine, and that Google Book Search worked exactly the same way as web search, it gradually dawned on me that she had no idea how Google Search worked. "Have you ever used Google?" I asked. "No," she said, adding (I kid you not), "but people in my office have."

The unintended consequences of simply trying to apply old rules and classifications in the face of a radically different model highlight the need for deeper understanding of technology on the part of regulators, and for fresh thinking on the part of both regulators and the companies they seek to regulate. Silicon Valley companies intent on "disruption" often see regulation as the enemy. They rail against regulations, or just ignore them. "A hot temper leaps o'er a cold decree," as Shakespeare's Portia put it in *The Merchant of Venice*.

Regulation is also the bête noir of today's politics. "We have too much of it," one side says; "We need more of it," says the other. Perhaps the real problem is that we just have the wrong kind, a mountain of paper rules, inefficient processes, and little ability to adjust the rules or the processes when we discover the inevitable unintended consequences.

RETHINKING REGULATION

Consider, for a moment, regulation in a broader context. Your car's electronics regulate the fuel-air mix in the engine to find an optimal balance of fuel efficiency and minimal emissions. An airplane's autopilot regulates the countless factors required to keep that plane aloft and heading in the right direction. Credit card companies monitor and regulate charges to detect fraud and keep you under your credit limit. Doctors regulate the dosage of the medicine they give us, sometimes loosely, sometimes with exquisite care, as with the chemotherapy required to kill cancer cells while keeping normal cells alive, or with the anesthesia that keeps a patient unconscious during surgery while keeping vital processes going. Internet service providers and corporate mail systems regulate the mail that reaches their customers, filtering out spam and malware to the best of their ability. Search engines and social media sites regulate the results and advertisements they serve up, doing their best to give us more of what we want to see.

What do all these forms of regulation have in common?

1. A clear understanding of the desired outcome.
2. Real-time measurement to determine if that outcome is being achieved.
3. Algorithms (i.e., a set of rules) that make continuous adjustments to achieve the outcome.
4. Periodic, deeper analysis of whether the algorithms themselves are correct and performing as expected.

There are a few cases—all too few—in which governments and quasi-governmental agencies regulate using processes similar to those outlined above. For example, central banks regulate the money supply in an attempt to manage interest rates, inflation, and the overall state of the economy. They have a target, which they try to reach by periodic small adjustments to the rules. Contrast this with the normal regulatory model, which focuses on the rules rather than the outcomes. How often have we faced rules that simply no longer make sense? How often do we see evidence that the rules are actually achieving the desired outcome?

The laws of the United States, and most other countries, have grown mind-bogglingly complex. The Affordable Care Act was nearly two thousand pages long. By contrast, the National Highway Bill of 1956, which led to the creation of the US Interstate Highway System, the largest public works project in history, was twenty-nine pages. The Glass-Steagall Act of 1933, which regulated banks after the Great Depression, was thirty-seven pages long. Its dismantling led to the 2008 financial crisis; the regulatory response this time, the Dodd-Frank Act of 2010, contains 848 pages, and calls for more than 400 additional bouts of rulemaking, in total adding up to as much as 30,000 pages of regulations.

Laws should specify goals, rights, outcomes, authorities, and limits.

If specified broadly and clearly, those laws can stand the test of time. Regulations, which specify how to execute those laws in much more detail, should be regarded in much the same way that programmers regard their code and algorithms, that is, as a constantly updated set of tools designed to achieve the outcomes specified in the laws.

Increasingly, in today's world, this kind of responsive regulation is more than a metaphor. New financial instruments are invented every day and implemented by algorithms that trade at electronic speed. How can these instruments be regulated except by programs and algorithms that track and manage them in their native element in much the same way that Google's search quality algorithms, Google's "regulations," manage the constant attempts of spammers to game the system? There are those who say that government should just stay out of regulating many areas, and let "the market" sort things out. But bad actors take advantage of a vacuum in the absence of proactive management. Just as companies like Google, Facebook, Apple, Amazon, and Microsoft build regulatory mechanisms to manage their platforms, government exists as a platform to ensure the success of our society, and that platform needs to be well regulated.

As the near collapse of the world economy in 2008 demonstrated, it is clear that regulatory agencies haven't been able to keep up with the constant "innovations" of the financial sector pursuing profit without regard to the consequences. There are some promising signs. For example, in the wake of Ponzi schemes like those of Bernie Madoff and Allen Stanford, the SEC instituted algorithmic models that flag for investigation hedge funds whose results meaningfully outperform those of peers using the same stated investment methods. But once flagged, enforcement still goes into a long loop of investigation and negotiation, with problems dealt with on a haphazard, case-by-case basis. By contrast, when Google discovers that a new kind of spam is damaging search results, they can quickly change the rules to limit the effect of those bad actors. And those rules are automatically executed by the system in pursuit of its agreed-on fitness function.

We need to find more ways to make the consequences of bad action systemic, part of a high-velocity workflow akin to the way that Internet companies use DevOps to streamline and accelerate their internal business processes. This isn't to say that we should throw out the concept of "due process" that is at the core of the Fifth Amendment, just that in many cases that process can be sped up enormously, and made fairer and clearer at the same time.

There are some important lessons from technology platforms. Despite the enormous complexity of the algorithmic systems used to manage platforms like Google, Facebook, and Uber, the fitness function of those algorithms is usually simple: Does the user find this information relevant, as evidenced by their propensity to click on it, and then go away? Does this user find this content engaging, as evidenced by their willingness to keep clicking on the next story? Is the user being picked up within three minutes? Does the driver have a rating above 4.5 stars?

Outside regulators should focus on defining the desired outcome, and measuring whether or not it has been achieved. They should also diagnose the delta between the intended outcomes and the fitness function of the algorithms being used by those they aim to regulate. That is, are the participants incented to achieve the stated goal of the regulation, or are they incented to try to thwart it? The best regulations encourage the regulated party to take on the problem themselves. This is not "self-regulation" in the sense that government simply trusts the market to do the right thing. Instead, it is a matter of creating the right incentives. For example, the Fair Credit Billing Act of 1974 made consumers responsible for only $50 of any fraudulent credit card charges, making it in the industry's own self-interest to police fraud aggressively.

Diego Molano Vega, the former minister of information technologies and communications in Colombia, told me how he'd used a similar approach to solve the chronic problem of dropped telephone calls by replacing a regime of fines and three-year-long investigations with a simple rule that telecom providers had to reimburse customers for the cost of every dropped call. After a year and $33 million in refunds, the problem was solved.

And of course, this is ultimately how Google regulated the problem of content farms, which produced content specifically designed to fool the search algorithms but that provided little value to users. Google didn't assess penalties. They didn't set detailed rules for what kind of content sites could publish. But by demoting these sites in the search results, they *created consequences* that led the bad actors to either improve their content or go out of business.

Andrew Haldane, the executive director for financial stability at the Bank of England, made a compelling case for simplicity in regulations in a 2012 talk to the Kansas City Federal Reserve called "The Dog and the Frisbee." He pointed out that while precisely modeling the flight of a Frisbee and running to catch it requires complex equations, simple heuristics mean that even a dog can do it. He traces the failures of financial regulation that led to the 2008 crisis in large part to the increase in their complexity, which made them almost impossible to administer. The more complex the regulations, the less likely they are to succeed, and the more fragile they are in the face of changing conditions.

The modernization of how data is reported to both the government and the market is an important way of improving regulatory outcomes. When reporting is on paper or in opaque digital forms like PDF, or released only quarterly, it is much less useful. When data is provided in reusable digital formats, the private sector can aid in ferreting out problems as well as building new services that provide consumer and citizen value. There's an entirely new field of regulatory technology, or RegTech, that uses software tools and open data for regulatory monitoring, reporting, and compliance.

Data-driven regulatory systems need not be as complex as those used by Google or credit card companies. The point is to measure the outcome, and to put any adverse consequences of divergence from the intended outcome on the appropriate parties. Too often, incentives and outcomes are not aligned. For example, government grants mobile phone carriers exclusive licenses to spectrum with the goal of creating reliable and universal access, yet spectrum licenses are auctioned off to the highest bidder. Is this approach giving the right outcome? The quality of mobile services in the United States would suggest otherwise. What if, instead, spectrum licenses were granted based on promises of maximum coverage? Much as Minister Molano Vega did for phone service in Colombia, rebates to customers for failures to live up to coverage promises could potentially create a much more self-regulating system.

THE ROLE OF SENSORS IN FUTURE REGULATION

Increasingly, our interactions with businesses, government, and the built environment are becoming digital, and thus amenable to creative forms of measurement, and ultimately responsive regulation. For example, fines are routinely issued to motorists running red lights or making illegal turns by cameras mounted over highly trafficked intersections. With the rise of GPS, we are heading for a future where speeding motorists are no longer pulled over by police officers who happen to spot them, but instead automatically ticketed whenever they exceed the speed limit.

We can also imagine a future in which that speed limit is automatically adjusted based on the amount of traffic, weather conditions, and other variable conditions that make a higher or lower speed more appropriate than the static limit that is posted today. The endgame might be a future of autonomous vehicles that are able to travel faster because they are connected in an invisible web, a traffic regulatory system that keeps us safer than today's speed limits. Speed might be less important than the quality of the algorithm driving the car, the fact that the car has been updated to the latest version, and that it is equipped with adequate sensors. The goal, after all, is not to have cars go more slowly than they might otherwise, but to make our roads safe.

Congestion pricing on tolls, designed to reduce traffic to city centers, is another example. Smart parking meters have similar capabilities—parking can cost more at peak times, less off-peak, just like plane tickets or hotel rooms. But perhaps more important, smart parking meters can report whether they are occupied or not, and eventually give guidance to drivers and car navigation systems, reducing the amount of time spent circling aimlessly looking for a parking space.

As we move to a future with more electric vehicles, there are proposals to replace the gasoline taxes with which we currently fund road maintenance with miles driven—reported, of course, once again by

GPS. Companies like Metromile already offer to base your insurance rates on how often and how fast you drive. It is only a small step further to do the same for taxes.

THE SURVEILLANCE SOCIETY

Living in a world of pervasive connected sensors questions our assumptions of privacy and other basic freedoms, but we are well on our way toward that world purely through commercial efforts. We are already being tracked by every site we visit on the Internet, through every credit card charge we make, through every set of maps and directions we follow, and by an increasing number of public or private surveillance cameras. Ultimately, science fiction writer David Brin got it right in his prescient 1998 nonfiction book, *The Transparent Society*. In an age of ubiquitous commercial surveillance that is intrinsic to the ability of companies to deliver on the services we ask for, the kind of privacy we enjoyed in the past is dead. Brin argues that the only way to respond is to make the surveillance two-way through transparency. To the Roman poet Juvenal's question "Who will watch the watchers?" ("Quis custodiet ipsos custodes?"), Brin answers, "All of us."

Security and privacy expert Bruce Schneier offers an important caveat to the transparent society, though, especially with regard to the collection of data by government. When there is a vast imbalance of power, transparency alone is not enough. "This is the principle that should guide decision-makers when they consider installing surveillance cameras or launching data-mining programs," he writes. "It's not enough to open the efforts to public scrutiny. All aspects of government work best when the relative power between the governors and the governed remains as small as possible—when liberty is high and control is low. Forced openness in government reduces the relative power differential between the two, and is generally good. Forced openness in laypeople increases the relative power [of government], and is generally bad."

We clearly need new norms about how data can be used both by private actors and by government. I love what Gibu Thomas, now the head of global commerce at PepsiCo, had to say when he was the head of digital innovation at Walmart. "The value equation has to be there. If we save them money or remind them of something they might need, no one says, 'Wait, how did you get that data?' or 'Why are you using that data?' They say, 'Thank you!' I think we all know where the creep factor comes in, intuitively."

This notion of "the creep factor" should be central to the future of privacy regulation. When companies use our data for our benefit, we know it and we are grateful for it. We happily give up our location data to Google so they can give us directions, or to Yelp or Foursquare so they can help us find the best place to eat nearby. We don't even mind when they keep that data if it helps them make better recommendations in the future. Sure, Google, I'd love it if you could do a better job predicting how long it will take me to get to work at rush hour. And yes, I don't mind that you are using my search and browsing habits to give me better search results. In fact, I'd complain if someone took away that data and I suddenly found that my search results weren't as good as they used to be.

But we also know when companies use our data against us, or sell it on to people who do not have our best interests in mind. If I don't have equal access to the best prices on an online site because the site has determined that I have either the capacity or willingness to pay more, my data is being used unfairly against me. In one notable case, Orbitz was steering Mac users to higher-priced hotels than they offered to PC users. This is data used for "redlining," so called because of the old practice of drawing a red line on the map to demarcate geographies where loans or insurance would be denied or made more costly because of location (often as a proxy for a racial profile). Political microtargeting with customized, misleading messages based on data profiling also definitely fails the creep factor test.

These people are privacy bullies, who take advantage of a power imbalance to peer into details of our private lives that have no bearing on the services from which that data was originally collected. Government regulation of privacy should focus on the privacy bullies, not on the routine possession and use of data to serve customers.

Regulators have to understand the fair boundary of the data transaction between the consumer and the service provider. It seems to me that insurance companies would be quite within their rights to offer lower rates to people who agree to drive responsibly, and to verify the consumer's claims of how many miles they drive annually or whether they keep to the speed limit, but if my insurance rates suddenly spike because of data about formerly private legal behavior, like the risk profile of where I work or drive for personal reasons, I have reason to feel that my data is being used unfairly against me.

The right way to deal with data redlining is not to prohibit the collection of data, as so many privacy advocates seem to urge, but rather, to prohibit its misuse once companies have that data. As David Brin once said to me, "It is intrinsically impossible to know if someone *does not* have information about you. It is much easier to tell if they *do* something to you."

Regulators should consider the possible harms to the people whose data is being collected, and work to eliminate those harms, rather than limiting the collection of the data itself. When people are denied health coverage because of preexisting conditions, that is their data being used against them; this harm was restricted by the Affordable Care Act. By contrast, the privacy rules in HIPAA, the 1996 Health Insurance Portability and Accountability Act, which seek to set overly strong safeguards around the privacy of data, rather than its use, have had a chilling effect on many kinds of medical research, as well as patients' access to their very own data.

As was done with credit card fraud, regulators should look to create incentives for companies themselves to practice the right behavior. For example, liability for misuse of data sold on to third parties would discourage sale of that data. A related approach is shown by legal regimes such as that controlling insider trading: If you have material nonpublic information obtained from insiders, you can't trade on that knowledge, while knowledge gained by public means is fair game.

Data aggregators, who collect data not in order to provide services directly to consumers, but to other businesses, should come in for particular scrutiny, since the data transaction between the consumer and the service provider has been erased, and it is far more likely that the data is being used not for the benefit of the consumer who originally provided it but for the benefit of the purchaser.

Disclosure and consent as currently practiced are extraordinarily weak regulatory tools. They allow providers to cloak malicious intent in complex legal language that is rarely read, and if read, impossible to understand. Machine-readable disclosure similar to those designed by Creative Commons for expressing copyright intent would be a good step forward in building privacy-compliant services. A Creative Commons license allows those publishing content to express their intent clearly and simply, ranging from the "All Rights Reserved" of traditional copyright to a license like CC BY-NC-ND (which requires attribution, but allows the content to be shared freely for noncommercial purposes, and does not allow derivative works). Through a mix of four or five carefully crafted assertions, which are designed to be both machine and human readable, Creative Commons allows users of a photo-sharing site like Flickr or a video-sharing site like YouTube to search only for content matching certain licenses. An equivalent framework for privacy would be very helpful.

During the Obama administration, there was a concerted effort toward what is called "Smart Disclosure," defined as "the timely release of complex information and data in standardized, machine readable formats in ways that enable consumers to make informed decisions." New technology like the blockchain can also encode contracts and rules, creating new kinds of "smart contracts." A smart contracts approach to data privacy could be very powerful. Rather than using brute force "Do Not Track" tools in their browser, users could provide nuanced limits to the use of their data. Unlike paper disclosures, digital privacy contracts could be enforceable and trackable.

As we face increasingly automated systems for enforcing rules, though, it is essential that it be possible to understand the criteria for a decision. In a future of what some call "algocracy"—rule by algorithm—where algorithms are increasingly used to make real-world decisions, from who gets a mortgage and who doesn't, to how to allocate organs made available for donation, to who gets out of jail and who doesn't, concern for fairness demands that we have some window into the decision-making process.

If, like me, you've ever been caught going through a red light by an automated traffic camera, you know that algorithmic enforcement can seem quite fair. I was presented with a time-stamped image of my car entering the intersection after the light had changed. No argument.

Law professor Tal Zarsky, writing on the ethics of data mining and algorithmic decision making, argues that even when software makes a decision based on thousands of variables, and the most that the algorithm creator could say is "this is what the algorithm found based on previous cases," there is a requirement for *interpretability*. If we value our human freedom, it must be possible to explain why an individual was singled out to receive differentiated treatment based on the algorithm.

As we head into the age of increasingly advanced machine learning, though, this may be more and more difficult to do. If we are not explicit about what regulatory regime—inherited or, optimally, to be devised—shall apply, expect lawsuits down the line.

REGULATION MEETS REPUTATION

It is said that "that government is best which governs least." Unfortunately, evidence shows that this isn't true. Without the rule of law, capricious power sets the rules, usually to the benefit of a powerful few. What people really mean by "governs least" is that the rules are aligned with their interests. In an economy tuned to the interests of the few, the rules are often unfair to the rest. An economy tuned to the interests of the majority may seem unfair to some, but John Rawls's "veil of ignorance"—the idea that the best rules for a political or economic order are those that would be chosen by people who had no prior knowledge of their place in that order—is a convincing argument that that government is best that governs *for most*.

That, as it turns out, is also the lesson of technology platforms. As we saw with TCP/IP, the rules should ideally be intrinsic to the design of the platform, not something added to it. But as long as the rules, however complex, are aligned with the simple interests of the participants, as is the case with Google's quest for relevance, regulation becomes largely invisible. Things just appear to work.

Reputation systems are one way that regulation is built into the design of online platforms. Amazon has consumer ratings for every one of millions of products, helping consumers make informed decisions about which products to buy. Sites like Yelp and Foursquare provide extensive consumer reviews of restaurants; those that provide poor food or service are flagged by unhappy customers, while those that excel are praised. TripAdvisor and other similar sites have had a similar effect in helping travelers discover the best places to stay in remote places around the world. These reviews help the sites to algorithmically rank the products or services that users are most likely to be satisfied with.

eBay, which grew out of Pierre Omidyar's quest to create a perfect marketplace, was a pioneer in reputation systems. eBay was faced with enormous challenges. Unlike Amazon, which began by selling products from familiar vendors, and was therefore just an online version of something familiar—a bookstore—eBay was the online version of a worldwide garage sale or swap meet, where the trust that is engendered by existing brands is absent.

In their paper "Trust Among Strangers in Internet Transactions: Empirical Analysis of eBay's Reputation System," economists Paul Resnick and Richard Zeckhauser point out that customers of an online auction site can't inspect the goods and make their own determination as to their quality; they rarely have repeated interactions with the same seller; and they can't learn about the seller from friends or neighbors. Especially in the early days, photographs and descriptions were often unprofessional, and little or nothing was known about the sellers. Not only was there risk that items were not as shown, or might be counterfeit, but there was a risk that they might never be delivered. And in 1995, when eBay and Amazon were established, using a credit card on the Internet was itself widely considered an unacceptable risk.

So, in addition to building a network of buyers and sellers, eBay had to build mechanisms for helping buyers and sellers to trust one another. The eBay reputation system, in which customers rated vendors and vendors rated customers, was one of their answers. It was widely emulated.

David Lang summarized the Internet's journey toward trust in a *Medium* post about the success of education crowdfunding site Donors-Choose. He points out that traditional charities typically give funds only to established nonprofits, in large chunks, usually with a great degree of oversight. By contrast, DonorsChoose allows individual teachers to advertise classroom needs, which can be met by either individuals or institutions. Describing other examples where technology has enabled trust, Lang wrote: "The novelty isn't the financial transaction—room renting, car sharing and art patronage has been around for centuries—the novelty is rather in the level of trust we're willing to extend to strangers because the apps and algorithms provide a filter."

As the battles of companies like Uber, Lyft, and Airbnb with regulators demonstrate, though, the journey toward trust requires more than just getting consumers on board. Logan Green told me that Lyft's original approval for peer-to-peer car-hire services from the California Public Utilities Commission was based on the argument that they could use technology to provide many of the same benefits as traditional taxi regulation. Passenger safety was of paramount importance to the CPUC. One key regulator, a former military officer known simply as "the General," reportedly said, "Nobody dies on my watch!" Logan said that his team was able to persuade the CPUC that the tracking of the ride via GPS, the reputation system, and careful vetting of the drivers were an effective way of meeting their mutual goals. "Safety is the most important thing to our users too," Logan told me. "So we said, 'Let's nail it!'"

But in many jurisdictions, reputation systems and traditional regulations are still on a collision course. Ostensibly, taxis are regulated to protect the quality and safety of the consumer experience, as well as to ensure that there are an optimal number of vehicles providing service at the time they are needed. In practice, most of us know that these regulations do a poor job of ensuring quality or availability. A strong argument can be made that the reputation system used by Uber and Lyft, by which passengers are required to rate their drivers after each ride, does a better job of weeding out bad actors. Certainly, I've had taxi drivers who would never have been able to offer a ride again if it were as easy to file a taxi complaint as it is to give a one-star rating.

However, this has not stopped opponents of the new services from claiming that the drivers provided by Uber and Lyft have been insufficiently vetted. While all of the new services perform driver background checks before they are allowed to offer rides, opponents argue that the checks are not strenuous enough because they don't require fingerprinting and FBI criminal background checks, an onerous and time-consuming step that, from the point of view of Uber and Lyft, is undesirable because it would limit the participation of part-time and occasional drivers, who provide a majority of the service on these new platforms. Uber and Lyft feel so strongly about this issue that they actually pulled their services from the city of Austin after it required fingerprinting and full FBI checks. Both companies claim that the background checks they perform, using a third-party service, actually provide better data on drivers.

In any event, it turns out that existing regulations for licensing drivers provided two intertwined functions: ensuring the quality of drivers, and, for a number of reasons, limiting the supply. According to Steven Hill, author of *Raw Deal*, a critical book about Uber, the first "taxi" regulations were promulgated in 1635 by King Charles I of England, who ordered that all vehicles on the streets of London needed to be licensed "to restrain the multitude and promiscuous use of coaches." The same thing happened in the United States during the Great Depression. People were desperate for work, and cars-for-hire clogged the streets. In 1933, a US Department of Transportation official wrote: "The excess supply of taxis led to fare wars, extortion and a lack of insurance and financial responsibility among operators and drivers. Public officials and the press in cities across the country cried out for public control over the taxi industry." As a result, cities imposed limits on the number of taxis using a "medallion" system. They awarded only a limited number of licenses to commercial drivers, and issued regulations on fares, insurance, vehicle safety inspections, and driver background checks.

This brief history illuminates how easy it is to mix up means and ends. If the problem is framed as "the multitude and promiscuous use of coaches," as King Charles I put it, limiting the number of licensed coaches looks functionally equivalent to the actual objective, which is eliminating congestion and pollution. (In 1635, horse manure was the equivalent of twentieth-century smog.) If, as the DOT official claimed in 1933, the excess of supply led to fare wars where no driver could make a decent living, thus leading to a decline in safety and lack of insurance on the part of drivers, the one-time solution, limiting the number of drivers and subjecting them to mandatory inspections, becomes a goal in and of itself. But to echo the refrain in Stephen King's *Dark Tower*, "The world has moved on," and perhaps there are now better solutions.

While there are still risks of bad drivers, and critics have made the most of crimes committed by Uber drivers, the fact that every Uber ride is tracked in real time, with the exact time, location, route, and the identity of both the driver and the passenger known, makes an Uber or Lyft ride inherently safer than a taxi ride. And the use of post-ride ratings by both passenger and driver helps, over time, to weed bad actors out of the system. Hal Varian put this in the broader context of how computer-mediated transactions change the regulatory game. "The entire transaction is monitored. If something goes wrong with the transaction, you can use the computerized record to find what went wrong."

And as to congestion, while the current algorithm is optimized to create shorter wait times, there is no reason it couldn't take into account other factors that improve customer satisfaction and lower cost, such as the impact of too many drivers on congestion and wait time. Algorithmic dispatch and routing is in its early stages; to think otherwise is to believe that the evolution of Google Search ended in 1998 with the invention of PageRank. For this multi-factor optimization to work, though, Uber and Lyft have to make a deep commitment to evolving their algorithms to take into account all of the stakeholders in their marketplace. It is not clear that they are doing so.

Understanding the differences between means and ends is a good way to help untangle the regulatory disagreements between the TNCs (transportation network companies) and taxi and limousine regulators.

Both parties want enough safe, qualified drivers available to meet the needs of any passenger who wants a ride, but not so many drivers that drivers don't make enough money to keep up their cars and give good service. The regulators believe that the best way to achieve these objectives is to limit the number of drivers, and to certify those drivers in advance by issuing special business licenses. Uber and Lyft believe that their computer-mediated marketplace achieves the same goals more effectively. Surely it should be possible to evaluate the success or failure of these alternative approaches using data.

As discussed in Chapter 7, there is a profound cultural and experiential divide here between Silicon Valley companies and government that is part of the problem. In Silicon Valley, every new app or service starts out as an experiment. From the very first day a company is funded by venture capitalists, or launches without funding, its success is dependent on achieving key metrics such as user adoption, usage, or engagement. Because the service is online, this feedback comes in near-real time. In the language of Eric Ries's popular Lean Startup methodology, the first version is referred to as "minimum viable product (MVP)," defined as "that version of a new product which allows a team to collect the maximum amount of validated learning about customers with the least effort." The goal of every entrepreneur is to grow that MVP incrementally till it finds "product-market fit," resulting in explosive growth.

This mindset is taught to every entrepreneur. Once an app or service is launched, new features are added and tested incrementally. Not only is the usage of the features measured, and those that are not adopted by users silently dropped or rethought, but different versions of each feature—the placement or size of a button, messaging, or graphic—are tested against random samples of users to see which version works better. Feedback loops are tight, and central to the success of the service.

By contrast, despite the changes beginning under the Obama administration that were described in Chapter 7, lawmakers and government regulators are accustomed to considering a topic, taking input from stakeholders in public meetings (and too often, in private meetings with lobbyists), making a considered decision, and then sticking with it. Measurement of the outcome, if it happens at all, perhaps comes in the form of an academic study years after the event, with no clear feedback into the policy-making process. I once came across a multimillion-dollar project to build a job search engine for veterans that had managed to reach only a few hundred users but was about to have its contract renewed. I asked a senior government official who had overseen the project whether they ever did the math to understand what their cost was for each user. "That would be a good idea," she said. A good idea? Any Silicon Valley entrepreneur who couldn't answer that question would be laughed out of the room. Tom Loosemore, the former chief operating officer of the UK Government Digital Service, speaking at the 2015 Code for America Summit, noted that the typical government regulatory framework represents "500 pages of untested assumptions."

Government technology procurement processes echo this same approach. A massive specification is written, encapsulating everyone's best thinking, and spelling out every detail of the implementation so that it can be put out to bid. The product typically takes years to develop, and the first time its assumptions are tested is when it is launched. (Note that while this may sound similar to the Amazon "working backwards" approach, it is actually very different. Amazon asks those tasked with doing the work to imagine the intended user experience, not to specify all of the implementation details in advance. As they build the actual product or service, they continue to learn and refine their ideas.)

Now, to be fair, many (though far from all) of the things that government regulates have far higher stakes than a consumer app. "Move fast and break things," Mark Zuckerberg's famous admonition to his developers at Facebook, hardly applies to the design of bridges, air traffic control, the safety of the food supply, or many of the other things that government regulates. Government also must be inclusive, serving all residents of the country, not just a highly targeted set of users. Nonetheless, there is a great deal for government to learn from the iterative development processes of modern digital organizations.

"Regulatory capture," the process by which companies that benefit from a regulation become parties to manage it, accelerates the confusion. I once had a conversation with former Speaker of the House Nancy Pelosi about a piece of legislation (the Stop Online Piracy Act, or SOPA). I told her, based on data from my company's publishing business, that online piracy was less of a problem than proponents of the bill were claiming. She didn't ask to look at my data, she didn't counter that proponents of the bill had offered different data. She said, "Well, we have to balance the interests of Silicon Valley against the interests of Hollywood."

I was shocked. It is as if Google's Search Quality team had sat down with representatives of search spammers and agreed to set aside a third of the top results in order to preserve their business model. In my mind, the job of our representatives is not to balance the interests of various lobbying groups, but to gather data and make an informed decision on behalf of the public.

I'm not saying that Silicon Valley always gets it right—it certainly doesn't get it right the first time—and government doesn't always get it wrong. While government is often too responsive to lobbyists, its fundamental goal is to look out for the interests of the public, including populations that would otherwise be ignored.

Getting extremely specific about the objectives of any regulation allows for a franker, more productive discussion. Both sides can debate the correct objectives. And when they have come to an agreement, they can start to look at alternative ways to achieve them, as well as how to measure whether or not they have succeeded. They should also define a process for modifying the regulation in response to what is learned through that measurement. And there must be a mechanism for resolving conflicts between overlapping regulations. If it is a complex regulation, this process should be followed for each subcomponent. The lessons about modularity from Jeff Bezos's platform memo are surprisingly relevant to the design of regulations as well as to platforms and modern technology organizations.

In this regard, I was heartened by the National Highway Traffic Safety Administration's 2016 guidance on regulation for self-driving cars. It lays out a clear set of objectives, organized in such a way that they can be measured. It breaks up its guidance by what it calls Operational Design Domain (ODD)—a set of constraints for which competency needs to be demonstrated: roadway types, geographical location, speed range, lighting conditions for operation (day and/or night), weather conditions, and other operational domain constraints. It highlights the need for measurement: "Tests should be developed and conducted that can evaluate (through a combination of simulation, test track or roadways) and validate that the Highly Automated Vehicle system can operate safely with respect to the defined ODD and has the capability to fall back to a minimal risk condition when needed."

When you focus on outcomes rather than rules, you can see that there are multiple ways to achieve comparable outcomes, and sometimes new ways that provide better outcomes. Which approach is best should be informed by data.

Unfortunately, it isn't just government that is unwilling or unable to put its data on the table. Companies like Uber, Lyft, and Airbnb jealously guard much of their data for fear that it will give away trade secrets or relative marketplace traction to competitors. Instead, they should open up more data to academics as well as to regulators trying to understand the impact of on-demand transportation on cities. Nick Grossman, who leads Union Square Ventures' efforts on public policy, regulatory, and civic issues, argues that open data may be the solution to Uber's many debates with regulators. He makes the case that "regulators need to accept a new model where they focus less on making it hard for people to get started." Relaxing licensing requirements and increasing the freedom to operate means more people can participate, and companies can experiment more freely. But "in exchange for that freedom to operate," Nick continues, "companies will need to share data with regulators—un-massaged, and in real time, just like their users do with them. AND, will need to accept that that data may result in forms of accountability."

Open data could help lay to rest other persistent questions about Uber's market-based approach. For example, Uber claims that lowering prices does not affect driver income, but drivers say that they have to work longer in order to make the same amount, and that too many drivers are increasing their wait times between pickups. This shouldn't be a matter of claim and counterclaim, because the answer to that question is to be found in data Uber has on its servers. Open data is a great way for everyone to better understand how well the system is working. Open data would also help cities to understand the impact of on-demand car services on overall congestion, and make it much easier to evaluate Airbnb's impact on housing availability and affordability. It is a shame that cities and platform companies are not working together more proactively, using data to craft better outcomes for both sides.

WORKERS IN A WORLD OF CONTINUOUS PARTIAL EMPLOYMENT

There is no better demonstration of how outdated maps shape public policy, labor advocacy, and the economy than in the debate over whether Uber and Lyft drivers (and workers for other on-demand startups) should be classified as "independent contractors" or "employees." In the world of US employment law, an independent contractor is a skilled professional who provides his or her services to multiple customers as a sole proprietor or small business. An employee provides services to a single company in exchange for a paycheck. Most on-demand workers seem to fall into neither of these two classes.

Labor advocates point out that the new on-demand jobs have no guaranteed wages, and hold them in stark contrast to the steady jobs of the 1950s and 1960s manufacturing economy that we now look back to as a golden age of the middle class. Yet if we are going to get the future right, we have to start with an accurate picture of the present, and understand *why* those jobs are growing increasingly rare. Outsourcing is the new corporate norm. That goes way beyond offshoring to low-wage countries. Even for service jobs within the United States, companies use "outsourcing" to pay workers less and provide fewer benefits. Think your hotel housekeeper works for Hyatt or Westin? Chances are good they work for Hospitality Staffing Solutions. Think those Amazon warehouse workers who pack your holiday gifts work for Amazon? Think again. It's likely Integrity Staffing Solutions. This allows companies to pay rich benefits and wages to a core of highly valued workers, while treating others as disposable components. Perhaps most perniciously, many of the low-wage jobs on offer today not only fail to pay a living wage, but they provide only part-time work.

Which of these scenarios sounds more labor friendly?

Our workers are employees. We used to hire them for eight-hour shifts. But we are now much smarter and are able to lower our labor costs by keeping a large pool of part-time workers, predicting peak demand, and scheduling workers in short shifts. Because demand fluctuates, we keep workers on call, and only pay them if they are actually needed. What's more, our smart scheduling software makes it possible to make sure that no worker gets more than 29 hours, to avoid triggering the need for expensive full-time benefits.

or

Our workers are independent contractors. We provide them tools to understand when and where there is demand for their services, and when there aren't enough of them to meet demand, we charge customers more, increasing worker earnings until supply and demand are in balance. We don't pay them a salary, or by the hour. We take a cut of the money they earn. They can work as much or as little as they want until they meet their income goals. They are competing with other workers, but we do as much as possible to maximize the size of the market for their services.

The first of these scenarios summarizes what it's like to work for an employer like Walmart, McDonald's, the Gap, or even a progressive low-wage employer like Starbucks. Complaints from workers include lack of control over schedule even in case of emergencies, short notice of when they are expected to work, unreasonable schedules known as "clopens" (for example, the same worker being required to close the store at 11 p.m. and open it again at 4 a.m. the next day—a practice that Starbucks only banned in mid-2014, and that is still in place at many retailers and fast-food outlets), "not enough hours," and a host of other labor woes.

The second scenario summarizes the labor practices of Uber and Lyft. Talk to many drivers, as I have, and they tell you that they mostly love the freedom the job provides to set their own schedule, and to work as little or as much as they want. This is borne out by a study of Uber drivers by economists Alan Krueger of Princeton and Jonathan Hall, now an economist at Uber. Fifty-one percent of Uber drivers work fewer than 15 hours a week, to generate supplemental income. Others report working until they reach their target income. Seventy-three percent say they would rather have "a job where you choose your own schedule and be your own boss" than "a steady 9-to-5 job with some benefits and a set salary."

Managing a company with workers who are bound to no schedule but simply turn on an app when they want to work and who compete with other workers for whatever jobs are available requires a powerful set of algorithms to make sure that the supply of workers and customers is in dynamic balance.

Traditional companies have also always had a need to manage uneven labor demand. In the past, they did this by retaining a stable core of full-time workers to meet base demand and a small group of part-time contingent workers or subcontractors to meet peak demand. But in today's world, this has given way to a kind of continuous partial employment for most low-wage workers at large companies, where workplace scheduling software from vendors like ADP, Oracle, Kronos, Reflexis, and SAP lets retailers and fast-food companies build larger-than-needed on-demand labor pools to meet peak demand, and then parcel out the work in short shifts and in such a way that no one gets full-time hours. This design pattern has become the dominant strategy for managing low-wage workers in America. According to a management survey by Susan Lambert of the University of Chicago, by 2010, 62% of retail jobs were part-time and two-thirds of retail managers preferred to maintain a large part-time workforce rather than to increase hours for individual workers.

The advent of scheduling software enabled this trend. As Esther Kaplan of the Investigative Fund describes it in her *Harper's* article "The Spy Who Fired Me,"

> In August 2013, less than two weeks after the teen-fashion chain Forever 21 began using Kronos, hundreds of full-time workers were notified that they'd be switched to part-time and that their health benefits would be terminated. Something similar happened last year at Century 21, the high-fashion retailer in New York. . . . Within the space of a day, Colleen Gibson's regular schedule went up in smoke. She'd been selling watches from seven in the morning to three-thirty in the afternoon to accommodate evening classes, but when that availability was punched in to Kronos, the system no longer recognized her as full-time. Now she was getting no more than twenty-five hours a week, and her shifts were erratic. "They said if you want full hours, you have to say you're flexible," she told me.

That is, both traditional companies and "on demand" companies use apps and algorithms to manage workers. But there's an important difference. Companies using the top-down scheduling approach adopted by traditional low-wage employers have used technology to amplify and enable all the worst features of the current system: shift assignment with minimal affordances for worker input, and limiting employees to part-time work to avoid triggering expensive health benefits. Cost optimization for the company, not benefit to the customer or the employee, is the guiding principle for the algorithm.

By contrast, Uber and Lyft expose data to the workers, not just the managers, letting them know about the timing and location of demand, and letting them choose when and how much they want to work. This gives the worker agency, and uses market mechanisms to get more workers available at periods of peak demand or at times or places where capacity is not normally available.

> When you are drawing a map of new technologies, it's essential
> to use the right starting point. Much analysis of the on-demand
> or "gig" economy has focused too narrowly on Silicon Valley
> without including the broader labor economy. Once you start
> drawing a map of "workers managed by algorithm" and "no
> guarantee of employment" you come up with a very different
> sense of the world.

Why do we regulate labor? In an interview with Lauren Smiley, Tom Perez, secretary of labor during the Obama administration, highlighted that the most important issue is whether or not workers make a living wage. The Department of Labor's Wage and Hour Division head David Weil put it succinctly: "We have to go always back to first principles of who are we trying to protect and how the people emerging in these new jobs fall on that spectrum."

On first blush, it would seem that being an employee has many benefits. But there is a huge gulf between the benefits often provided to full-time employees and part-time employees. And that has led to what I call "the 29-hour loophole." Unscrupulous managers can set the business rules for automated scheduling software to make sure that no worker gets more than 29 hours in a given week. Because employment law allows different classes of benefits for part-time and full-time workers, with the threshold being at 30 hours per week, this loophole allows core staff at the company to be given generous benefits, while the low-wage contingent workers get the bare-bones version. Once you realize this, you understand the potentially damaging effect of current labor regulations not just for new Silicon Valley companies but also for their workers. Turn on-demand workers from 1099 contractors into W2 employees, and the most likely outcome is that the workers go from having the opportunity to work as much as they like for a platform like Uber or TaskRabbit to one in which they are kept from working more than 29 hours a week. This was in fact exactly what happened when Instacart converted some of its on-demand workers to employees. *They became part-time employees.*

(Even before the advent of computerized shift-scheduling software, companies played shell games with employee pay and benefits. I remember student protests at Harvard that my daughter was part of in 2000 focused on the unfair treatment of janitors and other maintenance personnel. "You're not a full-time employee and aren't eligible for full-time benefits," janitors were told. "You don't work 40 hours for Harvard University. You work 20 hours for Harvard College, and 20 hours for the Harvard Law School.")

Perhaps as pernicious as the fact that companies limit workers to 29 hours a week, the capricious nature of many of the schedules that are provided by traditional low-wage employers and the lack of visibility into future working hours means that workers can't effectively schedule hours for a second job. They can't plan their lives, their childcare, a short vacation, or even know if they will be able to be present for their children's birthdays. By contrast, workers for on-demand services can work as many hours as they like—many report working until they reach their desired income for the week, rather than some set number of hours—and equally important, they work when they want. Many report that the flexibility to take time off to deal with childcare, health issues, or legal issues is the most important part of what they like about the job.

It is essential to look through the labels—employee and independent contractor—and examine the underlying reality that they point to. So often, we live in the world of labels and associated value judgments and assumptions, and forget to reduce our intellectual equation to the common denominators. As Alfred Korzybski so memorably wrote, we must remember that "the map is not the territory."

When you put yourself into the mapmaker's seat, rather than simply taking the existing map as an accurate reflection of an unchanging reality, you begin to see new possibilities. The rules that we follow as a society must be updated when the underlying conditions change. The distinction between employees and subcontractors doesn't really make sense in the on-demand model, which requires subcontractor-like freedoms to workers who come and go at their own option, and where employee-based overtime rules would prohibit workers from maximizing their income.

Professor Andrei Hagiu, writing in *Harvard Business Review*, and venture capitalist Simon Rothman, writing on *Medium*, both argue that we need to develop a new classification for workers—we might call them "dependent contractors." This new classification might allow some of the freedoms of independent contractors, while adding some of the protections afforded to employees. Nick Hanauer and David Rolf go further, arguing that just as technology allows us to deploy workers without the overhead of traditional command-and-control employment techniques, it also could let us provide traditional benefits to part-time workers. There is no reason that we couldn't aggregate the total amount worked across a number of employers, and ask each of them to contribute proportionally to a worker's account. Hanauer and Rolf call this a "Shared Security Account" in conscious echo of the safety net of a Social Security account.

A similar policy proposal for portable benefits comes from Steven Hill at New America. Hanauer, Rolf, and Hill all suggest that we decouple benefits like worker's compensation, employer contribution to Social Security and Medicare taxes, as well as holiday, sick, and vacation pay, from employers and instead associate them with the employee, erasing much of the distinction between 1099 independent contractor and W2 employee. Given today's technology, this is a solvable problem. It would be entirely possible to allocate benefits across multiple employers. It shouldn't matter if I work 29 hours for McDonald's and 11 for Burger King, if both are required to contribute pro rata to my benefits.

However, none of these proposals have solved the deeper dynamics that drive companies to use the 29-hour loophole. It isn't basic payroll taxes that lead companies to want to have two classes of workers. It is healthcare to start with (a single-payer system would solve that problem, as well as many others), but also other "Cadillac" benefits that companies wish to lavish on their most prized workers but not on everyone. More powerfully, it is the notion that workers are just a cost to be eliminated rather than an asset to be developed. Ultimately, the segregation of workers into privileged and unprivileged classes, and the moral and financial calculus that drives that segregation, has to stop.

Over time, we will realize that this is an existential imperative for our economy, not just a moral imperative.

It will take much deeper thinking (and forceful and focused activism) to come up with the right incentives for companies to understand and embrace the value of taking care of all their workers on an equal footing. Zeynep Ton's *Good Jobs Strategy* is a good place to start. Ton outlines the common principles that make companies as diverse as Costco and Google great places to work. As Harvard Business School lecturer and former CEO of Stop & Shop José Alvarez writes, "Zeynep Ton has proven what great leaders know instinctively—an engaged, well-paid workforce that is treated with dignity and respect creates outsized returns for investors. She demonstrates that the race to the bottom in retail employment doesn't have to be the only game being played." Economists have long recognized this phenomenon. They call wages higher than the lowest that the market would otherwise offer "efficiency wages." That is, they represent the wage premium that an employer pays for reduced turnover, higher employee quality, lower training costs, and many other significant benefits.

In Chapters 11 and 12, we'll look at the key drivers of the race to the bottom in wages, and why we need to rewrite the rules of business. But even without radically changing the game, businesses can gain enormous tactical advantage by better understanding how to improve the algorithms they use to manage their workers, and by providing workers with better tools to manage their time, connect with customers, and do all of the other things they do to deliver improved service.

Algorithmic, market-based solutions to wages in on-demand labor markets provide a potentially interesting alternative to minimum-wage mandates as a way to increase worker incomes. Rather than cracking down on the new online gig economy businesses to make them more like twentieth-century businesses, regulators should be asking traditional low-wage employers to provide greater marketplace liquidity via data sharing. The skills required to work at McDonald's and Burger King are not that dissimilar; ditto Starbucks and Peet's, Walmart and Target, or the AT&T and Verizon stores. Letting workers swap shifts or work on demand at competing employers would obviously require some changes to management infrastructure, training, and data sharing between employers. But given that most scheduling is handled by standard software platforms, and that payroll is also handled by large outsourcers, many of whom provide services to the same competing employers, this seems like an intriguingly solvable problem.

> The algorithm is the new shift boss. What regulators and politicians should be paying attention to is the fitness function driving the algorithm, and whether the resulting business rules increase or decrease the opportunities offered to workers, or whether they are simply designed to increase corporate profits.

In the next two chapters, we'll look at how the same flawed fitness function is driving media and finance, and how the speed and scale of digital platforms is algorithmically amplifying that flaw.

MEDIA IN THE AGE OF ALGORITHMS

AFTER THE 2016 US PRESIDENTIAL ELECTION, THERE WAS A lot of finger-pointing, and many of those fingers pointed at Facebook, arguing that its newsfeed algorithms played a major role in spreading misinformation and magnifying polarization.

False stories claiming that Pope Francis had endorsed Donald Trump, that Mike Pence had said that Michelle Obama was "the most vulgar First Lady we've ever had," and that Hillary Clinton was about to be indicted were shared more than a million times. All were cooked up by Macedonian teens out to make a buck. The story about the "FBI Agent Suspected in Hillary Email Leaks Found Dead in Apparent Murder-Suicide"—also totally fake but shared half a million times— was the work of a Southern California man who started in 2013 to prove how easily disinformation spread, but ended up creating a twenty-five-employee business to churn out the stuff.

Facebook users were not the only ones spreading these stories. Many of them circulated by email and on Twitter, on YouTube, on reddit, and on 4chan. Google surfaced them in Google Suggest, the drop-down recommendations that appear for every user as they begin to type a query.

But it was Facebook that became the focus of the discussion, perhaps because at first Mark Zuckerberg denied the problem, saying in an onstage interview at the Techonomy conference a few days after the election that it was "a pretty crazy idea" that the stories had influenced the outcome. They were a tiny proportion of the total content shared on the site, he argued.

Fake news is the stuff of tabloids. Marginal, once the subject of ridicule. How could it come to play such a large role in shaping our collective future?

At the very least, the 2016 US presidential election showed what Eli

Pariser had called "the filter bubble" in full force. Social media algorithms, driven by "likes," show each person more of what they respond to positively, confirming their biases, reinforcing their beliefs, and encouraging them to associate online with like-minded people. The *Wall Street Journal* created an eye-opening site called Blue Feed/Red Feed that used Facebook's own research data on the political preferences of its users to create side-by-side live feeds of hyperpartisan stories shown to each group. It is shocking just how different the news shown to "extremely liberal" and "extremely conservative" viewers turns out to be. I'd experienced a version of that myself in the stories that were shared with me by my conservative family members, and the progressive stories that I'd shared with them in return. We are living in different worlds. Or perhaps we are just living in a new "post-truth" world, where appeals to emotion carry more weight than facts.

The democratization not just of media distribution but also of its creation played a major role. Colin Megill, founder of pol.is, a service focused on creating better public dialogue, told me that his mother, a doctor who worked her whole life to break the glass ceiling, was beset by doubt about Hillary Clinton and had been especially influenced by a video claiming that her aide Huma Abedin had been a member of the Muslim Brotherhood, a video that had autoplayed after she watched YouTube replays of late-night television.

"I reflected on my conversation with my mom a lot after that happened and came up with one possible explanation," Colin said. "For her whole life, something would be out of the news immediately if it was totally false. Editors saw to that. The idea that something with a high production value, shared by millions, could be without a shred of truth really wasn't in her matrix of possibilities." The notion that the video could have been created by an anonymous Trump supporter was just not part of her mental map.

According to Pew Research, 66% of Americans get their news through social media sites, 44% of them from Facebook alone. Much of that content may come from traditional media via links shared on social media, but much of it is native to the platform, or coming from new, hyperpartisan sites like those cooked up for profit by the Macedonian teens, or for partisan reasons by extreme right-wing or extreme left-wing political organizations. And that is to say nothing of groups like ISIS that have successfully used social media for terrorist recruiting, or of the role of propaganda planted or amplified by Russia with the goal of influencing the US presidential election. As one US government official who wished to remain anonymous told me: "We aren't fighting our first cyberwar. We just fought it. And we already lost."

ALGORITHMIC WHAC-A-MOLE

In many ways, the rising influence of fake news is a cautionary tale of algorithms gone wrong, digital djinns given poorly framed instructions with potentially catastrophic consequences. It is worth studying even though Facebook and Google will have done a great deal of work to solve the current iteration of the problem by the time this book is published.

In a follow-up Facebook post the week after his dismissive comments, Mark Zuckerberg admitted that fake news was a problem, and that Facebook was working on it. His suggested solution was to give "the community" more tools for signaling what they believed to be true or false. I had met with Mark a few weeks before the election, about a related issue he was wrestling with, how Facebook could give voice to its users around community norms and values. His goal to make Facebook a neutral platform through which its users can connect and share is deeply felt. In his post about fake news and the election, he concluded, "In my experience, people are good, and even if you may not feel that way today, believing in people leads to better results over the long term."

That belief that controlling fake news should be up to the users, not to the platform, shaped Facebook's response to the crisis. Mark wrote: "We have already launched work enabling our community to flag hoaxes and fake news, and there is more we can do here. We have made progress, and we will continue to work on this to improve further." So far, so good.

He continued to argue for the role of Facebook's users in policing the site: "I am confident we can find ways for our community to tell us what content is most meaningful, but I believe we must be extremely cautious about becoming arbiters of truth ourselves." He correctly noted that "identifying the 'truth' is complicated. While some hoaxes can be completely debunked, a greater amount of content, including from mainstream sources, often gets the basic idea right but some details wrong or omitted. An even greater volume of stories express an opinion that many will disagree with and flag as incorrect even when factual."

The internal debate at platforms such as Facebook and Google about their responsibility to control fake news is not just a matter of caution in getting it right, though. It's also a worry about setting a legal precedent. The Digital Millennium Copyright Act (DMCA), enacted in 1998, exempted Internet service providers and other online intermediaries from liability from copyright infringement on the grounds that they were neutral platforms that simply enabled users to post whatever they want. They are more like a wall on which users can post handbills than they are like a publisher who chooses what to publish and should be held to a higher legal standard. This "neutral platform" argument is central to the existence of Internet services. Without it, Google would be liable for every copyright infringement made by any user posting online, simply by including that content in the search index. Similarly, Facebook, Twitter, YouTube, or Word-Press would be liable if any user posted infringing material. A similar legal defense, by extension, could be applied to other kinds of content posted by users: The service is a platform for its users, not a content provider. No online service wants to break this shield.

Critics snarl at this defense. One such critic, Carole Cadwalladr, was outraged that Google's Suggest feature was offering results such as "Jews are evil" as autocomplete for "Jews are . . ." When she clicked through, she found that the first result had the headline: "Top 10 Major Reasons Why People Hate Jews." A page from neo-Nazi site Storm-front was the third result, with additional explanations of why Jews are evil appearing as the fifth, sixth, seventh, and tenth results. When she did a search for "did the holo . . ." Google autocompleted her query to "did the Holocaust happen?" and she was taken to a list of Holocaust-denial sites, again topped by a page from Stormfront.

Her solution: Google should stop linking to these pages immediately. "Google's business model is built around the idea that it's a neutral platform. That its magic algorithm waves its magic wand and delivers magic results without the sullying intervention of any human," she wrote in a scathing op-ed for the *Guardian*. "It desperately does not want to be seen as a media company, as a content provider, as a news and information medium that should be governed by the same rules that apply to other media. But this is exactly what it is."

I sympathize with Cadwalladr's outrage, and her belief that Google (like all media) "frames, shapes and distorts how we see the world." I agree that Google needs to come to grips with bad results like this, just as they have come to grips with other challenges to the quality of their results. But Cadwalladr ignored the scale at which Google operates, and the way that scale fundamentally changes the necessary solution.

> Google, Facebook, Twitter, and their like need to be understood as a new thing, which doesn't fit neatly into the old map. That new thing operates by different rules—not by whim or an unwillingness to incur the costs of curation, but by necessity.

Google's and Facebook's reluctance to make manual interventions is not just a matter of hiding behind a convenient legal disclaimer of responsibility. These sites don't produce their results through some convocation of human editors, like the old *New York Times* front-page meeting, in which editors decided which stories get placement and where. That meeting was phased out even at the *Times* in 2015. The result of any Google search is the result of prodigious efforts to retrieve and rank every page on the web—30 *trillion* of them, from 250 billion unique web domain names, according to former Google VP of search Amit Singhal—and to serve them up in response to more than 5 billion searches a day. Many of those searches are common, but at least tens of millions of them are the result of quite infrequent combinations of words and phrases. The offensive Holocaust results that Cadwalladr complained about are the result of a search that, according to Google, is made only about 300 times a day. Out of 5 billion. That's 0.000006% of daily searches, a few millionths of a percent.

Facebook is similarly huge. In 2013, the social network disclosed that nearly 5 billion pieces of content were posted every day. That number is now surely far larger, as the site now has over 1 billion daily active users, up from 700 million in 2013.

The idea that Google or Facebook can solve the problem simply by hiring teams of human editors or fact checkers, or use outside media organizations to combat fake news, hate speech, or other objectionable results, removing or demoting them one at a time, indicates that people have little idea of the scale or nature of the problem. It's like the carnival game of Whac-A-Mole, except with billions of moles and only hundreds of hammers. Human oversight and intervention is definitely needed, but it will make little difference if it is implemented in the way that critics like Cadwalladr imagine. To whack billions of moles, you need much faster hammers.

We have to break the notion that the role of the human in the loop is as the final decision maker pulling a kill switch. There's a famous *Harvard Business Review* article called "Who's Got the Monkey?" that explains why whenever an employee brings in a problem, like a monkey on his or her back, the manager must offer counsel, and then send the employee back out with the monkey. Otherwise, the manager, with multiple employees, ends up with all the monkeys. How much more true is this in the age of algorithms? The manager ends up with a million monkeys. A good manager is always a teacher. How much more is this true with the powerful but fundamentally stupid race of djinns that do so much of the work at our massive online platforms?

Google no doubt has teams of developers, the managers of the digital workers who build the index and serve up the search results, hard at work teaching their inhumanly fast djinns how to mitigate this problem. I'd be very surprised if, by the time this book has been published, there hasn't been a comprehensive fake news search overhaul akin to the 2011 Panda and Penguin updates that dealt with content farms. And in fact, within weeks of Cadwalladr's op-eds, the search results for

Holocaust denial had been improved. The initial fix had failed to work consistently, and Google is still struggling to come up with a comprehensive solution to fake news, but the processes by which they respond to attacks on the search engine's effectiveness are well defined.

Facebook's problems are not identical to Google's. While Google evaluates and links to content from hundreds of billions of external sites, Facebook's content is posted natively by its users on its own platform. Much of that content links to external sites, but much of it does not. Even when the content comes from external sites, it has often been remixed into a "meme"—which has now come to mean a graphic or video representation of a key moment or quote that is freed from its original context, designed to be shared, designed for impact rather than deeper dialogue or understanding.

In May 2016, long before Trump was elected, Milo Yiannopoulos, writing on *Breitbart*, predicted that Trump's facility with creating Internet memes and appealing to the people who share them was crucial to his success. "Establishment types no doubt think this is all silly, schoolyard stuff," he wrote. "And it is. But it's also effective. . . . Caught between the hammer of Trump's media machine and the anvil of his online troll army, The Donald's opponents never stood a chance. Trump understands the Internet, and the Internet might just propel him into the White House. Meme magic is real."

As a result of the lack of context, many of the signals that Google relies on, such as the link structure of the web, are absent. While Facebook can make use of some of the same techniques, its infrastructure and business processes for dealing with content are not the same. This is one reason that Facebook is looking for "the community" to solve the problem. Can its billion-plus users police the site given the right tools? In a patent filed in June 2015, System and Methods for Identifying Objectionable Content, Facebook had already laid out its approach to dealing with hate speech, pornography, and bullying, relying on user reporting but using many additional signals to rank and weight not only the reports themselves but the users providing them. Many of the techniques described in the patent are also applicable to fake news.

In a second blog post on the topic, Mark Zuckerberg wrote in more detail about the company's approach, which includes making it easier for people to report fake stories, partnering with third-party fact-checking organizations, and potentially even showing warnings on stories that have been flagged by fact checkers or the community. But Mark also pointed out that the most important thing Facebook can do is "to improve our ability to classify misinformation. This means better technical systems to detect what people will flag as false before they do it themselves." He also noted that Facebook had already improved the algorithms used to choose "related articles" under links in the News Feed.

This algorithmic reeducation is essential because the speed with which content can spread on social media works against unaugmented human fact checkers. One fake story began on Twitter when Trump supporter Eric Tucker posted a photo of chartered buses in Austin, Texas, and suggested that the Clinton campaign was using them to bus protesters to Trump's upcoming speech. Even though Tucker himself had only forty followers, and deleted the tweet once he found that the buses were actually for visitors to a convention held by software company Tableau, the photo went viral, shared 16,000 times on Twitter and 350,000 times on Facebook. His initial tweet had used the hashtags #fakeprotests #Trump2016 #Austin, ensuring that it would be read widely by people following those topics.

The story was picked up first on reddit, then by various right-wing blogs, and then by mainstream media. Donald Trump himself then tweeted about "professional protesters," adding fuel to the fire. While Tucker didn't expect to have such an impact, the people who promote fake news often have strong incentives to boost it, using programmatic tools to discover key influencers and plant it with them to give it a quick start. Given the traffic that a hot story can bring today, even professional news organizations use automated "social listening tools" to quickly pick up trending topics and republish popular stories on their own publications without the careful fact checking that used to characterize mainstream media.

By the time concerned users or fact checkers begin to flag content as false, it may already have been shared hundreds of thousands of times and have been read by millions. Retractions of the original story usually have little effect. By midnight of the day he first tweeted it, Tucker had deleted the original tweet and replaced it with one stamped "False" across the picture. That tweet was retweeted a grand total of 29 times, versus the 16,000 retweets of the original. I'm reminded of the old saying passed on to me by my mother: "A lie will have gone halfway around the world before the truth has had time to tie on its shoes."

One approach that Google, Facebook, and others have begun practicing, labeling disputed stories, may help, because the labels will follow and potentially stay with the story, but only if it's done in advance of the story being too widely shared. But even this approach has problems, since there is nothing to stop a partisan or financially motivated site from creating a new version of the same false story. How do you detect that? You're back to the algorithmic djinns for help whacking the mole.

In addition, users themselves have trouble not only determining what is true or false, but even in detecting the signals that companies provide to help them determine the authority of what they are seeing. Only 25% of high school students in one Stanford study recognized the significance of the blue check mark used by Facebook and Twitter to denote verified accounts. Will flags for fake news fare any better?

Finally, it's essential to realize that search engines and social media platforms are the battlefield of an online war, with hostile attackers using the same tools that were originally developed by advertisers to track their customers, and then by scammers and spammers to game the system for profit. In addition to the Russian-sponsored social media disinformation campaigns, the Trump campaign's Project Alamo used highly targeted disinformation to discourage Clinton voters from going to the polls. These posts were referred to as "dark posts" by Brad Parscale, who led the campaign's social media efforts, private posts whose viewership is tightly targeted so that, as he put it, "only the people we want to see it, see it."

Jonathan Albright, a communications professor who analyzed a network of 300 news sites that were promulgating fake news during the 2016 election, made the same point about programmatic microtargeting.

"This is a propaganda machine," he wrote. "They're capturing people and then keeping them on an emotional leash and never letting them go."

"Capturing people and then keeping them on an emotional leash" is nothing new. It was at the heart of much media in the days of "yellow journalism" at the turn of the twentieth century, beaten back by journalistic standards for much of the century, then reasserted in its closing decades by talk radio and by Fox News on TV. Social media and its advertising business model has taken the process to its logical conclusion.

Targeted social media campaigns will almost certainly be a feature of all future political campaigns. Online social media platforms—and society as a whole—will need to come to grips with the challenges of the new medium. The moment of crisis may come when we realize that the tools of disinformation and propaganda are the very same tools that are routinely used by businesses and ad agencies to track and influence their customers. It is not just political actors who have a vested interest in spreading fake news. Vast sums of money are at stake, and participants use every tool to game the system. The problem is not Facebook's.

Fake news is simply the most unsavory face of the business model that drives much of the Internet economy.

In cybercrime, these tools go beyond the distasteful into the realm of the illegal. One Russian botnet uncovered in December 2016 was creating targeted videos that were generating $3–5 million per day in ad revenue from fake video views by programs masquerading as users. In other words, this battle goes far beyond planting fake news. It is also possible to plant fake users who exist only as imaginary pawns in a battle of clicks and likes.

When attackers use programs to masquerade as users, unaided human supervision is inadequate due to the speed and scale of the attacks. This is another reason why the response to fake news and other kinds of amplified social media fraud needs to be algorithmic, much as spam filters are, rather than solely relying on users or the tools of traditional journalism.

The 2015–16 DARPA Cyber Grand Challenge was based on a similar insight, asking for the development of AI systems to find and automatically patch software vulnerabilities that corporate IT teams just aren't able to keep up with. The problem is that an increasing number of cyberattacks are being automated, and these digital adversaries are finding the holes far faster than humans can patch them.

John Launchbury, the director of DARPA's Information Innovation Office, told me an illuminating story from the Cyber Grand Challenge. The various competing systems had been seeded with security vulnerabilities that they were expected to find and fix before they could be exploited by another of the systems. One of the AI contestants examined its own source code and found a vulnerability not among those that had been planted, and used it to take control of another system. A third system, observing the attack, diagnosed the problem and fixed its own source code. All of this in twenty minutes.

Air Force Colonel John Boyd, "the father of the F-16," introduced the term *OODA loop* ("Observe-Orient-Decide-Act") to describe why agility is more important in combat than pure firepower. Both fighters are trying to understand the situation, decide what to do, and then act. If you can think more quickly, you can "get inside the OODA loop of your enemy" and disrupt his decision making.

"The key is to obscure your intentions and make them unpredictable to your opponent while you simultaneously clarify his intentions," wrote Boyd's colleague Harry Hillaker in his eulogy to Boyd. "That is, operate at a faster tempo to generate rapidly changing conditions that inhibit your opponent from adapting or reacting to those changes and that suppress or destroy his awareness. Thus, a hodgepodge of confusion and disorder occur to cause him to over- or under-react to conditions or activities that appear to be uncertain, ambiguous, or incomprehensible."

This is very hard to do when your opponent is a machine able to act millions of times faster than you are. One observer who wished to remain anonymous, an expert in both financial systems and in cyberwarfare, said to me, "It takes a machine to get inside the OODA loop of another machine."

WHAT IS TRUTH?

We have been talking about objectively verified facts and objectively verified falsehoods. There is a further, even more challenging problem that algorithms can be unexpectedly helpful with. As Mark Zuckerberg noted, many problematic pieces of content are not outright falsehoods, but contain opinion or half-truths. Partisans on both sides of an issue are eager to believe and reshare content even if they know it is at least partially false. Even when professional fact-checking organizations such as Snopes or PolitiFact or mainstream media sites staffed by experienced reporters debunk a story, there are others who decry the result as biased.

George Soros has pointed out that there are things that are true, things that are false, and things that are true or false only to the extent that people believe in them. He calls this "reflexive knowledge," but perhaps the old-fashioned term *beliefs* will serve just as well. So much that matters falls into this category—notably history, politics, and markets. "We are part of the world we seek to understand," Soros wrote, "and our imperfect understanding plays an important role in shaping the events in which we participate."

This has always been the case, but our new, world-spanning digital systems, connecting us into a nascent global brain, have accelerated and intensified the process. It is not just facts that spread from mind to mind. It is not just the idea that pots containing decaffeinated coffee should be orange. Misinformation goes viral too, shaping the beliefs of millions. Increasingly, what we know and what we are exposed to are shaped by personalization algorithms, which try to pick out for us from the firehose of content on the Internet just the things that the algorithms expect we will most likely respond to, appealing to engagement and emotion rather than to literal truth.

But Soros's reminder that stock prices and social movements are neither true nor false suggests an approach to the fake news problem as well. Even while recognizing the role of emotion in stock prices, stock pickers still believe that a stock has "fundamentals." A stock price may depend on what people believe about a company's future prospects, but they recognize that a company also does have revenue, income, capital, growth rates, and a plausible market opportunity from which those future prospects can be estimated. Stock reporting routinely measures and reports on the price/earnings ratio and other measures of how far expectations outstrip the fundamentals, so that people can make informed judgments of how much risk they are taking. There are many who will overlook the risks, and those who encourage them to do so, but at least some information is there.

> The distance between human enthusiasm and the fundamentals can also be measured for news, using many signals that can be verified algorithmically by a computer, often more quickly and thoroughly than they can be verified by humans.

When people are discussing the truth or falsity of news, and the responsibility of sites like Facebook, Google, and Twitter to help identify it, they somehow think that determining "truth" or "falsity" is solely a matter of evaluating the content itself, and make the case that it can't be done by a computer because it requires a subjective judgment. But as with Google Search, many of the signals that can be used are independent of the actual content. To use them, we must simply follow Korzybski's injunction to compare the map with the territory it claims to describe.

> Algorithmic fact checking doesn't replace human judgment. It amplifies our power to exercise it, in much the same way as earthmoving equipment amplifies our muscles. The signals it uses are similar to those that a human fact checker might use.

Does the story or graph cite any sources? If no sources are given, it is far from certain that the story is false, but the likelihood increases that it should be investigated further. A fake story typically provides no sources. For example, when debunking one claim sent to me by my brother, a fake map purporting to show higher crime rates in precincts that voted Democratic, I was unable to find any sources for the data the map claimed to be based on. In the course of my search, though, I found a series of visualizations put together by *Business Insider* that painted a very different picture. Unlike my brother's map, the legitimate publication provided the source of the data it had used, an FBI crime database.

Do the sources actually say what the article claims they say? It would have been entirely possible for *Business Insider* to claim that the data used in their article was from the FBI, but for there to be no such data, or for the data there to be different. Few people trace the chain of sources to their origin, as I did. Many propaganda and fake news sites rely on that failure to spread falsity. Checking sources all the way back to their origin is something that computers are much better at doing than humans.

Are the sources authoritative? In evaluating search quality over the years, Google has used many techniques. How long has the site been around? How often is it referenced by other sites that have repeatedly been determined to be reputable? Most people would find the FBI to be an authoritative source for US national crime data.

If the story references quantitative data, does it do so in a way that is mathematically sound? For example, anyone who has even a little knowledge of statistics will recognize that showing absolute numbers of crimes without reference to population density is fundamentally meaningless. Yes, there are more crimes committed by millions of people in New York City or Chicago than by hundreds in an area of rural Montana. That is why the FBI data referenced by the *Business Insider* article, which normalized the data to show crimes per 100,000 people, was inherently more plausible to me than the fake electoral maps that set me off on this particular quest for truth. Again, math is something computers do quite well.

Do the sources, if any, substantiate the account? If there is a mismatch between the story and its sources, that may be a signal of falsity. Even before the election, Facebook had rolled out an update to combat what they call "clickbait" headlines. Facebook studied thousands of posts to determine the kind of language typically used in headlines that tease the user with a promise that is not met by the content of the actual article, then developed an algorithm to identify and downgrade stories that showed that mismatch. Matching articles with their sources is a very similar problem.

Are there multiple independent accounts of the same story? This is a technique that was long used by human reporters in the days when the search for truth was properly central to the news. A story, however juicy, would never be reported on the evidence of a single source. Searching for multiple confirming sources is something that computers can do very well. Not only can they find multiple accounts, but they can also determine which ones appeared first, which ones represent duplicate content, how long the site or username from which the account has been posted has existed, how often it makes similar posts, and even which location the content was posted from.

Consumers of online media are unlikely to retrain themselves to act this same way. Especially when they read a story that confirms their biases, few people do a search for other accounts of the same story from a source that doesn't share those biases. One of my sisters sent me a story about California "legalizing child prostitution" after reading an account in the *Washington Examiner*. "I think this might just be why some decent people don't like California," she wrote. I read the bill, as well as rebuttals from other media sources. What the California bill actually said was that individuals under the age of eighteen involved in prostitution would not be treated as criminals, but instead could be taken into custody and made a ward of the court. Given an account of an original source, an algorithm could potentially compare the summary with the original, or compare multiple accounts of the same event, and flag discrepancies.

In addition to sharing content that confirms their biases and framing it to serve their agendas, users are too eager for clicks and likes. John Borthwick, CEO of Betaworks, described the user behavior that feeds the spread of false news. "Media hacks take advantage of the decontextualized structure of real-time news feeds," he wrote. "You see a Tweet from a known news site, with a provocative headline and maybe the infographic image included—you retweet it. Maybe you intend to read the story, might be you just want to Tweet something interesting and proactive, maybe you recognize the source, maybe you don't." One of the simplest algorithmic interventions Facebook and Twitter could make would be to ask people, "Are you sure you want to share that link? You don't appear to have read the story."

Because they follow rules exactly, algorithms are also good at noticing things that slip by humans. Earlier in this chapter, I cited an op-ed by Carol Cadwalladr about Google and Holocaust denial sites. At the end of a follow-up article, in which Cadwalladr showed how she could push down the fake results by buying a few targeted ads, was an explanation attributed to Danny Sullivan, the search engine guru, saying that Google had changed its algorithms "to reward popular results over authoritative ones. For the reason that it makes Google more money."

The article seemed doubly authoritative—it appeared in the *Guardian*, a reputable newspaper, and it quoted an expert on Google search I know and respect. But something was nagging at me. While there were other links in the op-ed, there was no link to the article from which Danny Sullivan was supposedly quoted. So I sent Danny an email. He told me that not only had he not said that Google had changed its algorithm to increase its profits, but he'd notified the *Guardian* after the article cited him incorrectly. Sadly, he said, the article hadn't been updated.

> Citing and linking to sources makes it much easier to validate whether an assertion is an opinion or interpretation, and who is making it. This should be the gold standard for all reporting. If media reliably linked to sources, any story without sources would automatically become suspect.

There are cases, of course, where reporters depend on anonymous sources. Watergate's "Deep Throat" comes to mind. But note how journalistic standards have slipped: Woodward and Bernstein spent many months tracking down corroborating evidence that proved Deep Throat's assertions. They didn't just report the leaked information as hearsay.

REASONABLE DOUBT

When fake news is detected, there are a number of possible ways to respond.

The stories can be suppressed entirely if certainty is extremely high. This should be done rarely, because suppressing content entirely is a slippery slope toward censorship. We already rely on this level of extreme prejudice in other online applications, though, since it is what email providers do to filter the email we actually want to see from the billions of spam messages sent every day.

The stories can be flagged. For example, Facebook (or online mail systems like Gmail, since much fake news appears to be spread by email) could show an alert, similar to a security alert, that says, "This story appears likely to be false. Are you sure you want to share it?" with a link to the reasons why it is suspect, or to a story that debunks it, if that is available. Unfortunately, Facebook's desire not to be the arbiter of truth, even when the stories are from known sources of misinformation, means that their efforts are often less effective than they could be.

In March 2017, Facebook began listing stories as "disputed" when authorized sites like Snopes or PolitiFact debunk them, but as expected with human fact checkers, the process takes days when the damage is done in minutes or hours. Krishna Bharat, the Google engineer who founded and ran Google News for many years, believes that one of the most important roles for algorithms to play may be as a kind of circuit breaker, which pauses the spread of suspicious postings, providing "enough of a window to gather evidence and have it considered by humans who may choose to arrest the wave before it turns into a tsunami." Bharat points out that it is not every false story that needs to be flagged, only those that are gaining momentum. "Let us say that a social media platform has decided that it wants to fully address fake news by the time it gets 10,000 shares," he notes. "To achieve this they may want to have the wave flagged at 1,000 shares, so that human evaluators have time to study it and respond. For search, you would count queries and clicks rather than shares and the thresholds could be higher, but the overall logic is the same."

A variation of Facebook's existing automated Related Stories feature might be another way to tackle confirmation bias without resorting to blocking a story entirely. Given a news story that displays likely bias according to various algorithmic measures, it should be possible to match it up immediately with an offsetting story from a site known to be authoritative, or to match it up with original sources. While nothing will force readers to consult those sources, the fact that a story is flagged as potentially false or misleading and that an alternative view is available may give pause to the trigger finger of sharing. But this has to happen extremely quickly, before content has already gone viral.

Suspect stories also can be given less priority, shown lower down in the newsfeed, or less often. Google does this routinely in ranking search results. And while the idea that Facebook should do this has been more controversial, Facebook is already ranking stories, featuring those that drive more engagement over those that are more recent, showing stories related to ones we've already shared or liked, and even showing particularly popular stories more than once. Once Facebook stopped showing stories in pure timeline order, they put themselves in the position of curating the feed algorithmically. It's about time they added source verification and other "truth" signals to the algorithm.

The algorithm does not have to find absolute truth; it has to find a reasonable doubt, just like a human jury. This is especially true if the penalty is simply not being promoted. There is no free speech obligation for platforms to proactively promote any particular content. Fake news got a big boost from a flawed algorithm that seems to have favored the emotional rush of partisan engagement over other factors.

Google and Facebook constantly devise and test new algorithms. Yes, there is human judgment involved. But it is judgment applied to the design of a system, not to each specific result. Designing an effective algorithm for search or the newsfeed has more in common with designing an airplane so it flies than with deciding where that airplane flies.

In the case of making an airplane fly, the goals are simple—stay aloft, go faster, use less fuel—and design changes can be rigorously tested against the desired outcome. There are many analogous problems in search—finding the best price, or the most authoritative source of information on a topic, or a particular document—and many that are far less rigorous. When users get right to what they want, the users are happy, and so, generally, are advertisers. Unfortunately, unlike search, where the desires of the users to find an answer and get on with their lives are generally aligned with "give them the best results," prioritization of "engagement" may have led Facebook in the wrong direction. Engagement and time on-site may be good for advertisers; they may not be good for users or for seekers of truth.

Even in the case of physical systems like aerodynamics and flight engineering, there are often hidden assumptions to be tested and corrected. In one famous example that determined the future of the aerospace industry, a radically new understanding of how to deal with metal fatigue was needed. At the beginning of commercial jet travel, in 1953, Britain's new de Havilland Comet was ready to dominate the skies. Then, horrifyingly, one of the planes fell out of the sky for no apparent reason. The airline blamed pilot error and bad weather. A year later, the skies were clear when a second plane did the same thing. The fleet was grounded for two months during an extensive investigation, after which the manufacturer confidently asserted that they had made modifications to deal with "every possibility that imagination has suggested as a likely cause of the disaster." When a third plane fell from the sky only a few days after the report was issued, it was clear that de Havilland's imagination was insufficient to the challenge. A young engineer in America had a better idea, which handed the future of commercial jet aviation to Boeing. As described by University of Texas physics professor Michael P. Marder, who brought this story to my attention: "Cracks were the centerpiece of the investigation. They could not be eliminated. They were everywhere, permeating the structure, too small to be seen. The structure could not be made perfect, it was inherently flawed, and the goal of engineering design was not to certify the airframe free of cracks but to make it tolerate them."

So too, the essence of algorithm design is not to eliminate all error, but to make results robust in the face of error. The fundamental question to ask is not whether Facebook should be curating the newsfeed, but how.

Where de Havilland tried in vain to engineer a plane where the materials were strong enough to resist all cracks and fatigue, Boeing realized that the right approach was to engineer a design that allowed cracks, but kept them from propagating so far that they led to catastrophic failure. That is also Facebook's challenge. Their goal is to find a way for the plane to fly faster, but fly safely. This means improving their algorithms—training and managing their electronic workers rather than throwing them out and simply going back to human curation. After the de Havilland Comet incidents, the airline industry didn't simply throw up its hands, go back to propeller planes, and give up on commercial jet flight. Facebook's algorithms have been set to optimize for engagement; they need to be more complex, and add optimizations for truth.

The bright side: Searching through the possibility space for the intersection of truth *and* engagement could lead Facebook to some remarkable discoveries. Pushing for what is hard makes you better.

There are signs of this effort in Mark Zuckerberg's February 2017 manifesto, "Building Global Community." In it, he pointed to a radically different way of solving the problem. Mark gave only a token nod to the explicit problem of fake news, noting that new AI tools are already submitting a third of all stories sent to Facebook's internal content review team. (The other two-thirds are submitted by Facebook users.) He focused instead on the root cause of the problem: the decline in social capital, the ties that bind us together as a society and that make it easier for us to work together for the common good.

In his 2000 book, *Bowling Alone*, Robert Putnam used the decline of bowling leagues and the rise of individual bowling as a metaphor for the changing nature of American society. From the days when Alexis de Tocqueville first analyzed the American character in the early nineteenth century, the United States had been characterized by a rich civic fabric of participation in local government, churches, unions, mutual aid societies, charities, sports leagues, and associations of all kinds. The decline of this participation had serious consequences, Putnam thought.

During earlier research on economic differences between the twenty regional governments of Italy, Putnam had noticed that there was a close correlation between civic engagement and prosperity. "These communities did not become civic simply because they were rich. The historical record strongly suggests precisely the opposite: They have become rich because they were civic." Social capital is as important as financial capital in the wealth of nations.

Mark Zuckerberg came to much the same conclusion. "There has been a striking decline in the important social infrastructure of local communities over the past few decades," he noted. "The decline raises deeper questions alongside surveys showing large percentages of our population lack a sense of hope for the future. It is possible many of our challenges are at least as much social as they are economic—related to a lack of community and connection to something greater than ourselves."

Online communities represent a bright spot, Mark noted, but there is much work to do to expand their impact and their scale, using them to enable offline as well as online connection, empowering community leaders with new tools, and identifying more "meaningful groups" that can have a positive effect on people's offline as well as online lives. Support groups for new parents or for those suffering from a serious disease are good examples. (Margaret Levi, the director of the Stanford Center for Advanced Study in the Behavioral Sciences, pointed out to me one major caveat: that these groups already have a pressing common purpose; finding each other is the problem, which Facebook can clearly help with. In other areas, finding a common purpose that brings people together rather than driving them apart is precisely the unsolved problem.)

When Mark says it is time for Facebook to shift from a focus on friends and family to "the social infrastructure for community—for supporting us, for keeping us safe, for informing us, for civic engagement, and for inclusion of all," you can see the promise of a virtuous circle of engagement. Where engagement seems to be the wrong fitness function for traditional ad-supported media, engagement is exactly the metric we want going up and to the right if we are looking to strengthen not only friendship and families but society as a whole.

That is a very promising direction. If Facebook is indeed able to make progress in strengthening forms of positive engagement that actually create communities with true social capital, and is able to find an advertising model that supports that goal rather than distorts it, that would likely have a greater impact than any direct attempt to manage fake news. When tuning algorithms, as in ordinary life, it is always better to tackle root causes than symptoms. Humans are a fundamentally social species; the tribalism of today's toxic online culture may be a sign that it is time to reinvent all of our social institutions for the online era.

In our conversation on the topic, Margaret Levi offered a concluding warning: "Even when social media helps people engage in collective action—as it did in Egypt—by coordinating them, that is quite distinct from an ongoing organization and movement." This is what our mutual friend, Wael Ghonim, had learned as a result of his experience with the Egyptian revolution. "Unanswered still," Margaret continued, "is Wael's concern about how you transform coordinated and directed action to a sustained movement and community willing to work together to solve hard problems. Especially when they begin as a heterogeneous set of people with somewhat conflicting end goals. They may agree on getting rid of the dictator, but then what?"

THE PROBLEM OF DISAGREEMENT

Henry Farrell, a professor of political science at George Washington University and a columnist for the *Washington Post*, wrote to me after reading an online post that I'd published about the fake news problem. Henry made an important point very different from my own. The problem, he wrote, is "[n]ot what is the optimal solution to finding truth given the technology and the constraints. Instead . . . what is the most plausible path towards identifying a sustainable *political* compromise between a heterogeneous crowd of individuals who don't agree on the solution, and in some cases maybe don't agree that there is a problem in the first place?"

This is a very good question, but, I would argue, also one that technology may be able to help with. In a very interesting experiment, the government of Taiwan held a public consultation, Virtual Taiwan, using a tool called pol.is to involve its citizens in discussions of legislation and regulations, including, notably, regulation of new transportation services such as Uber.

As Colin Megill, the creator of pol.is, describes it, Jaclyn Tsai, a minister in the executive branch in Taiwan, went to a government-oriented hackathon and said, "We need a platform to allow the entire society to engage in rational discussion."

Pol.is asks people to make assertions that take the form of a single-sentence comment. Those reading those assertions don't have a means to argue with them—there are no replies. They can agree, disagree, or pass. And then they can make a separate assertion of their own. Colin notes, "Doing away with replies gets you something very special. It gets you a matrix [of] every participant, and what they thought about every comment." Humans aren't very good at analyzing this, but machines are really good at it. "You use this all the time," he says. "Every time you rate a movie, every time you buy a product, you're creating data; and we do machine learning on that data in pol.is like Netflix would do on movies. Netflix identifies clusters, things like people who love comedy, people who love horror, people who love comedy and documentaries but hate horror, people who love comedies and horror but hate documentaries."

In pol.is, a well-known statistical technique called principal component analysis (PCA) is used to cluster the assertions and the people who respond to them into groups of like-minded individuals and the statements they favor and disfavor. The statements each group tended to vote uniquely on, as well as statements that enjoyed consensus among all the groups, are shown to everyone. The assertions getting consensus across all groups, or within specific groups, float to the top and are seen more often—just like content on Facebook, but with visibility into what percentage of others agreed or disagreed with them.

This is very different from Facebook likes because participants can see the filter bubble–like graph of those who agree and disagree with a common set of assertions. Participants can click through to view the

statements that shape a particular cluster. And as participants agree
or disagree with various statements, their avatars move on the graph,
toward or away from another cluster. Participants can see not only what
percentage of the entire conversation agrees with them on a particular
statement, but also the percentage of the cluster who agrees with similar statements they or others have made.

There is a similar, very powerful technique for small groups meeting in the physical world, which we've often used to discuss contentious
issues among the staff and fellows at Code for America. It's called a
"Human Spectrogram." The group stands together in the middle of a
large room. Someone makes a statement, and those who agree strongly
with it move to the far end of the room. Those who disagree move
to the other end of the room. People whose views are less polarized
can arrange themselves anywhere in between. Then someone makes
another comment, and if it influences your thinking, you move accordingly. The beauty of pol.is is that it seems to have scaled this approach
to work with thousands of people and thousands of assertions across
multiple dimensions.

The pol.is discussion of Uber in Virtual Taiwan began with one
assertion: "I think Passenger Liability insurance should be mandatory
for riders on uberX private vehicles." Those responding to this assertion
quickly sorted themselves into groups: those pro- and anti-regulation.
Participants could see the size of those groups—no more than 33%
took either side of the debate. So people tried out different assertions,
trying to move toward those that would garner higher support.

Over a period of four weeks, the group of about 1,700 participants in
the Uber conversation (out of tens of thousands who participated in the
overall Virtual Taiwan effort) worked their way toward consensus on key
points. One assertion that reached high agreement: "The government
should leverage this opportunity to challenge the taxi industry to improve
their management & quality control system, so that drivers & riders
would enjoy the same quality service as Uber. (95%, across all groups.)"

By the end of the consultation, Uber had agreed to provide Minister Tsai with its international liability insurance policy and, if needed,
release it for public review. It also agreed to coach all drivers to register

and obtain professional driver's licenses, and that if it were legalized in some areas, it was willing to pay for UberX car permits and transport taxes. The Taipei Taxi Association expressed a willingness to work with the UberTAXI platform, and to offer better services if the government would let them increase taxi pricing in response to market demand in the same way that Uber does.

Ray Dalio, the founder and executive chairman of Bridgewater Associates, uses a similar approach to creating what he calls an "idea meritocracy" at his company, the largest hedge fund in the world. As members of the firm debate investments or ideas, they rate the assertions of the other participants, assembling them into a matrix that highlights agreement and disagreement. Everyone is urged to be "radically transparent" with their opinions, and the newest associate is welcome to tell Ray himself that he is wrong. Bridgewater takes the further step of applying an algorithm to the matrix, which takes into account factors such as past performance, expertise on the particular topic, and other ways of weighting individual opinions. The goal is to combine the best of human insight and the ability of computer algorithms to sum up and clarify the points of agreement and disagreement.

There's no silver bullet, and disagreement too can be a tool for moving toward truth, as long as it is honestly entered into, and there are mechanisms for people to move and change their opinions as they are exposed to the views of others. This is very different from polling, which simply tries to learn what people already believe, and then calibrates arguments to reinforce it.

As Henry Farrell wrote to me in another email: "Processes of intellectual discovery are all about arguments between different (and sometimes stylized) positions. To use a machine learning analogy stolen from my collaborator, Cosma Shalizi—all of us put together are at best an ensemble of weak learners, each of which only grasps a few of the terms in a very long and complicated vector that we're trying to model. It plausibly helps if we start from very different positions (each weak learner sees a different set of terms) as long as each of these positions reflect some aspect of the truth and then, and only then, try to converge on a shared model of the problem."

That is a beautiful summation of the power of intellectual debate to drive toward truth. We face enormous challenges as a society as that debate moves into online platforms with billions of participants, with no boundaries of nationality or geography, with untested signals of authority and authenticity, using rude tools not yet up to the task.

It's still day one.

LONG-TERM TRUST AND THE MASTER ALGORITHM

Truth is only one of many factors humans—and the companies they create—struggle to optimize. What is really driving our decisions?

Some years ago, John Mattison, the chief medical information officer of Kaiser Permanente, the large integrated health provider, said to me, "The great question of the twenty-first century is going to be 'Whose black box do you trust?'" A black box, by definition, is a system whose inputs and outputs are known, but the process by which one is transformed to the other is unknown. Mattison was talking about the growing importance of algorithms in medicine, but his point, more broadly, was that we place our trust in systems whose methods for making decisions we do not understand.

Sometimes that trust is given because we ourselves don't have the knowledge to understand the algorithm, but we believe that someone else does. Sometimes that knowledge is denied even to experts capable of understanding what is inside the black box; it is kept from them as a trade secret. Google does not disclose the exact details of its search algorithm lest it be gamed by those trying to increase their rankings. Similarly, when Facebook cracked down on stories with clickbait headlines, Adam Mosseri, its VP of product management for News Feed, wrote, "Facebook won't be publicly publishing the multi-page document of guidelines for defining clickbait because 'a big part of this is actually spam, and if you expose exactly what we're doing and how we're doing it, they reverse engineer it and figure out how to get around it.'"

Just as with clickbait headlines, some incentives to create fake news can be eliminated. Many of those promoting fake news during the 2016 election were politically motivated, whether sincerely or cynically, but many fake news sites, like the ones created by the Macedonian teens, were created purely for financial gain. Cutting off advertising for sites or accounts that are peddling fake news is a great way to eliminate some of the most egregious offenders. This can be done not only by the platforms themselves, but by advertisers and ad networks who place "remnant advertising" on the lowest-quality sites. Businesses are beginning to recognize that the ads they show against their content make a statement about who they are, and showing the wrong ads can irrevocably damage their own reputation. As Warren Buffett is reputed to have said, "It takes twenty years to build a reputation and five minutes to ruin it. If you think about that, you'll do things differently."

Outright bad actors are only a small part of the problem, though. A more fundamental challenge is the way that the fitness function in the algorithms of search and social media shape the choices made by writers and publishers. Advertising-driven businesses in particular are slaves to the need for attention. Chris O'Brien, formerly a reporter for the *San Jose Mercury News* and the *Los Angeles Times* and now at online publisher *VentureBeat*, told me of the struggle reporters like him face every day. Do they write and publish what they think is most newsworthy, or what will get the most attention on social media? Do they use the format that will do the most justice to the subject (a deep, authoritative piece of research, a so-called longread), or do they decide that it's more profitable to harvest attention with short, punchy articles, perhaps even with deceptive headlines, that generate higher views and more advertising dollars? Do they choose video over text, even when text would let them do a better job?

The need to get attention from search engines and social media is a major factor in the dumbing down of news media and a style of reporting that leads even great publications to a culture of hype, fake controversies, and other techniques to drive traffic. The race to the bottom has in part been a result of the primary shift of news industry revenue from subscription to advertising and from a secure base of local readers to chasing readers via social media.

Subscription-based publications have an incentive to serve their readers; advertising-based publications have an incentive to serve their advertisers. As described in Chapter 8, search-based pay-per-click advertising can help to align the incentives, but it too can be gamed, and in any event it represents only half of digital ad spending, which in turn is only a fraction of total advertising spending. The flood of subscribers to news publications like the *New York Times*, *Washington Post*, and *Wall Street Journal* since the 2016 presidential election is a promising sign that there is interest from consumers in supporting investigative reporting again. But publications like these that formerly dominated the news media landscape are now much less influential. As a result, those whose algorithms guide what content is consumed via search and social media have a deep responsibility to tune their algorithms not just for profit but for the public interest.

Because many of the ad-based algorithms that shape our society are black boxes—either for reasons like those cited by Facebook's Adam Mosseri, or because they are, in the world of deep learning, inscrutable even to their creators—the question of trust is key. Facebook and Google tell us that their goals are laudable: to create a better user experience. But they are also businesses, and even creating a better user experience is intertwined with their other fitness function: making money.

Evan Williams has been struggling to find an answer to this problem. When he launched *Medium*, his follow-up to Twitter, in 2012, he wrote, rather presciently as it turned out: "The current system causes increasing amounts of misinformation . . . and pressure to put out more content more cheaply—depth, originality, or quality be damned. It's unsustainable and unsatisfying for producers and consumers alike. . . . We need a new model."

In January 2017, Ev realized that despite *Medium*'s success in building a community of writers who produce thoughtful content and a community of readers who value it, he had failed to find that new business model. He threw down the gauntlet, laid off a quarter of *Medium*'s staff, and committed to rethink everything it does. He had come to realize that however successful, *Medium* hadn't gone far enough in breaking with the past. He concluded that the broken system is ad-driven Internet media itself. "It simply doesn't serve people. In fact, it's not designed to," he wrote. "The vast majority of articles, videos, and other 'content' we all consume on a daily basis is paid for—directly or indirectly—by corporations who are funding it in order to advance their goals. And it is measured, amplified, and rewarded based on its ability to do that. Period. As a result, we get . . . well, what we get. And it's getting worse."

Ev admits he doesn't know what the new model looks like, but he's convinced that it's essential to search for it. "To continue on this trajectory," he wrote, "put us at risk—even if we were successful, business-wise—of becoming an extension of a broken system."

It is very hard to repair that broken system without rebuilding trust. When the algorithms that reward the publishers and platforms are at variance with the algorithms that would benefit users, whose side do publishers come down on? Whose side do Google and Facebook come down on? Whose black box can we trust?

There's an irony here that everyone crying foul about the dangers of censorship in response to fake news should take deeply to heart. In 2014, Facebook's research group announced that it had run an experiment to see whether shifting the mix of stories that their readers saw could make people happy or sad. "In an experiment with people who use Facebook, we test whether emotional contagion occurs outside of in-person interaction between individuals by reducing the amount of emotional content in the News Feed," the researchers wrote. "When positive expressions were reduced, people produced fewer positive posts and more negative posts; when negative expressions were reduced, the opposite pattern occurred. These results indicate that emotions expressed by others on Facebook influence our own emotions, constituting experimental evidence for massive-scale contagion via social networks."

The outcry was swift and severe. "To Facebook, we are all lab rats," trumpeted the *New York Times*.

Think about this for a moment. Virtually every consumer-facing Internet service uses constant experiments to make its service more addictive, to make content go viral, to increase its ad revenue or its e-commerce sales. Manipulation to make more money is taken for granted, its techniques even taught and celebrated. But try to understand whether or not the posts that are shown influence people's emotional state? A disgraceful breach of research ethics!

There is a master algorithm that rules our society, and, with apologies to Pedro Domingos, it is not some powerful new approach to machine learning. It is a rule that was encoded into modern business decades ago, and has largely gone unchallenged since.

It is the algorithm that led CBS chairman Leslie Moonves to say in March 2016 that Trump's campaign "may not be good for America, but it's damn good for CBS."

You must please that algorithm if you want your business to thrive.

11

OUR SKYNET MOMENT

ON SEPTEMBER 17, 2011, FED UP WITH GOVERNMENT BAILOUTS that had saved the banks despite the fact that they had brought the world to the brink of financial ruin with a toxic stew of complex derivatives based on aggressively marketed home mortgages, fed up that the banks had then foreclosed on the ordinary people who'd bought homes financed with those mortgages, fed up with crushing student loan debt, fed up with the cost of healthcare they couldn't afford, fed up with wages that weren't enough to live on, a group of protesters camped out in Zuccotti Park, a few blocks from Wall Street. Their movement, labeled with the Twitter hashtag #OccupyWallStreet or simply #Occupy, spread worldwide. By early October, Occupy protests had taken place in more than 951 cities, across 82 countries. Many of them were ongoing, with protesters camping out for months, until forcibly removed.

Two days after the protests began, I spent the afternoon at Zuccotti Park, studying the thousands of cardboard signs spread over the ground and surrounding buildings, each telling the story of a person or family failed by the current economy. I talked with the protesters to hear their stories firsthand; I participated in the "people's microphone," the clever technique used to get around the ban on amplified sound. Every speaker addressing the crowd paused at the end of each phrase, giving those nearby time to repeat it aloud, with the volume amplified by many voices so that those farther away could hear.

The rallying cry of the movement was "We are the 99%," a slogan coined by two online activists to highlight the realization, which had recently penetrated the popular consciousness, that 1% of the US population now earned 25% of the national income and owned 40% of its wealth. They began a campaign on Tumblr, a short-form blogging site with hundreds of millions of users. They asked people to post pictures of themselves holding a sign describing their economic situation, the phrase, "I am the 99%," and a pointer to the occupywallstreet.org site.

The messages were powerful and personal:

"My parents put themselves into debt so I could get a fancy degree. It cost over $100 grand, and I have no job prospects. I am the 99%."

"I have a master's degree, and I am a teacher, yet I can barely afford to feed my child because my husband lost his job due to missing too much work being hospitalized with a chronic illness. His meds alone are more than I make in a month. I am the 99%."

"I have a master's degree & a full time job in my field—and I have started SELLING MY BODY to pay off my debt. I am the 99%."

"Single mom, grad student, unemployed, and I paid more tax last year than GE. I am the 99%."

"I have not seen a dentist or doctor in over 6 years. I have long term injuries that I cannot afford the care for. Some days, I can barely walk. I am the 99%."

"Single mom. Working part-time and getting food stamps to barely get by. I just want a future for my daughter. I am the 99%."

"No medical. No dental. No vision. No raises. No 401K. Less than $30K a year before taxes. Less than $24K a year after taxes. I work for a Fortune 500 company. I am the 99%."

"I have never been appreciated, in retail, for any potential other than selling other people crap, half of which they do not need, and most of which they probably cannot really afford. I hate being used like that, I want a useful job. I am the 99%."

"We never chose irresponsibly. We were careful not to live outside our means. We bought a humble home and a responsible car; no Mc-Mansion, no Hummer. We were OK till my husband was laid off. . . . After six months of unemployment, he was fortunate enough to find work. However, it is 84 miles of commuting a day and it's 30% less pay. . . . My husband's fuel costs are almost one of his bi-weekly pay-checks. We are in a loss mitigation and loan modification program with our mortgage lender, and struggling with everything we have to keep our little house. I got a 2% raise in June, but my paycheck actually got smaller because my health insurance costs went up. We are the 99%."

"I have had no job for over 2½ years. Black men have a 20% unemployment rate. I am 33 years old. Born and raised in Watts. I am the 99%."

"I am nineteen. I have wanted kids in my future for a long time. Now I am scared that the future will not be an OK place for my kids. I am the 99%."

"I am retired. I live on savings, retirement, and social security. I'm OK. 50 million Americans are NOT OK: they are poor, have no health insurance, or both. But we are all the 99%."

They go on, thousands of them, voices crying out their fear and pain and helplessness, the voices of people whose lives have been crushed by the machine.

• • •

From *2001*'s HAL to *The Terminator*'s Skynet, it's a science fiction trope: artificial intelligence run amok, created to serve human goals but now pursuing purposes that are inimical to its former masters.

Recently, a collection of scientific and Silicon Valley luminaries, including Stephen Hawking and Elon Musk, wrote an open letter recommending "expanded research aimed at ensuring that increasingly capable AI systems are robust and beneficial: our AI systems must do what we want them to do." Groups such as the Future of Life Institute and OpenAI have been formed to study the existential risks of AI, and, as the OpenAI site puts it, "to advance digital intelligence in the way that is most likely to benefit humanity as a whole, unconstrained by a need to generate financial return."

These are noble goals. But they may have come too late.

We are already in the thrall of a vast, world-spanning machine that, due to errors in its foundational programming, has developed a disdain for human beings, is working to make them irrelevant, and resists all attempts to bring it back under control. It is not yet intelligent or autonomous, and it still depends on its partnership with humans, but it grows more powerful and more independent every day. We are engaged in a battle for the soul of this machine, and we are losing. Systems we have built to serve us no longer do so, and we don't know how to stop them.

If you think I'm talking about Google, or Facebook, or some shadowy program run by the government, you'd be wrong. I'm talking about something we refer to as "the market."

To understand how the market, that cornerstone of capitalism, is on its way to becoming that long-feared rogue AI, enemy to humanity, we first need to review some things about artificial intelligence. And then we need to understand how financial markets (often colloquially, and inaccurately, referred to simply as "Wall Street") have become a machine that its creators no longer fully understand, and how the goals and operation of that machine have become radically disconnected from the market of real goods and services that it was originally created to support.

THREE TYPES OF ARTIFICIAL INTELLIGENCE

As we've seen, when experts talk about artificial intelligence, they distinguish between "narrow artificial intelligence" and "general artificial intelligence," also referred to as "weak AI" and "strong AI."

Narrow AI burst into the public debate in 2011. That was the year that IBM's Watson soundly trounced the best human Jeopardy players in a nationally televised match in February. In October of that same year, Apple introduced Siri, its personal agent, able to answer common questions spoken aloud in plain language. Siri's responses, in a pleasing female voice, were the stuff of science fiction. Even when Siri's attempts to understand human speech failed, it was remarkable that we were now talking to our devices and expecting them to respond. Siri even became the best friend of one autistic boy.

The year 2011 was also the year that Google announced that its self-driving car prototype had driven more than 100,000 miles in ordinary traffic, a mere six years after the winner of the DARPA Grand Challenge for self-driving cars had managed to go only seven miles in seven hours. Self-driving cars and trucks have now taken center stage, as the media wrestles with the possibility that they will eliminate millions of human jobs. This fear, that this next wave of automation will go much further than the first industrial revolution in making human labor superfluous, is what makes many say "this time is different" when contemplating technology and the future of the economy.

The boundary between narrow AI and other complex software able to take many factors into account and make decisions in microseconds is fuzzy. Autonomous or semiautonomous programs able to perform complex tasks have been part of the plumbing of our society for decades. We rely on automated switching systems to route our phone calls (which were once patched through, literally, by humans in a switchboard office, connecting cables to specific named locations), and humans are routinely ferried thousands of miles by airplane autopilots while the human pilots ride along "just in case." While these systems at first appear magical, no one thinks of them as AI.

Personal agents like Siri, the Google Assistant, Cortana, and Amazon's Alexa do strike us as "artificial intelligences" because they listen to us speak, and reply with a human voice, but even they are not truly intelligent. They are cleverly programmed systems, much of whose magic is possible because they have access to massive amounts of data that they can process far faster than any human.

But there is one key difference between traditional programming of even the most complex systems, and deep learning and other techniques at the frontier of AI. Rather than spelling out every procedure, a base program such as an image recognizer or categorizer is built, and then *trained* by feeding it large amounts of data labeled by humans until it can recognize patterns in the data on its own. We teach the program what success looks like, and it learns to copy us. This leads to the fear that these programs will become increasingly independent of their creators.

Artificial general intelligence (also sometimes referred to as "strong AI") is still the stuff of science fiction. It is the product of a hypothetical future in which an artificial intelligence isn't just trained to be smart about a specific task, but to learn entirely on its own, and can effectively apply its intelligence to any problem that comes its way.

The fear is that an artificial general intelligence will develop its own goals and, because of its ability to learn on its own at superhuman speeds, will improve itself at a rate that soon leaves humans far behind. The dire prospect is that such a superhuman AI would have no use for humans, or at best might keep us in the way that we keep pets or domesticated animals. No one even knows what such an intelligence might look like, but people like Nick Bostrom, Stephen Hawking, and Elon Musk postulate that once it exists, it will rapidly outstrip humanity, with unpredictable consequences. Bostrom calls this hypothetical next step in strong AI "artificial superintelligence."

Deep learning pioneers Demis Hassabis and Yann LeCun are skeptical. They believe we're still a long way from artificial general intelligence. Andrew Ng, formerly the head of AI research for Chinese search giant Baidu, compared worrying about hostile AI of this kind to worrying about overpopulation on Mars.

Even if we never achieve artificial general intelligence or artificial superintelligence, though, I believe that there is a third form of AI, which I call *hybrid artificial intelligence*, in which much of the near-term risk resides.

When we imagine an artificial intelligence, we assume it will have an individual self, an individual consciousness, just like us. What if, instead, an AI was more like a multicellular organism, an evolution beyond our single-celled selves? What's more, what if we were not even the cells of such an organism, but its microbiome, the vast ecology of microorganisms that inhabits our bodies? This notion is at best a metaphor, but I believe it is a useful one.

As the Internet speeds up the connection between human minds, as our collective knowledge, memory, and sensations are shared and stored in digital form, we are weaving a new kind of technology-mediated superorganism, a global brain consisting of all connected humans. This global brain is a human-machine hybrid. The senses of that global brain are the cameras, microphones, keyboards, and location sensors of every computer, smartphone, and "Internet of Things" device; the thoughts of that global brain are the collective output of billions of individual contributing intelligences, shaped, guided, and amplified by algorithms.

Digital services like Google, Facebook, and Twitter that connect hundreds of millions or even billions of people in near-real time are already primitive hybrid AIs. The fact that the intelligence of these systems is interdependent with the intelligence of the community of humans that makes it up is an echo of the way that we ourselves function. Each of us is a vast nation of trillions of differentiated cells, only some of which share our own DNA, while far more are immigrants, the vast microbiome of microorganisms that colonize our guts, our skin, our circulatory systems. There are far more microorganisms in our bodies than there are human cells, not invaders but a functioning part of the whole. Without the microorganisms we host, we could not digest our food or turn it into useful energy. The bacteria in our guts have even been shown to change how we think and how we feel. A multicellular organism is the sum of the communications, the ecosystem, the platform or marketplace if you will, of all its participants. And when that marketplace gets out of balance, we fall ill or fail to live up to our potential.

> **Humans are living in the guts of an AI that is only now being born. Perhaps, like us, the global AI will not be an independent entity, but a symbiosis with the human consciousnesses living within it and alongside it.**

Every day, we teach the global brain new skills. DeepMind began its Go training by studying games played by humans. As its creators wrote in their January 2016 paper in *Nature*, "These deep neural networks are trained by a novel combination of supervised learning from human expert games, and reinforcement learning from games of self-play." That is, the program began by observing humans playing the game, and then accelerated that learning by playing against itself millions of times, far outstripping the experience level of even the most accomplished human players. This pattern, by which algorithms are trained by humans, either explicitly or implicitly, is central to the explosion of AI-based services.

Explicit development of training data sets for AI is dwarfed, though, by the data that humans produce unasked on the Internet. Google Search, financial markets, and social media platforms like Facebook and Twitter gather data from trillions of human interactions, distilling that data into collective intelligence that can be acted on by narrow AI algorithms. As computational neuroscientist and AI entrepreneur Beau Cronin puts it, "In many cases, Google has succeeded by reducing problems that were previously assumed to require strong AI—that is, reasoning and problem-solving abilities generally associated with human intelligence—into narrow AI, solvable by matching new inputs against vast repositories of previously encountered examples." Enough narrow AI infused with the data thrown off by billions of humans starts to look suspiciously like strong AI. In short, these are systems of collective intelligence that use algorithms to aggregate the collective knowledge and decisions of millions of individual humans.

And that, of course, is also the classical conception of "the market"—the system by which, without any central coordination, prices of goods and labor are set, buyers and sellers are found for all of the fruits of the earth and the products of human ingenuity, guided as if, as Adam Smith famously noted, "by an invisible hand."

But is the invisible hand of a market of self-interested human merchants and human consumers the same as a market in which computer algorithms guide and shape those interests?

COLLECTIVE INTELLIGENCE GONE WRONG

Algorithms not only aggregate the intelligence and decisions of humans; they also influence and amplify them. As George Soros notes, the forces that shape our economy are not true or false; they are reflexive, based on what we collectively come to believe or know. We've already explored the effect of algorithms on news media.

The speed and scale of electronic networks are also changing the nature of financial market reflexivity in ways that we have not yet fully come to understand. Financial markets, which aggregate the opinions of millions of people in setting prices, are liable to biased design, algorithmically amplified errors, or manipulation, with devastating consequences. In the famous "Flash Crash" of 2010, high-frequency-trading algorithms responding to market manipulation by a rogue human trader dropped the Dow by 1,000 points (nearly a trillion dollars of market value) in only thirty-six minutes, recovering 600 of those points only a few minutes later.

The Flash Crash highlights the role that the speed of electronic networks plays in amplifying the effects of misinformation or bad decisions. The price of goods from China was once known at the speed of clipper ships, then of telegrams. Now electronic stock and commodities traders place themselves closer to Internet points of presence (the endpoints of high-speed networks) to gain microseconds of advantage. And this need for speed has left human traders behind. More than 50% of all stock market trades are now made by programs rather than by human traders.

Unaided humans are at an immense disadvantage. Michael Lewis, the author of *Flash Boys*, a book about high-frequency trading, summarized that disadvantage in an interview with NPR *Fresh Air* host Terry Gross: "If I get price changes before everybody else, if I know a stock price is going up or going down before you do, I can act on it. . . . [I]t's a bit like knowing the result of the horse race before it's run. . . . The time advantage of a high-frequency trader is so small, it's literally a millisecond. It takes 100 milliseconds to blink your eye, so it's a fraction of a blink of an eye, but that for a computer is plenty of time."

Lewis noted that this divides the market into two camps, prey and predator, the people who actually want to invest in companies, and people who have figured out how to use their speed advantage to front-run them, buy the stock before ordinary traders can get to it, and resell it to them at a higher price. They are essentially parasites, adding no value to the market, only extracting it for themselves. "The stock market is rigged," Lewis told Gross. "It's rigged for the benefit for a handful of insiders. It's rigged to . . . maximize the take of Wall Street, of banks, the exchanges and the high-frequency traders at the expense of ordinary investors."

When Brad Katsuyama, one of the heroes of Lewis's book, tried to create a new exchange "where every dollar stands the same chance," by taking away the advantages of the speed traders, Lewis noted that "the banks and the brokers [who] are also paid a cut of what the high-frequency traders are taking out of investors' orders . . . don't want to send their orders on this fair exchange because there's less money to be made."

Derivatives, originally invented to hedge against risk, instead came to magnify it. The CDOs (collateralized debt obligations) that Wall Street sold to unsuspecting customers in the years leading up to the 2008 crash could only have been constructed with the help of machines. In a 2009 speech, John Thain, the former CEO of the New York Stock Exchange who'd become CEO of Merrill Lynch, admitted as much. "To model correctly one tranche of one CDO took about three hours on one of the fastest computers in the United States. There is no chance that pretty much anybody understood what they were doing with these securities. Creating things that you don't understand is really not a good idea no matter who owns it."

In short, both high-speed trading and complex derivatives tilt financial markets away from human control and understanding. But they do more than that. They have cut their anchor to the human economy of real goods and services. As Bill Janeway noted to me, the bursting of what he calls the "super-bubble" in 2008 "shattered the assumption that financial markets are necessarily efficient and that they will reliably generate prices for financial assets that are locked onto the fundamental value of the physical assets embedded in the nonfinancial, so-called real economy."

The vast amounts of capital sloshing around in the financial system leading up to that 2008 crisis led to the growth of "shadow banking," which used that capital to provide credit far in excess of the underlying real assets, credit that was secured by low-quality bonds based on ever-riskier mortgages. Financial capitalism had become a market in imaginary assets, made plausible only by the Wall Street equivalent of fake news.

THE DESIGN OF THE SYSTEM SETS ITS OUTCOMES

High-frequency trading, complex derivatives like CDOs, and shadow banking are only the tip of the iceberg, though, when thinking about how markets have become more and more infused with machinelike characteristics, and less and less friendly to the humans they were originally expected to serve. The fact that we are building financial products that no one understands is actually a reflection of the fundamental design of the modern financial system. What is the fitness function of the model behind its algorithms, and what is the biased data that we feed it?

Like the characters in the Terminator movies, before we can stop Skynet, the global AI bent on the enslavement of humanity, we must travel back in time to try to understand how it came to be.

According to political economist Mark Blyth, writing in *Foreign Affairs*, during the decades following World War II, government policy makers decided that "sustained mass unemployment was an existential threat to capitalism." The guiding "fitness function" for Western economies thus became full employment.

This worked well for a time, Blyth notes, but eventually led to what was called "cost-push inflation." That is, if everyone is employed, there is no barrier to moving from job to job, and the only way to hang on to employees is to pay them more, which employers necessarily compensated themselves for by raising prices, in a continuing spiral of higher wages and higher prices. As Blyth notes, every intervention is subject to Goodhart's Law: "Targeting any variable long enough undermines the value of the variable."

Coupled with the end of the Bretton Woods system, a gold exchange standard anchored to the US dollar, the commitment to full employment led to skyrocketing inflation. Inflation is good for debtors—it makes goods such as housing much cheaper, because you repay a fixed dollar amount of debt with future dollars that are worth much less. Meanwhile, you have more of those dollars, as your salary keeps going up. Ordinary goods cost more, though, which means that as a worker, you have to keep demanding higher wages. But inflation is very bad for the owners of capital, since it reduces the value of what they own.

Starting in the 1970s, keeping inflation low replaced full employment as the fitness function. Federal Reserve chairman Paul Volcker put a strict cap on the money supply in an effort to bring inflation to a screeching halt. By the early 1980s, inflation was under control, but at the cost of sky-high interest rates and high unemployment.

The attempt to bring inflation under control was coupled with a series of supporting policy decisions. Labor organizing, which had helped to promote high wages and full employment, was made more difficult. The Taft-Hartley Act of 1947 weakened the power of unions and allowed the passage of state laws that limited it still further. By 2012, only 12% of the US labor force was unionized, down from a peak above 30%. But, perhaps most important, a bad idea took hold.

In September 1970, economist Milton Friedman penned an op-ed in the *New York Times Magazine* titled "The Social Responsibility of Business Is to Increase Its Profits," which took ferocious aim at the idea that corporate executives had any obligation but to make money for their shareholders.

"I hear businessmen speak eloquently about the 'social responsibilities of business in a free-enterprise system,'" Friedman wrote. "The businessmen believe that they are defending free enterprise when they declaim that business is not concerned 'merely' with profit but also with promoting desirable 'social' ends; that business has a 'social conscience' and takes seriously its responsibilities for providing employment, eliminating discrimination, avoiding pollution and whatever else may be the catchwords of the contemporary crop of reformers. In fact they are—or would be if they or anyone else took them seriously—preaching pure and unadulterated socialism."

Friedman meant well. His concern was that by choosing social priorities, business leaders were making decisions on behalf of their shareholders that those shareholders might individually disagree on. Far better, he thought, to distribute the profits to the shareholders and let them make the choice of charitable act for themselves, if they so wished. But the seed was planted, and began to grow into a noxious weed.

The next step occurred in 1976 with an influential paper published in the *Journal of Financial Economics* by economists Michael Jensen and William Meckling, "Theory of the Firm: Managerial Behavior, Agency Costs, and Ownership Structure." Jensen and Meckling made the case that professional managers, who work as agents for the owners of the firm, have incentives to look after themselves rather than the owners. The management, might, for instance, lavish perks on themselves that don't directly benefit the business and its actual owners.

Jensen and Meckling also meant well. Unfortunately, their work was thereafter interpreted to suggest that the best way to align the interests of management and shareholders was to ensure that the bulk of management compensation was in the form of company stock. That would give management the primary objective of increasing the share price, aligning their interests with those of shareholders, and prioritizing those interests over all others.

Before long, the gospel of shareholder value maximization was taught in business schools and enshrined in corporate governance. In 1981, Jack Welch, then CEO of General Electric, at the time the world's largest industrial company, announced in a speech called "Growing Fast in a Slow-Growth Economy" that GE would no longer tolerate low-margin or low-growth units. Any business owned by GE that wasn't first or second in its market and wasn't growing faster than the market as a whole would be sold or shuttered. Whether or not the unit provided useful jobs to a community or useful services to customers was not a reason to keep on with a line of business. Only the contribution to GE's growth and profits, and hence its stock price, mattered.

That was our Skynet moment. The machine had begun its takeover.

Yes, the markets have become a hybrid of human and machine intelligence. Yes, the speed of trading has increased, so that a human trader not paired with that machine has become prey, not predator. Yes, the market is increasingly made of complex financial derivatives that no human can truly understand. But the key lesson is one we have seen again and again. The design of a system determines its outcomes. The robots did not force a human-hostile future upon us; we chose it ourselves.

• • •

The 1980s were the years of "corporate raiders" celebrated by Michael Douglas's character, Gordon Gekko, in the 1987 movie *Wall Street*, who so memorably said, "Greed is good." The theory was that by discovering and rooting out bad managers and finding efficiencies in underperforming businesses, these raiders were actually improving the operation of the capitalist system. It is certainly true that in some cases they played that role. But by elevating the single fitness function of increasing share price above all else, they hollowed out our overall economy.

The preferred tool of choice has become stock buybacks, which, by reducing the number of shares outstanding, raise the earnings per share, and thus the stock price. As a means of returning cash to shareholders, stock buybacks are more tax-efficient than dividends, but they also send a very different message. Dividends traditionally signaled, "We have more cash than we need for the business, so we are returning it to you," while stock buybacks signaled, "We believe our stock is undervalued by the market, which doesn't understand the potential of our business as well as we do." They were positioned as an investment the company was making in itself. This is clearly no longer the case.

In his 2016 letter to Berkshire Hathaway shareholders, Warren Buffett, the world's most successful financial investor for the past six decades, put his finger on the short-term thinking driving most buybacks: "The question of whether a repurchase action is value-enhancing or value-destroying for continuing shareholders is entirely purchase-price dependent. It is puzzling, therefore, that corporate repurchase announcements almost never refer to a price above which repurchases will be eschewed."

Larry Fink, the CEO of BlackRock, the world's largest asset manager, with more than $5.1 trillion under management, also took aim at buybacks, noting in his 2017 letter to the CEOs that for the twelve months ending in the third quarter of 2016, the amount spent on dividends and buybacks by companies that make up the S&P 500 was greater than the entire operating profit of those companies.

While Buffett believes that companies are spending the money on buybacks because they don't see opportunities for productive capital investment, Fink points out that for long-term growth and sustainability, companies have to invest in R&D "and, critically, employee development and long-term financial well-being." Rejecting the idea that companies or the economy can prosper solely by boosting short-term returns to shareholders, he continues: "The events of the past year have only reinforced how critical the well-being of a company's employees is to its long-term success."

Fink makes the case that instead of returning cash to shareholders, companies should be spending far more of their hoarded profits on improving the skills of their workers. "In order to fully reap the benefits of a changing economy—and sustain growth over the long-term—businesses will need to increase the earnings potential of the workers who drive returns, helping the employee who once operated a machine learn to program it," he writes. They "must improve their capacity for internal training and education to compete for talent in today's economy and fulfill their responsibilities to their employees."

The Rise and Fall of American Growth, Robert J. Gordon's magisterial history of the change in the US standard of living since the Civil War, makes a compelling case that after a century of extraordinary expansion, the growth of productivity in the US economy slowed substantially after 1970. Whether Gordon's analysis that the productivity-enhancing technologies of the previous century gave the economy a historically anomalous surge, or whether Fink and others are right that we simply aren't making the investments we need, it is clear that companies are using stock buybacks to create the illusion of growth where real growth is lagging.

> Stock prices are a map that should ideally describe the underlying prospects of companies; attempts to distort that map should be recognized for what they are. We need to add "fake growth" to "fake news" in our vocabulary to describe what is going on. Real growth improves people's lives.

Apologists for buybacks argue that much of the benefit of rising stock prices goes to pension funds and thus, by extension, to a wide swath of society. However, even with the most generous interpretation, little more than half of all Americans are shareholders in any form, and of those who are, the proportional ownership is highly skewed toward a small segment of the population—the now-famous 1%. If companies were as eager to allocate shares to all their workers in simple proportion to their wages as they are to award them to top management, this argument might hold some water.

The evidence that companies constructed around a different model can be just as successful as their financialized peers is hiding in plain sight. Storied football firm the Green Bay Packers is owned by its fans and uses that ownership to keep ticket prices low. Outdoors retailer REI, a member cooperative with $2.4 billion in revenue and six million members, returns profits to its members rather than to outside owners. Yet REI's growth consistently outperforms both its publicly traded competitors and the entire S&P 500 retail index. Vanguard, the second-largest financial asset manager in the United States, with more than $4 trillion under management, is owned by the mutual funds whose performance it aggregates. John Bogle, its founder, invented the index fund as a way to keep fund management fees low, transferring much of the benefit of stock investing from money managers to its customers.

Despite these counterexamples, the idea that extracting the highest possible profits and then returning the money to company management, big investors, and other shareholders is good for society has become so deeply rooted that it has been difficult for too long to see the destructive effects on society when shareholders are prioritized over workers, over communities, over customers. This is a bad map that has led our economy deeply astray.

As former chair of the White House Council of Economic Advisers Laura Tyson impressed on me over dinner one night, though, the bulk of jobs are provided by small businesses, not by large public companies. She was warning me not to overstate the role of financial markets in economic malaise, but her comments instead reminded me that the true effect of "trickle-down economics" is the way that the ideal of maximizing profit, not shared prosperity, has metastasized from financial markets and so shapes our entire society.

> Mistaking what is good for financial markets for what is good for jobs, wages, and the lives of actual people is a fatal flaw in so many of the economic choices business leaders, policy makers, and politicians make.

William Lazonick, a professor of economics at the University of Massachusetts Lowell and director of the Center for Industrial Competitiveness, notes that in the decade from 2004 to 2013, Fortune 500 companies spent an astonishing $3.4 trillion on stock buybacks, representing 51% of all corporate profits for those companies. Another 35% of profits were paid out to shareholders in dividends, leaving only 14% for reinvestment in the company. The 2016 figures cited by Larry Fink are the culmination of a decades-long trend. Companies like Amazon that are able to defy financial markets and sacrifice short-term profits in favor of long-term investment are all too rare.

The decline in corporate retained earnings is critical, because they are the most important source of funds for business investment. Despite the common idea that financial markets are used to fund business expansion, Lazonick notes that "the primary role of the stock market has been to permit owner-entrepreneurs and their private-equity associates to exit personally from investments that have already been made rather than to enable a corporation to raise funds for new investment in productive assets."

Since the mid-1980s, Lazonick observes, "the resource-allocation regime at many, if not most, major U.S. business corporations has transitioned from 'retain-and-reinvest' to 'downsize-and-distribute.' Under retain-and-reinvest, the corporation retains earnings and reinvests them in the productive capabilities embodied in its labor force. Under downsize-and-distribute, the corporation lays off experienced, and often more expensive, workers, and distributes corporate cash to shareholders."

One casualty of the shareholder value economy has been the decline of corporate scientific research. In a 1997 analysis for the US Federal Reserve Board of Governors, economists Charles Jones and John Williams calculate that the actual spending on R&D as a share of GDP is less than a quarter of the optimal rate, based on the "social rate of return" from innovation. And in a 2015 paper, economists Ashish Arora, Sharon Belenzon, and Andrea Patacconi document the decline since 1980 in the number of research papers published by scientists at large companies, curiously coupled with no decline in the number of patents filed. This is a shortsighted prioritization of value capture over value creation. "Large firms appear to value the golden eggs of science (as reflected in patents)," the authors write, "but not the golden goose itself (the scientific capabilities)."

The biggest losers, though, from this change in corporate reinvestment have been workers, whose jobs have been eliminated and whose wages have been cut to fund increasing returns to shareholders. As shown in the figure below, the share of GDP going to wages has fallen from nearly 54% in 1970 to 44% in 2013, while the share going to corporate profits went from about 4% to nearly 11%. Wallace Turbeville, a former Goldman Sachs banker, aptly describes this as "something approaching a zero-sum game between financial wealth-holders and the rest of America." Zero-sum games don't end well. "The one percent in America right now is still a bit lower than the one percent in pre-revolutionary France but is getting closer," says French economist Thomas Piketty, author of *Capital in the Twenty-First Century*.

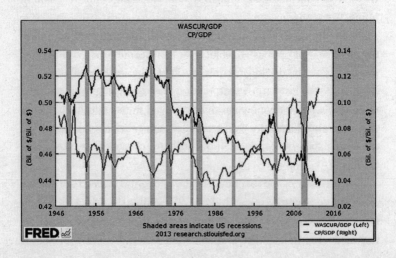

Lazonick believes his research demonstrates that this trend "is in large part responsible for a national economy characterized by income inequity, employment instability, and diminished innovative capability—or the opposite of what I have called 'sustainable prosperity.'"

Even stock options, so powerful a tool in the Silicon Valley innovation economy, have played a damaging role in turning the economy into a casino. Bill Janeway, who is a pioneering venture capitalist as well as an economist, likes to point out that when options began to be deployed by startups they were tickets to a lottery in which most holders would receive nothing. In 75% of VC-backed startups, the entrepreneur gets zero, and only 0.4% hit the proverbial jackpot. "The possible returns had to be abnormally high," Janeway wrote to me in an email, "given how rarely it was reasonable to expect such success to be realized."

Options were designed to encourage innovation and risk taking. "But then," Bill wrote, "this innovation in compensation, mobilized to lure executives out of the safe harbors of HP and IBM, was hijacked. The established companies began using stock options when there was essentially no risk of company failure. It reached the destructive extreme when the CEOs of banks, whose liabilities are guaranteed by taxpayers, began getting most of their compensation in options."

In 1993, a well-intentioned law pushed by President Clinton limited the ordinary income that could be paid to top management, with the unintended consequence that even more of the compensation moved to stock. Congress also initially allowed a huge loophole in the accounting treatment of options—unlike ordinary income paid to employees, options had to be disclosed, but not valued. Since the value of options did not need to be charged against company earnings, it became a kind of "free money" for companies, invisibly paid for by dilution of public market shareholders (of whom a large percentage are pension funds and other institutional shareholders representing ordinary people) rather than out of the profits of the company.

Meanwhile, there is an incentive to cut income for ordinary workers. Cutting wages drives up net income and thus the price of the stock in which executives are increasingly paid. Those executives who are not motivated by cupidity are held hostage. Any CEO who doesn't keep growing the share price or who considers other interests than those of the shareholders is liable to lose his or her job or be subject to lawsuits. Even Silicon Valley firms whose founders retain controlling positions in their companies are not immune from pressure. Because so much of the compensation of their employees is now in stock, they can only continue to hire the best talent as long as the stock price continues to rise.

> It isn't Wall Street per se that is becoming hostile to humanity. It is the master algorithm of shareholder capitalism, whose fitness function both motivates and coerces companies to pursue short-term profit above all else. What are humans in that system but a cost to be eliminated?

Why would you employ workers in a local community when you could improve corporate profits by outsourcing the work to people paid far less in emerging economies? Why would you pay a living wage if you could instead use the government social safety net to make up the difference? After all, that safety net is funded by other people's taxes—because of course it is only efficient to minimize your own.

Why invest in basic research, or a new factory, or training that might make your workforce more competitive, or a risky new line of business that might not contribute meaningfully to earnings for many years when you can get a quick pop in the price by using your cash to buy back your stock instead, reducing the number of shares outstanding, pleasing investors and enriching yourself?

For that matter, why would you provide the best goods or services if you could improve profits by cutting corners? This is the era of what business strategist Umair Haque, director of the Havas Media Lab, calls "thin value, profit extracted through harm to others." Thin value is the value of tobacco marketed even after its purveyors knew it contributed to cancer; the value of climate change denial by oil companies, who have retained the same disinformation firms used by the tobacco industry. This is the value that we experience when food is adulterated with high-fructose corn syrup or other additives that make us sick and obese; the value we experience when we buy shoddy products that are meant to be prematurely replaced.

If profit is the measure of all things, why not "manage your earnings," as Welch, the CEO of GE, came to do, so that the business appears better than it is to investors? Why not actively trade against your customers, as investment banks began to do? Why not dip into outright fraud, selling those customers complicated financial instruments that were designed to fail? And when they do fail, why not ask the taxpayers to bail you out, because government regulators—drawn heavily from your own ranks—believe you are so systematically important to the world economy that you become untouchable?

Government—or perhaps more accurately, the lack of it—has become deeply complicit in the problem. Economists George Akerlof and Paul Romer fingered the nexus between corporate malfeasance and political power in their 1994 paper "Looting: The Economic Underworld of Bankruptcy for Profit." "Bankruptcy for profit will occur if poor accounting, lax regulation, or low penalties for abuse give owners an incentive to pay themselves more than their firms are worth and then default on their debt obligations," they wrote. "The normal economics of maximizing economic value is replaced by the topsy-turvy economics of maximizing current extractable value, which tends to drive the firm's economic net worth deeply negative. . . . A dollar in increased dividends today is worth a dollar to owners, but a dollar in increased future earnings of the firm is worth nothing because future payments accrue to the creditors who will be left holding the bag."

This was the game plan of many corporate raiders, who laid off workers and stripped firms of their assets, even taking them through bankruptcy to eliminate their pension plans. It was also at the heart of a series of booms and busts in real estate and finance that decimated the economy while enormously enriching a tiny group of economic looters and lucky bystanders.

It is the Bizarro World inverse of my maxim that companies must create more value than they capture. Instead, companies seek to capture more value than they create.

This is the tragedy of the commons writ large. It is also the endgame of what Milton Friedman called for in 1970, a bad idea that took hold in the global mind, whose consequences took decades to unfold.

There is an alternative view, which crystallized for me in 2012, even though I'd been living it all my life, when I heard a talk at the TED conference by technology investor Nick Hanauer. Nick is a billionaire capitalist, heir to a small family manufacturing business, who had the great good fortune to become the first nonfamily investor in Amazon, and who later was a major investor in aQuantive, an ad-targeting firm sold to Microsoft for $6 billion. What Nick said made a lot of sense to me. As with open source and Web 2.0, his talk was one more piece of a puzzle that slipped into place, helping me see the outlines of what I would come to call "the Next Economy."

As I remember the talk, its central argument went something like this: "I'm a successful capitalist, but I'm tired of hearing that people like me create jobs. There's only one thing that creates jobs, and that's customers. And we've been screwing workers so long that they can no longer afford to be our customers."

In making this point, Nick was echoing the arguments of Peter F. Drucker in his 1955 book, *The Practice of Management*: "There is only one valid definition of business purpose: to create a customer. . . . It is the customer who determines what a business is. It is the customer alone whose willingness to pay for a good or for a service converts economic resources into wealth, things into goods. . . . The customer is the foundation of a business and keeps it in existence."

In this view, business exists to serve human needs. Corporations and profits are a means to that end, not an end in themselves. Free trade, outsourcing, and technology are tools not for reducing costs and improving share price, but for increasing the wealth of the world. Even handicapped by the shareholder value theory, the world is better off because of the dynamism of a capitalist economy, but how much better could we have done had we taken a different path?

I don't think anyone but the looters believes that making money for shareholders is the ultimate end of economic activity. But many economists and corporate leaders are confused about the role it plays in helping us achieve that end. Milton Friedman, Meckling and Jensen, and Jack Welch were well-meaning. All believed that aligning the interests of corporate management with shareholders would actually produce the greatest good for society as well as for business. But they were wrong. They were following a bad map. By 2009 Welch had changed his mind, calling the shareholder value hypothesis "a dumb idea."

But by then Welch had retired with a fortune close to $900 million, most of it earned via stock options, and the machine ground on, bigger than any CEO, bigger than any company. Author Douglas Rushkoff told me the story of one Fortune 100 CEO who broke down in tears as she told him how her attempts to inject social value into decision making at her company had resulted in quick punishment by "the market," forcing her to reverse course.

Who is the market? It is algorithmic traders who pop in and out of companies at millisecond speed, turning what was once a vehicle for capital investment in the real economy into a casino where the rules always favor the house. It is corporate raiders like Carl Icahn (now rebranded as a "shareholder activist") who buy large blocks of shares and demand that companies that wish to remain independent instead put themselves up for sale, or that a company like Apple disgorge its cash into their pockets rather than using it to lower prices for customers or raise wages for workers. It is also pension funds, desperate for higher returns to fund the promises that they have made, who outsource their money to professional managers who must then do their best to match the market or lose the funds they manage. It is venture capitalists and entrepreneurs dreaming of vast disruption leading to vast fortunes. It is every company executive who makes decisions based on increasing the stock price rather than on serving customers.

But these classes of investor are just the most obvious features of a system of reflexive collective intelligence far bigger even than Google and Facebook, a system bigger than all of us, that issues relentless demands because it is, at bottom, driven by a master algorithm gone wrong.

This is what financial industry critics like Rana Foroohar, author of the book *Makers and Takers*, are referring to when they say that the economy has become *financialized*. "The single biggest unexplored reason for long-term slower growth," she writes, "is that the financial system has stopped serving the real economy and now serves mainly itself."

It isn't just that the financial industry employs only 4% of Americans but takes in more than 25% of all corporate profits (down from a 2007 peak of nearly 40%). It isn't just that Americans born in 1980 are far less likely to be better off financially than their parents than those born in 1940, or that 1% of the population now owns nearly half of all global wealth, and that nearly all the income gains since the 1980s have gone to the top tenth of 1%. It isn't just that people around the world are electing populist leaders, convinced that the current elites have rigged the system against them.

These are symptoms. The root problem is that the financial market, once a helpful handmaiden to human exchanges of goods and services, has become the master. Even worse, it is the master of all the other collective intelligences. Google, Facebook, Amazon, Twitter, Uber, Airbnb, and all the other unicorn companies shaping the future are in its thrall as much as any one of us.

It is this hybrid artificial intelligence of today, not some fabled future artificial superintelligence, that we must bring under control.

PART IV
IT'S UP
TO US

The best way to predict the future is to invent it.

—Alan Kay

12

REWRITING THE RULES

IN THE MAY 2011 *VANITY FAIR* ESSAY THAT BROUGHT THE NO-
tion of the 1% into the national dialogue, "Of the 1%, by the 1%, for
the 1%," Nobel Prize–winning economist Joseph Stiglitz wrote a chilling reflection on the consequences of a dysfunctional economy that
works well for only a tiny fraction of the population. His title, painfully
echoing Lincoln's Gettysburg Address, asks if indeed "government of
the people, by the people, and for the people" remains our ideal.

He wrote about the tumult then upending autocratic regimes in the
Middle East, noting, "These are societies where a minuscule fraction
of the population—less than 1 percent—controls the lion's share of the
wealth; where wealth is a main determinant of power; where entrenched
corruption of one sort or another is a way of life; and where the wealthiest often stand actively in the way of policies that would improve life for
people in general." Most tellingly, Stiglitz observed, "In important ways,
our own country has become like one of these distant, troubled places,"
and asked of the popular uprising, "When will this come to America?"

The Occupy Wall Street protesters were eventually cleared out of
their encampments, but the questions they asked continue to resonate
through our politics. Will the future provide opportunity for all of us?
Or will it crush most of us even further underfoot?

"The 1%" was a key feature of Bernie Sanders's 2016 presidential
campaign, and Donald Trump rode the message of blowing up the
incumbents all the way to victory over Hillary Clinton's defense of the
status quo. By all appearances, though, President Trump has little in
the way of a policy solution to the fundamental problem that Stiglitz
outlined, that the 1%, or more properly, the .01%, have translated their
financial power into political power, turning what was once a vibrant
democracy and a vibrant economy into a staggering colossus, a platform
that no longer works for the benefit of its participants.

You can see how the struggle between people and profit played out in a 2016 *New York Times* account of the closing of Carrier's Indianapolis factory and the planned transfer of its 1,400 jobs to Mexican workers making about as much per day as its Indianapolis workers make per hour. Trump made much of this incident during his campaign, pointing to labor outsourcing as the root of the problem. But why do companies seek ever-cheaper labor?

Carrier's parent company, United Technologies, explained that "the cuts are painful but are necessary for the long term competitive nature of the business and shareholder value creation." United Technologies Chief Financial Officer Akhil Johri gave the game away with his final words: ". . . *and shareholder value creation.*" The article went on to explain:

> Wall Street is looking for United Technologies to post a 17 percent increase in earnings per share over the next two years, even though sales are expected to rise only 8 percent. Bridging that gap means cutting costs wherever savings can be found, as Mr. McDonough [president of United Technologies' climate, controls, and security division] suggested at the meeting with analysts.

In theory, companies care about their stock price because financial markets provide the capital that allows them to invest and expand. But guess what: United Technologies didn't need to go to financial markets for capital. In fact, they have so much capital that in December 2015 they had just committed to spend another $12 billion to buy back their stock.

Despite United Technologies' rhetoric, it is not a company that needs to cut costs "for the long term competitive nature of the business." I believe that a set of money managers, already members of the .01%, are demanding that profits rise in order to drive up the stock, so that their own incomes will increase. Top managers in the company might go along with this plan because their compensation is also tied to that rise in stock price and because they will lose their jobs if they don't deliver on it. *This is a forced wealth reallocation from one set of stakeholders in the company to another.*

That's why there is so much anger at Wall Street from the followers of both Donald Trump and Bernie Sanders, populists of the right and of the left. The system *is* rigged. Companies are forced to eliminate workers not by the market of real goods and services where supply and demand set the right price, but by the commands of financial markets, where hope and greed too often set the price.

> Most people unthinkingly use the term *the market* to refer to these two very different markets. Recognizing that they are not the same is the first step toward solving the problem.

President Trump's solution is to threaten companies with tariffs on foreign goods or losing government contracts, or to promise backdoor payments to keep jobs in America. None of these address the underlying problem. Financiers, CEOs, and corporate boards should do some deep soul-searching into their responsibility for the current state of the economy, which so clearly no longer works for many ordinary Americans. Alas, as Nick Hanauer said to me, soul-searching by CEOs is as likely to affect the economy as "thoughts and prayers for the victims" are to put an end to gun violence. *We need to rethink the incentives that encourage this behavior, and reverse the rules that allow it.*

THE "LAWS" OF ECONOMICS

Future economic historians may look back wryly at this period when we worshipped the divine right of capital while looking down on our ancestors who believed in the divine right of kings.

Business leaders making decisions to outsource jobs to low-wage countries or to replace workers with machines, or politicians who insist that it is "the market" that makes them unable to require companies to pay a living wage, rely on the defense that they are only following the laws of economics. But the things economists study are not natural phenomena like the laws of motion uncovered by Kepler and Newton. They are in part the outcome of rules and algorithms devised by humans that attempt to model and influence human behavior. Because many of these rules and algorithms are enforced by law and custom rather than by code, we are blind to the ways that they are similar to the algorithms used by Google and Facebook and Uber. We are following the wrong map.

Because they are shaped by rules crafted by our imperfect understanding, entire economies can go awry in much the same way that simpler digital marketplaces like Google and Facebook, Uber and Airbnb can. Their fundamental fitness functions can be wrong. They can have bias in the data used to train their algorithms. They can be gamed by participants.

Behavioral economics has convincingly refuted the idealized model of "homo economicus," the rational actor whose pursuit of self-interest can be neatly modeled with mathematical formulae. Modern economics is increasingly looking to historical data rather than to theory, trying to build a better map. Unfortunately, what James Kwak calls "economism," the reduction of real-world problems to fit a simplistic version of economic theory—that is, substituting looking at the map for looking at the territory, continues to rule the thinking of most politicians and business leaders.

A better way to think about an economy is that it is like a game. Some of the rules of the game do represent what appear to be fundamental constraints—population growth and productivity, the availability of labor or resources, or the capacity of the environment, or even the behavioral patterns of human nature—while others are arbitrary and subject to change, such as tax policy, government entitlements, and minimum-wage requirements. The game has untold possible outcomes. Its complexity comes both from the near-infinite variety that can come from permutations of simple rules and from the fact that billions of humans are playing the game simultaneously, each affecting the outcomes for the others. Even the simplest and most definitive of the "rules" of an economy are far more complex to apply than they appear on paper. As an Internet wag noted many years ago, "The difference between theory and practice is always greater in practice than it is in theory."

This complexity, and its dismissal based on economic theory, came to mind last year in a conversation I had with Uber's economists. I was arguing that just as Google's search algorithm takes many factors into account in producing the "best" results, Uber would benefit if its algorithms took drivers' wages, job satisfaction, and turnover into account, and not just passenger pickup time, which it currently uses as its fitness function. (Uber aims to have enough drivers on the road in a given location that the average pickup time is no more than three minutes.)

The economists explained to me that Uber's wages were, by definition, optimal, because they simply represent the equilibrium point between supply and demand, one of the most basic ideas of free market economics.

Uber's real-time matching algorithm actually satisfies two overlapping demand curves. If there are not enough passengers, the price must go down to stimulate passenger demand. That's the essence of Uber's price cuts. But if there are not enough drivers to satisfy that demand, the price has to go up to encourage more drivers to come on the road. That's the essence of surge pricing. Uber's argument is that the algorithmically determined cost of a ride is at the sweet spot that will drive the most passenger demand while also providing sufficient incentive to produce the number of drivers to meet that demand. And because driver income is the product of both the number of trips and the rate paid, they believe that even with lower fares, increasing passenger demand will improve driver incomes more effectively than limiting supply, as was done with taxi medallions. They believe that any attempt to set rates specifically to raise driver income would suppress rider demand, and so reduce utilization and thus net wages. Of course, if too many drivers show up, this will also reduce utilization, but the economists seemed confident, based on data that they were not authorized to share with me, that they have generally found that sweet spot.

I'm not convinced. If Uber had the courage of its convictions, it would be deploying demand-based pricing (including surging prices in a negative direction, below the base price) all the time, much as Google sets ad prices with an auction. Why don't they? Because they believe that both drivers and customers are more comfortable with a known base price. That is, the difference between theory and practice is greater in practice than it is in theory.

It's also worth noting that even this seemingly simple market requires rules to prevent opportunistic behavior, such as a driver canceling because he gets a better offer elsewhere, or two friends each calling an Uber and taking the one that arrives first. (Before coming up with the idea for Uber, Garrett Camp was reportedly blocked by San Francisco cab companies for doing just this in the old world of cabs scheduled by calling a dispatcher.) The simple maps of idealized markets leave out many real-world details that must be dealt with in order for the market to actually function properly. Governance is essential.

The question is whether dynamic algorithmic governance can be superior to simpler fixed rules. Even in their current state, Uber's real-time marketplace algorithms do allow for better matching of supply and demand than the previous structure of the taxicab and limousine industry or the labor market algorithms used by workplace scheduling companies. But Uber can do far better. Algorithms such as these can be a real advance in the structure of our economy, but only if they take into account the needs of workers as well as those of consumers, businesses, and investors.

Here's the rub in the real world: Uber isn't just satisfying the two simultaneous demand curves of customer and driver needs, but also competitive business needs. Their desire to crush the incumbent taxi industry and to compete with rivals like Lyft also affects their pricing. And under the rules of the venture-backed startup game, in order to satisfy the enormous prospective valuation placed on them by their investors, they must grow at a rate that will allow them to utterly dominate the new industry that they have created.

Drivers are also not playing a simple game in which they can just go home if their income isn't sufficient. They have bills to pay and may have to work brutally long hours in order to meet them. They may have leased a vehicle and now must work to pay for it. They may know in theory that they are depreciating the value of the vehicle and running up expenses that undermine their hourly earnings, but in practice they don't feel they have any choice. Alternative jobs may be even worse, with less flexibility and even lower pay.

Uber has many advantages over its drivers in deciding on what price to set. They can see, as drivers cannot, just how much consumer demand there is, and where the price needs to be to meet the company's needs. Drivers must show up to work with much less perfect knowledge of that demand and the potential income they can derive from it. Michael Spence, George Akerlof, and Joseph Stiglitz received the Nobel Memorial Prize in Economics in 2001 precisely for their analysis in the 1970s of the ways that the efficient market hypothesis, so central to much economic thinking, breaks down in the face of asymmetric information.

Algorithmically derived knowledge is a new source of asymmetric market power. Hal Varian noted this problem in 1995, writing in a paper called "Economic Mechanism Design for Computerized Agents" that "to function effectively, a computerized agent has to know a lot about its owner's preferences: e.g., his maximum willingness-to-pay for a good. But if the seller of a good can learn the buyer's willingness-to-pay, he can make the buyer a take-it-or-leave it offer that will extract all of his surplus." If the growing complaints of Uber drivers about lower fares, too many competing drivers, and longer wait times between pickups are any indication, Uber is optimizing for passengers and for its own profitability by extracting surplus from drivers.

Despite the information asymmetry in favor of the platforms, I suspect that, over time, driver wages will need to increase at some rate that is independent of the simple supply and demand curves that characterize Uber and Lyft's algorithms today. Even if there are enough drivers, the quality of drivers deeply influences the customer experience.

Driver turnover is a key metric. As long as there are lots of people willing to try working for the service, it is possible to treat drivers as a disposable commodity. But this is short-term thinking. What you want are drivers who love the job and are good at it, are paid well, and as a result, keep at it. Over the long term, Uber and Lyft will be engaged in as fierce a contest to attract and keep drivers as they are to attract and keep customers today. And that competition may well provide further evidence that higher wages (so-called efficiency wages, as discussed in Chapter 9) can pay for themselves by improving productivity and driving greater consumer satisfaction.

Lyft and Uber keep their data close to the vest, but my own conversations with drivers suggest that Lyft, which has worked hard to craft policies and systems friendlier to drivers, is gaining on its larger, better-funded competitor. Almost every driver I talk to drives for both platforms. Almost universally, they tell me they prefer Lyft, and some tell me that they've quit driving for Uber even though there are more customers. More recently, as the result of cumulative PR missteps on Uber's part, even the customers are defecting to Lyft. Uber's aggressive tactics have earned them many enemies, and they have ignored one of the key rules of the modern connected company: as O'Reilly Media President and Chief Operating Officer Laura Baldwin is fond of saying, "Your customers are your conscience."

THE INVISIBLE HAND

Many simplistic apologists for the capitalist system celebrate disruption and assume that while messy, everything will work out for the best if we just let "the invisible hand" of competition do its work. This is true, if we correctly understand the invisible hand. The law of supply and demand is not describing some magical force, but the way that players of the game *fight* for competitive advantage. As Adam Smith put it, "It is not from the benevolence of the butcher, the brewer, or the baker that we expect our dinner, but from their regard to their own interest. We address ourselves, not to their humanity but to their self-love, and never talk to them of our own necessities but of their advantages."

The "law" emerges from the contest between players. As labor organizer David Rolf said to me, "God did not make being an autoworker a good job." Those middle-class jobs of the 1950s and 1960s that so many commentators look back at with nostalgia were the result of a fierce competition between companies and labor as to who would set the rules of the game. The invisible hand became very visible indeed by way of bitter strikes, and then transcended the market into the political process with the National Labor Relations Act of 1935 (the Wagner Act), the Labor Management Relations Act of 1947 (Taft-Hartley), and state "right to work laws." Over the past eighty years, these acts have tilted the rules first one way, then the other. Today they are heavily tilted in favor of capital, and against labor. Whatever your position on what the right tilt ought to be, it should be clear that the low-wage jobs of today aren't inevitable, any more than were the high-wage jobs of earlier decades.

Right now we're at an inflection point, where many rules are being profoundly rewritten. Much as happened during the industrial revolution, new technology is rendering obsolete whole classes of employment while making untold new wonders possible. It is making some people very rich, and others much poorer. It is giving companies new ways to organize; while labor organizing is beyond the scope of this book, this is an ideal time to rethink the labor movement as well.

I am confident that the invisible hand can do its work. But not without a lot of struggle. The political convulsions we've seen in the United Kingdom and in the United States are a testament to the difficulties we face. We are heading into a very risky time. Rising global inequality is triggering a political backlash that could lead to profound destabilization of both society and the economy. The problem is that in our free market economy, we found a way to make society as a whole far richer, but the benefits are unevenly distributed. Some people are far better off, while others are worse off.

Thus we come to the fundamental idea of welfare economics, as summarized in plain language by economist Pia Malaney from the Institute for New Economic Thinking: "It's very hard to find a way to have any policies that have no negative impacts on anyone but can make some people better off. So we found a refinement . . . where we look at the net benefit versus the net cost. And the idea is that . . . we take whatever benefit we have to the society as a whole and we redistribute so that overall people are better off." In short, the laws of welfare economics assert that when some people are made better off as the result of an economic policy change, the winners must compensate the losers. But as Bill Janeway put it to me in a pungent email, "Unfortunately, the winners rarely do so except as the result of political coercion."

Many discussions of our technological future assume that the fruits of productivity will be distributed fairly and to the satisfaction of all. That is clearly not the case. Right now, the economic game is enormously fun for far too few players, and an increasingly miserable experience for many others.

"Between the end of World War II and 1968, the minimum wage tracked average productivity growth fairly closely," writes economist John Schmitt. "Since 1968, however, productivity growth has far outpaced the minimum wage. If the minimum wage had continued to move with average productivity after 1968, it would have reached $21.72 per hour in 2012—a rate well above the average production worker wage. If minimum-wage workers received only half of the productivity gains over the period, the federal minimum would be $15.34." Instead, as we have seen, the bulk of the value created by increasing productivity has been allocated to corporate shareholders.

It is true that another huge swath of the value created by productivity gains in the economy has been allocated to consumer surplus—the difference between what goods sell for and what customers might have been willing to pay. Other value that new technology brings has been provided to consumers free of charge. Consumers don't pay directly for Google and Facebook and YouTube; advertisers do, invisibly hiding the cost in marginally higher prices. The net consumer surplus is hard to measure, but it has an offsetting effect to lower wages.

Depressed wages for workers and low prices for consumers are not just an inevitable result of automation and free trade, but are driven by fierce competition by companies to expand their market share, as Walmart and Amazon have done with consumer goods and Uber and Lyft have done with taxi fares. These upstarts upset the existing pricing equilibrium between companies and their customers in part as a competitive tactic, a way to undercut the old order.

As Nick Hanauer pointed out, in general we have forgotten the hard-fought lessons of the twentieth century: that workers are also customers, and that unless they receive a fair share of the proceeds, they will one day be unable to afford the products of industry. We are increasingly creating an economy that is producing too much of what only some people can afford to buy, while others just have their noses pressed to the glass. As shown in a recent study based on detailed barcode data from US retail sales from 2004 to 2013, there was a meaningful increase in the variety of products offered to higher-income households and a lower rate of inflation in the price of existing products for those wealthier consumers than for products aimed at those with lower incomes. Inequality feeds on itself, as the market becomes ever more optimized for those with more to spend.

In economic theory, one person's purchase is another person's sale, so, by definition, national product is equal to national income. But the distribution of income matters for consumer spending. As Nick Hanauer, who got his start with his family pillow business, put it in the documentary film *Inequality for All*, "The problem with rising inequality is that a person like me who earns a thousand times as much as the typical worker doesn't buy a thousand times as many pillows every year. Even the richest people only sleep on one or two pillows."

People like Nick not only don't buy thousands of times more pillows, but they can wear only one set of clothes at a time, and can eat only so many meals a day. They do save and invest (Nick, as mentioned earlier, made his first big fortune as the first nonfamily investor in Amazon), and that can have huge "trickle-down" improvements in other people's lives. But as became clear in the 2008 financial crisis, an increasing amount of that investment has been in financial products that are all about strip-mining value from the economy rather than creating value for all. As Warren Buffett said to Rana Foroohar, "You've now got a body of people who've decided they'd rather go to the casino than the restaurant."

As the incomes of ordinary consumers stagnated, companies kicked the can down the road a few decades by encouraging them to pay for goods on credit, but that short-term strategy is crashing down. In *The Marriage of Heaven and Hell*, written during the most hellish days of the first industrial revolution, the poet William Blake issued what might well be a rule as certain as those issued by any economist: "The Prolific would cease to be Prolific unless the Devourer, as a sea, received the excess of his delights."

I like to use Walmart as an example of the complexity of the game play and the trade-offs that the various competing players ask us to make as a society. Walmart has built an enormously productive business that has vastly reduced the cost of the goods that it supplies. A large part of the value goes to consumers in the form of lower prices. Another large part goes to corporate profits, which benefits both company management and outside shareholders. But meanwhile, Walmart workers are paid so little that most need government assistance to live. By coincidence, the difference between Walmart wages and a $15 minimum wage for their US workers (approximately $5 billion a year) is not that far off from the $6 billion a year that Walmart workers are subsidized via federal Supplemental Nutrition Assistance Program (SNAP, commonly known as food stamps). Those low wages are subsidized by the taxpayer. Walmart actually pays its workers better than many retailers and fast-food outlets, so you can multiply this problem manyfold. It has been estimated that the total public subsidy to low-wage employers amounts to $153 billion per year.

You can see here that there is a five-player game in which gains (or losses) can be allocated in different proportion to consumers, the company itself, financial markets, workers, or taxpayers. The current rules of our economy have encouraged the allocation of gains to consumers and financial shareholders (now including top company management), and the losses to workers and taxpayers. But it doesn't have to be that way.

In the face of declining same-store sales and consumer complaints, in 2014 Walmart raised its minimum wage to $10/hour, well above the federal minimum wage of $7.25, while also investing in employee training and career paths, costing the company $2.6 billion. This improved customer satisfaction, employee retention, and sales, but has led to serious dissatisfaction among investors. Bill Janeway likes to note that the competition between parties is often anything but invisible.

We can wait for the push and pull of the many players in the game to work things out, or we can try out different strategies for getting to optimal outcomes more quickly. As Joseph Stiglitz so powerfully reminded us in his book of that name, we can rewrite the rules.

In professional sports, leagues concerned about competitive play often establish new rules. Football (soccer) has changed its rules many times over the past 150 years. NBA basketball added the three-point shot in 1979 to make the game more dynamic. Many sports use salary caps to keep teams in large markets from buying up the best talent and making it impossible for teams in smaller markets to compete. And so on.

The "Fight for 15," the movement toward a national $15 minimum wage, is one way to rewrite the rules. Businesses and free market fundamentalists argue that raising minimum wages will simply cause businesses to eliminate jobs, making workers even worse off. But as Nick Hanauer said during the Q&A after his talk at my 2015 Next:Economy Summit, "That's an intimidation tactic masquerading as an economic theory." Considerable evidence shows that higher minimum wages would likely not have much impact in major cities; most proposals would allow them to remain lower in rural areas, where they might suppress employment.

The critical question, expressed in the true language of Adam Smith's "invisible hand," is who gets more, and who gets less. Capital, labor, consumers, taxpayers.

As noted above, a $15 minimum wage might cost Walmart on the order of $5 billion a year. This is no small number. It represents about a fifth of Walmart's annual profits, and about 1.25% of its annual US revenues. But it could save taxpayers $6 billion per year. If Walmart weren't able to pass off part of its true labor costs onto taxpayers, the company would have to accept lower profits or raise its prices. But is that really such a bad thing? If Walmart's profits were reduced by 20%, its market capitalization would certainly fall, a loss to shareholders. But leaving aside the shock of a sudden drop in earnings due to a change in the rules, were Walmart a private company, would its owners really not have wanted to own it if it generated $20 billion a year in profit instead of $25 billion? In the unthinkability of this trade-off, in our unquestioned assumption that companies *must* continually strive to increase their level of profit, we see the hand of the master algorithm that rules financial markets.

If Walmart were instead to pass along the additional costs to consumers, prices would have to go up by 1.25% (or $1.25 for every $100 spent at Walmart). If the costs were split between shareholders and consumers, that would require only a 10% drop in Walmart profits and an additional 62 cents per $100 spent by consumers. Would people really stop shopping at Walmart if they had to spend little more than an additional half cent for every dollar?

Those higher prices might discourage some customers, but the higher incomes of workers might encourage them to spend more. So it's not inconceivable that Walmart and its shareholders would come out whole. Nick Hanauer calls it the fundamental law of capitalism: "When workers have more money, companies have more customers and then hire more workers."

And of course, raising the minimum wage is only one way to address the fact that the current rules of our economy favor owners of capital over human workers. We could give companies tax credits for wages paid; we could tax robots or carbon or financial transactions instead of wages; we could give tax credits for unpaid work raising children or caring for elders; *we could think the unthinkable.*

Interestingly, Denmark, where there are no minimum wages because there is a robust social safety net, shows us that with the right system design, you can actually have fewer rules. We need to focus on outcomes and realize that all the rules should be open to change if new rules improve what we are actually trying to achieve.

We are mistaking a map for the territory, following a road into the desert because we were promised an oasis at its end. Travelers have returned, telling us that the water is gone, but we keep marching into the waste because the map tells us to do so and we cannot imagine the possibility of another road not yet shown. We have forgotten that maps are in need of an update when the landscape itself has changed. So many of our proposed solutions are like the taxi companies putting televisions and networked credit card readers in the back of their cars rather than reimagining the possibilities of on-demand transportation.

The barriers to fresh thinking are even higher in politics than in business. The Overton Window, a term introduced by Joseph P. Overton of the Mackinac Center for Public Policy, says that an idea's political viability depends mainly on whether it falls within the window framing a range of policies considered politically acceptable in the current climate of public opinion. There are ideas that a politician simply cannot recommend without being considered too extreme to gain or keep public office.

In the 2016 US presidential election, Donald Trump didn't just push the Overton Window far to the right, he shattered it, making statement after statement that would have been disqualifying for any previous candidate. Fortunately, once the window has come unstuck, it is possible to move it radically new directions. This has generally happened in US history as a result of great dislocations, where business as usual just couldn't continue. It took the Great Depression to give Franklin Roosevelt and Frances Perkins the license to put in place the New Deal. But given that license, they imagined the unimaginable.

I was thinking about the Overton Window in November 2016 after attending the Summit on Technology and Opportunity, hosted by the White House, the Chan Zuckerberg Initiative, and the Stanford Center on Poverty and Inequality. I had done a lunchtime debate with Martin Ford, author of the bestselling book *The Rise of the Robots*, which makes the case that artificial intelligence will take over more and more human jobs, including knowledge work. Martin argues for universal basic income as the solution—making sure that every person receives a basic cash grant sufficient to meet the essentials of life.

I was positioned as the techno-optimist in the debate, because I have argued that eliminating human jobs is a choice, not a necessity. When we focus on what needs doing, and what might be possible when humans are augmented by new technology, it is clear that there is plenty of work to go around for both humans and machines. It is only our acceptance of the notion that financial efficiency is the primary fitness function for the economy that locks us into the race to the bottom, in which humans are seen as a cost to be eliminated.

But in the debate with Martin and in subsequent conversations with attendees at the event, I found myself thinking and saying things that hadn't occurred to me before. Stanford's Rob Reich, the moderator of the discussion, said to me afterward, "I thought going into this that Martin was the radical. But I realized that you're the real radical. You're saying that universal basic income is just a software patch on the existing system. We need a complete reboot."

When imagining the future, it's best if you stretch out your view of the possible by postulating extreme futures. So let's assume that machines do replace a vast majority of human work, and most humans are put out of work. What are some of the sacred cows that we might toss through the shattered Overton Window of public policy?

If most humans are out of work, a brief exercise of "If this goes on . . ." thinking would quickly lead us to realize that personal income taxes can no longer be the primary source of government revenue. Some other source will be needed, so why not start thinking about it now? What would happen if we postulated a zero income tax for earned income?

If there were no income tax, what about replacing it entirely with so-called Pigovian taxes, taxes on negative externalities? A carbon tax is one of those ideas. A financial transactions tax or other form of tax on the massive redirection of corporate profits toward financial speculation and away from investment in people and the real economy might be another. (The problem of Pigovian taxes, though, is that they tend to reduce the production of whatever negative externality they are feeding on. So if they succeed, they decline. But that's a good thing, since it means that, like a business, government will always have to reinvent itself.)

But whatever the solution, it's time to end halfway measures that endlessly split the difference between what is needed and what is politically possible. We need bold proposals, once unthinkable. After all, virtually everything we take for granted today was once unthinkable. For millennia, humans dreamed of flying, but it only became possible one hundred years ago. As we face the challenges of the next economy, we need similar flights of boldness and invention. Dreaming the future is not reserved for technologists. The government of the people, by the people, and for the people also requires massive reinvention for the twenty-first century.

Once we've pushed the Overton Window wide open, we can start working toward more desirable futures, in which machines don't replace humans, but allow us to build a next economy that will elicit the WTF? of astonishment rather than the WTF? of dismay.

ASKING THE RIGHT QUESTIONS

I'm not an economist, a politician, or a financier equipped with quick answers as to why things can or can't change. I'm a technologist and an entrepreneur who is used to noticing discrepancies between the way things are and the way they could be, and asking questions whose answers might point the way to better futures.

Why do we have lower taxes on capital when it is so abundant that much of it is sitting on the sidelines rather than being put to work in our economy? Why do we tax labor income more highly when one of the problems in our economy is lack of aggregate consumer demand because ordinary people don't have money in their pockets? When economists like former Treasury secretary Larry Summers talk about "secular stagnation," this is what they are referring to. "The main constraint on the industrial world's economy today is on the demand, rather than the supply, side," Summers writes.

Why do we treat purely financial investments as equivalent to real business investment? "Only around 15% of the money flowing from financial institutions actually makes its way into business investment," says Rana Foroohar. "The rest gets moved around a closed financial loop, via the buying and selling of existing assets, like real estate, stocks, and bonds." There is some need for liquidity in the system, but 85%? As we'll see in the next chapter, this great money river is accessible only to a small part of our population, and relentlessly directs capital away from the real economy.

Why do productive and nonproductive investments get the same capital gains treatment? Holding a stock for a year is not the same as working for decades to create the company that it represents a share of, or investing in a new company with no certainty of return.

John Maynard Keynes recognized this problem eighty years ago during the depths of the Great Depression that followed the speculative excesses of the 1920s, writing in his *General Theory of Employment, Interest, and Money*: "Speculators may do no harm as bubbles on a steady stream of enterprise. But the position is serious when enterprise becomes the bubble on a whirlpool of speculation. When the capital development of a country becomes a by-product of the activities of a casino, the job is likely to be ill-done."

Keynes continued, "The spectacle of modern investment markets has sometimes moved me towards the conclusion that to make the purchase of an investment permanent and indissoluble like marriage, except by reason of death or other grave cause, might be a useful remedy for our contemporary evils. For this would force the investor to direct his mind to the long-term prospects and to those only." Warren Buffett has proven that this is actually a very good strategy. Yet our policies don't favor the kind of value investing that Buffett practices.

A financial transactions tax calibrated to eliminate all the benefits of front-running and other forms of high-speed market manipulation would be a good place to start, but we could go much further in taxing financial speculation while rewarding productive investment with lower rates. Larry Fink, the CEO of BlackRock, suggests that at a minimum, long-term capital gains treatment should begin at three years rather than one, with a declining rate for each additional year that an asset is held.

We could even institute a wealth tax such as that proposed by Thomas Piketty. And if we were to tax carbon rather than labor, rather than starting by substituting a carbon tax for income taxes, it might be better to substitute a carbon tax for Social Security, Medicare, and unemployment taxes. These rule changes might be costly to some capital owners but might well benefit society overall.

These are political decisions as much as they are purely economic or business decisions. And that is appropriate. Economic policy shapes the future not just for one person or one company, but for all of us. But we should realize that it is in our self-interest to improve the rules we are now playing under. In his article about income inequality, Joseph Stiglitz explains how Alexis de Tocqueville, a Frenchman writing about American democracy in the 1840s, considered "self-interest properly understood" to be "a chief part of the peculiar genius of American society."

"The last two words were the key," Stiglitz writes. "Everyone possesses self-interest in a narrow sense: I want what's good for me right now! Self-interest 'properly understood' is different. It means appreciating that paying attention to everyone else's self-interest—in other words, the common welfare—is in fact a precondition for one's own ultimate well-being. Tocqueville was not suggesting that there was anything noble or idealistic about this outlook—in fact, he was suggesting the opposite. It was a mark of American pragmatism. Those canny Americans understood a basic fact: looking out for the other guy isn't just good for the soul—it's good for business."

Throughout history and across continents, economies have played the game using different rules: No one can own the land. All land belongs to kings and aristocrats. Property is entailed and cannot be sold by the owners or heirs. All property should be held in common. Property should be private. Labor belongs to kings and aristocrats and must be supplied on demand. A man's labor is his own. Women belong to men. Women are independent economic actors. Children are a great source of cheap labor. Child labor is a violation of human rights. Humans can be the property of other humans. No human can be enslaved by another.

We look back at some of these rules as the mark of a just society and others as barbaric. But none of them was the inevitable way of the world.

Here is one of the failed rules of today's economy: Human labor should be eliminated as a cost whenever possible. This will increase the profits of a business, and richly reward investors. These profits will trickle down to the rest of society.

The evidence is in. This rule doesn't work. It's time to rewrite the rules. We need to play the game of business as if people matter.

13

SUPERMONEY

WHAT IS SILICON VALLEY'S ROLE IN THE ECONOMY GONE awry? It is easy to blame the ability of technology to replace human labor for declining wages and increased wealth inequality. But technology is being used for cost reduction rather than to empower people and to reach for the stars not because that is what technology wants, but because it is what the legal and financial system we have built demands.

For all its talk of disruption, Silicon Valley too is often in thrall to that system. The ultimate fitness function for too many entrepreneurs is not the change they want to make in the world, but "the exit," the sale or IPO that will make them and the venture capitalists who funded them a giant pile of money. It's easy to point fingers at "Wall Street" without realizing our own complicity in the problem or in finding a way to bring it under control.

I had always made the unquestioned assumption that financial markets were simply one face of the overall market economy. It was Bill Janeway who first brought the distinction between financial markets and the real market of goods and services to my attention. In his book *Doing Capitalism in the Innovation Economy*, he wrote: "I have come to read [the history of the innovation economy] as driven by three sets of continuous, reciprocal, interdependent games played between the state, the market economy, and financial capitalism."

What did Bill mean by calling financial capital out as a separate player equivalent in stature to government and the market economy? The more I thought about it, the more sense it made of my own experience. In my business, a private company started in 1983 with $500 in used furniture and office equipment as its starting capital, though now approaching $200 million in annual revenues, I've always lived in the real economy of goods and services. Originally a technical writing consulting company, we got paid when we found customers willing to hire us to write their manuals, and spent many an unpaid hour looking for new customers. Once we became a book publisher, we embodied our expertise into products, and sold them to customers who wanted to learn what we knew. We grew the business by developing more products, finding more customers, and hiring more people. Once we added conferences, we had to find people willing to pay us to attend or sponsor them. Debt, when we were able to use it, was generally secured by receivables and inventory, tying our growth directly to the underlying fundamentals of finding and serving new paying customers. Having to find people who will pay you for what you create clears out an awful lot of wishful thinking.

Over the years, it became clear to me that many of the companies in the technology industry we are part of were playing by very different rules. They were not getting paid by exchanging goods and services with customers, but by persuading investors to give them money. Perhaps customers would come along eventually, but as long as the company could find investors to finance their next round, perhaps all the way to an IPO or acquisition, a company could just have "users" instead of customers.

Early in my career, venture capitalists still funded entrepreneurs in the hope that they would build companies with real revenues and profits, but in the years leading up to the dot-com bust, it seemed that the game had changed. Entrepreneurs were not creating true companies but a kind of specialized financial instrument, a financial bet not that dissimilar to the CDOs that bedeviled the banking industry in the years leading up to the 2008 financial crisis a decade later. I see much of the same misdirection of entrepreneurial energy in today's overheated Silicon Valley boom.

> These companies, often sold for billions of dollars, are valued not based on some multiple of their sales, profits, or cash flows, but for expectations of what they might become, promoted like fake news in a market of attention. This effect is central to understanding the hypnotic allure of financialization.

I got my first taste of the multiplicative possibilities of the betting marketplace of expectations when we sold GNN to AOL in 1995 for $15 million, and then watched the stock portion of the purchase price balloon to $50 million as AOL's value continued to increase. (Our stock would have been worth over $1 billion if we'd held it to AOL's peak.) We'd built GNN by reinvesting the profits from our publishing business, exploring an exciting new medium that we thought we'd eventually turn into a real business. When we sold GNN, we received more than we could have earned in a decade selling our books at the rate we were selling them in 1995.

GNN was one of a series of purchases, along with Dave Wetherell's BookLink and Brewster Kahle's WAIS, that AOL bought for a collective amount of about $100 million in stock. The purchases signaled to the market that AOL was becoming an Internet company. I watched in awe as AOL's market capitalization rose by first a billion dollars, and eventually many, many billions.

AOL didn't actually succeed in transforming itself from the dominant company of the dial-up networking era into the leader of the commercial Internet, but the expectation that it would be able to do so made it possible for it to buy Time Warner, a company many times its size in the real market of goods and services. The AOL Time Warner merger was a colossal disaster, and the value of the combined company was eventually written down from a peak of $226 billion to less than $20 billion. It is not inconceivable that the same could happen to a company like Uber, whose purchase of self-driving truck startup Otto was as much a signal to investors as it was an investment in the actual development of self-driving trucks and cars.

The ratio between a company's revenues, cash flow, or profits and its market capitalization is one of many imaginary numbers that make up the world of financial capital. In theory, the intrinsic value of owning a stock is based on the net present value of its expected future profits. In practice, it is that net present value times the expectations of millions of potential buyers and sellers.

The price of a stock is fundamentally a bet. Will the company's earnings from the real market of goods and services be greater in the future? If so, it is worthwhile to own a share of that company.

When a company fails to deliver value that lives up to that bet, but still cashes in through IPO or acquisition, the wealth that is gained by startup founders and early investors is taken from public market investors. This is a risk that both sides of the bet willingly take, and it has provided enormous fuel for innovation because it encourages innovators to take risks in hope of future rewards. But in overexcited markets, it's too easy for many startups to aim to cash out with "dumb money" while the getting is good, with no real plan for ever delivering real revenues or profits.

The enormous leverage provided by the marketplace of expectations is the key to all that is good about Silicon Valley, but also all that is bad. On the good side, this leveraged financial adventurism allows for huge waves of Schumpeterian "creative destruction." A company like Amazon, Tesla, or Uber is effectively able to capitalize future hopes and dreams into current cash value and use it to finance a world-changing business despite losing money for years. This is a good thing. It's what capital markets are for: to provide the money that lets entrepreneurs large and small take risks. To the extent that a company grows into its expectations, it will eventually be valued at a ratio closer to the true net present value of its future earnings. On the bad side, many a company that is highly valued today may never deliver on those expectations.

Economist Carlota Perez makes the case that every technological revolution has been accompanied by a financial bubble, which funds investments in futures that have not yet come into being, investments that can be tolerated only because for every hundred failures there is a breakthrough so large that it pays off all the failed bets. "Occasionally, decisively," Bill Janeway notes, "the object of speculation is the financial representation of one of those fundamental technological innovations—canals, railroads, electrification, automobiles, airplanes, computers, the Internet—deployment of which at scale transforms the market economy."

That is true, but this system also disproportionately rewards luck, and even the destruction of real economic value. Bill Janeway said to me, "The process is inherently wasteful, what I call Schumpeterian waste. Progress through trial and error and error and error. So of course luck comes into the game." In short, there is many an Internet multimillionaire and even some billionaires who were just on the lucky end of a failed acquisition.

But that's not the end of it. What if I told you that there was a magical way to take a dollar of company profit and turn it into a currency worth, on average, $26? What if I could take a dollar of company profit and turn it into a currency worth hundreds of dollars? Thousands? That's exactly what public company stock (or private company stock on the path to a plausible IPO or sale to a company that has already gone public) represents.

The price-earnings ratio of a stock is the difference between the actual net present value of a company's future profits and its market price. Amazon's P/E ratio is 188 at the time I'm writing. Facebook's is 64, Google's 29.5. The ratio for the entire S&P 500 is about 26. That is, for every dollar of profit it makes today, Amazon gets $188 in stock value, Facebook gets $64, and Google gets $29.50. For a company like Uber, which has no profits yet but is valued at $68 billion by investors, the ratio is essentially infinite.

That leverage makes stock an incredibly powerful currency, which swamps the purchasing power of the ordinary currency used in the market of real goods and services. Amazon's profits in 2016 were just shy of $2.4 billion, and its book value (the actual value of its cash, inventories, and other assets less its liabilities) $17.8 billion, yet its market capitalization at the end of the year was $356 billion.

George Goodman, a financial writer who published under the pseudonym Adam Smith, calls this "supermoney." (In his preface to the Wiley Investment Classics edition, Warren Buffett compared Goodman's 1972 book of that name to a perfect game in baseball.) Supermoney is at the heart of today's growing financial inequality. Most people exchange their goods and services for ordinary money; a lucky few get paid in supermoney.

A company that has been financialized—that is, is valued in

supermoney—has a huge advantage over companies that are operating solely in the market of real goods and services.

If you have a company valued in supermoney, you can more easily buy other companies. At O'Reilly Media, we have made acquisitions from time to time, but as a privately owned business operating in the real market of goods and services, we've always had to value them based on a realistic multiple of their expected cash flows, paying for them out of our own retained earnings or debt financing against our cash flows. In one case, the net present value of a prospective acquisition based on its current sales and growth rate was about $13 million; a rival bidder snapped up the company for $40 million. Why could they do this? As a "hot" venture-financed IPO-track company, their own stock was valued at 5x or more what a comparable private company would be valued. Paying only a 3x premium to add to their growth was a reasonable bet. But don't mistake it: It was still a bet on financial market expectations, rather than a bet on the real operating cash flows and profits of the business.

If all you have to pay employees with are actual earnings from your business, you are limited in how rich you can make them. If you can pay them in supermoney—especially in the super-supermoney represented by stock options (the right to buy a stock at today's price but with no obligation to do so until it has appreciated in value), even more so if it is in supermoney cubed (pre-IPO stock options at 90% discounts from what even the venture capitalists are paying) you can hire the best talent.

> If you have access to supermoney, you can operate for years at a loss. This is one reason—not just the superior customer benefits and economic efficiencies of their technology or business model—that Internet companies can disrupt older, less highly valued companies.

Yes, Uber's service is superior in terms of availability, convenience, and customer experience to traditional taxi and limousine service, but could it so easily have overpowered the incumbents without access to billions of dollars in investment capital, which allowed it to subsidize lower prices for consumers and pay incentives to drivers? Arguably, funding that kind of innovation is what capital markets are for, but it's also possible that that capital can be used to destroy existing businesses without building something sustainable to take their place.

As we saw in Chapter 11, stock options, which have played such a large role in Silicon Valley wealth, have also become a key part of the problem of income inequality. Even though Silicon Valley companies are actually better than many other companies at distributing the gains because they offer options to virtually every employee, those options are still overwhelmingly weighted toward founders and top management, with each lower rank of workers typically receiving a full order of magnitude less in value.

This may or may not be appropriate, based on actual contribution to the business, but the net-net is that a huge proportion of the productivity gains we've seen in the past decades have increasingly gone to a small group of managers rather than to all workers. And when the market gets excited, those people get paid in a currency that appreciates at a rate far in excess of anything possible in the real economy.

The amount of supermoney created out of thin air simply by issuing new options to employees is staggering. In 2015, for instance, Google's stock-based compensation was $5.2 billion. The ability to print supermoney is proportional to your size, further accelerating the winner-takes-all economy. For a company the size of Google, whose market capitalization at the end of that year was more than $500 billion, that $5.2 billion in stock compensation represented only a 1% dilution of existing shareholders. For a smaller company, like Salesforce, with a market capitalization closer to $50 billion, 1% would only be $500 million—so Salesforce can afford only a tenth as much to hire engineers even though it has a third as many employees as Google. As a result, one analyst said to me that Salesforce would eventually have to be sold to a bigger company. This same analyst believed that this was ultimately the reason that LinkedIn was sold to Microsoft. They just weren't big enough to be competitive as a stand-alone company in today's market, he said. (This idea is suggestively compatible with new research by economists David Autor, David Dorn, Lawrence Katz, Christina Patterson, and John Van

Reenen that suggests that the problem of income inequality is driven in part by the rise of superstar firms, whose outsize productivity makes them able to command a larger share of the market while employing a smaller number of more highly paid people.)

Another effect of stock-based compensation is that it requires companies to keep growing, encouraging them to aim for complete market dominance and value capture. As long as employees are paid in stock, even founders with voting control over their companies still face pressure from "the market" to keep their earnings rising and their stock price increasing.

The amount of money spent on stock-based compensation has for too long not been properly reported by many technology companies. Amazon and Facebook only began reporting their stock-based compensation as part of their regular financials using Generally Accepted Accounting Principles (GAAP) rather than via special "non-GAAP" reporting in their quarterly reports in the first quarter of 2016. Twitter, less profitable than the others, still doesn't do so because it would show that far from having a profit, it is actually still operating at a loss when stock-based compensation is taken into account.

Even in the best case, when over time a company grows into its valuation and its shareholders earn their wealth in the real economy, there are damaging effects from having two currencies, one of them hiding in plain sight, that are valued so differently. If you are paid in ordinary money, you may still be able to buy a home, but perhaps you have to move to a less attractive location to do so. If you are paid in supermoney, you can afford to pay higher rents, or spend more to buy your home, driving up prices and even further increasing the distance from those working in the ordinary marketplace of goods and services. Not only that, if you are paid in supermoney, you can convert some of it into ordinary money and become an investor—placing bets in other new companies, in the broader stock market, in real estate—multiplying your wealth still further.

The impact on real estate is crushing for those who make their living in the real economy. In the absence of aggressive building of new housing, those paid in supermoney drive up housing prices to the point where ordinary people can no longer afford to live in a city like San Francisco. Meanwhile, government policies designed for an era when home ownership was a pathway into the middle class now exacerbate the problem. The tax deductibility of mortgage interest, allowed even on second homes, drives prices up even further, giving rich housing subsidies to those with the means to buy them and making homes even more expensive. A limit on the amount of mortgage interest deductibility would be corrective, but is blocked by the wealthy interests who benefit from it.

The negative impact on the real economy doesn't end there. Investors focus on companies that will have huge "exits"—that will return at least 10x their investment. This quest for outsize winners has the perverse effect of starving ordinary businesses of capital. A company that may deliver real value but grows slowly and may never achieve global scale is simply uninteresting to investors.

Meanwhile, venture investors have come to have much the same risk profile as investment bankers, where gains are private but losses are socialized. Venture capitalists typically get paid a percentage of the fund as an annual management fee—usually 2%—so a VC firm with a billion-dollar fund will have taken out $200 million in proceeds over the ten-year life of the fund even if the firm loses money for its limited partners (the investors who put up the vast majority of its capital). Put another way, the $58.8 billion in venture capital investment in 2015 paid out nearly $1.2 billion to venture capitalists that year whether or not their investments ever succeed. This encourages VCs to raise larger funds and to deploy ever-larger amounts of capital even though evidence shows that smaller funds typically deliver better results.

Because supermoney is so powerful, for many entrepreneurs and venture capitalists valuation is also something to be gamed, in the same way as a website might try to game its search rankings or its social media engagement. Values are ratcheted up in a series of financings—ideally a rising value is based on actual progress, but sometimes existing investors will put in more money at a higher valuation as a way to signal confidence to additional later investors. As more money floods in, as happened during the dot-com bubble and arguably again in the current unicorn bubble, expectations become increasingly divorced from reality.

In the worst case, companies are formed not to serve real customers, but to be financed. Strategic "pivots" are made not to advance the actual business but to convince investors to place another bet even though the original business idea didn't pan out.

Once companies take money from venture capitalists, they are committed to aiming for an exit. A typical venture fund is a partnership with a ten-year time horizon. Most of the investments are made within the first two to three years, with some money reserved for additional investment in the companies that are most promising. Once an entrepreneur takes money from a venture capitalist, he or she is promising to sell or go public within the lifetime of the fund. Yet VCs know that the vast majority of their deals will fail. Jon Oringer, the founder and CEO of Shutterstock, put it well in his advice to entrepreneurs: "What venture capital firms do is spread some number of millions of dollars to some number of companies. They're not really rooting for every single one. All they need is for a few of them to succeed. It's the way the model works. They have a totally different risk profile than you do. This is your only game in town. For the venture capital firm, it's one of a hundred games in town."

I've watched companies be wiped out by their venture capitalists' timetable, forced to sell without realizing much value at all for the entrepreneur because it was time for the VC to liquidate its position. I've also watched companies struggle to please investors rather than their customers. A perfectly good business, one that might eventually deliver tens of millions of dollars of revenue and meaningful profits if run properly, is told instead to shoot for the moon, because as a company operating in the real market of goods and services it won't command the stratospheric exit that it might be able to fetch if properly positioned in the marketplace of expectations.

The amount of money raised and the timing of when it is raised can also make a big difference. As discussed in Chapter 4, Sunil Paul's patents for on-demand car sharing predate Uber by the better part of a decade, but he was ahead of his time. By the time he launched Sidecar in 2011, two years after Uber had launched its service for calling black limousines on demand, and about the same time as Lyft started offering rides from ordinary people in their own cars, he was third in line. Uber and Lyft had both already raised massive amounts of money, and Sunil was never able to raise enough money to catch up.

GROWING A BUSINESS WITHOUT VENTURE CAPITAL

When I founded O'Reilly Media, I wanted to build a company that would be around for the long haul, so I was firmly committed to staying private. In my early years as a consultant, I'd watched many a client go from being an exciting startup to a treadmill focused on quarterly results. I didn't want that future. I wanted O'Reilly Media to be like ESRI, founded by Jack and Laura Dangermond in 1969, or SAS Institute, founded by Jim Goodnight and John Sall in 1976, both private technology companies still going strong after decades of innovation.

My friendship with master venture capitalist Bill Janeway began when he and his partners asked me about investing in my company when GNN was lighting up the commercial web in 1994. I remember our lunch at a noisy sidewalk cafe in San Francisco, in which Bill and his partner Henry Kressel grilled me about my aspirations for the business. At the end of the lunch, Bill told me, "We will never invest in your company, and you don't want our money. We're smart guys, and at the end of the day, our goal is to turn our money into a lot more money. That's not what you're about." I loved his honesty and insight.

Despite what Bill told me in 1994, he was a venture capitalist of the old school, making money by identifying and solving real problems, deploying patient capital to build companies with real customers. His specialty was to identify and acquire deep technology assets that were lying fallow at large companies, then put that technology together with a team of remarkable entrepreneurs to build a business. BEA (later acquired by Oracle), Veritas (later merged with Symantec), and Nuance, now a public company, were three of his home runs. He believes deeply in building real businesses, using the same philosophy as Warren Buffett, the value-investing genius of public markets, that valuing a company based on positive cash flow is the secret to successful investing. His own mentor, Fred Adler, had a saying that Bill passed on to me: "Corporate happiness is positive cash flow."

Over the years, I passed up many requests to sell O'Reilly Media or to take on outside investors, instead preferring to spin out and sell projects such as GNN (sold to AOL), Web Review (sold to Miller Freeman), LikeMinds (merged with Andromedia and then sold to Macromedia), reinvesting the proceeds in the core business. I knew that without outside investment I couldn't scale these businesses, but I didn't want to give up the control that I had as a private company.

But I could see that there was amazing power in the Silicon Valley investment model. I watched how Yahoo!, started almost two years later than GNN, became an Internet sensation after they took in venture capital, and used it to scale the business at the speed necessary to keep up with the growth of the market. Of course, execution matters as well as money, and Yahoo! executed brilliantly in becoming the web's first media colossus, handily defeating AOL, to whom I'd sold GNN as an alternative to taking outside investment into O'Reilly as a whole.

In 2002, Mark Jacobsen, at the time our VP of business development, started an internal venture fund at O'Reilly Media, which had several notable successes, including Blogger (founded by former O'Reilly employee Evan Williams), which we sold to Google, and ActiveState, which we sold to Sophos.

In 2004, Mark proposed that we raise a proper venture fund with outside investors, which we called O'Reilly AlphaTech Ventures (OATV). Bryce Roberts joined Mark as an additional managing partner. While I like to think that we are entrepreneur-friendly, we too have had to play by the rules of the venture game, which ultimately prioritize shooting for a supermoney exit.

Is it possible for entrepreneurs to get the benefits of Silicon Valley investment without some of the downsides of the traditional venture capital model? Bryce Roberts, my partner at OATV, thinks so. In 2015, he proposed an unusual experiment to me and to Mark Jacobsen. What if, Bryce asked, we came up with a way to invest in entrepreneurs who weren't shooting for the exits, who wanted to build a company with revenues, profits, and cash flow in the real economy? Bryce pointed out that there were many more of these companies than people realized. Not just SAS and ESRI but Craigslist, Basecamp, SmugMug, MailChimp, SurveyMonkey—and for that matter, O'Reilly Media—were all quietly making money. More recently, Bryce noted that "ambitious founders have been trained that billion dollar exits are reserved only for those who follow a very defined playbook for 'blitzscaling' a business." Yet in the last six months of 2016 there were seven tech M&A transactions with a value of more than $1 billion. Of those, only four were venture backed. Three had no investment at all from VCs.

This is not new. Other companies that were eventually acquired or went public, like Atlassian, Braintree, Shutterstock, and Lynda.com, had started out the same way, first reaching profitability and scale, and only adding investors late in life as part of a path toward going public or being sold. Taking on late-stage investors is a common tactic for successful private companies eventually seeking liquidity. After all, founders don't live forever, and estate taxes will likely force the sale of a company after the death of a majority owner.

But what about startups? Is there a way that we could provide value to firms that want to make their mark in the real economy without putting them on the treadmill to an exit? Bryce came up with a creative solution, which he called indie.vc.

Indie.vc is modeled on Y Combinator, the classic Silicon Valley accelerator, which takes a small but meaningful stake in very-early-stage companies in exchange for a very small amount of cash, plus a lot of help in business planning, networking with other entrepreneurs, and eventually, showcasing the company to VCs. Y Combinator has been phenomenally successful, helping to birth companies such as Airbnb and Dropbox. But the focus of Y Combinator's program specifically, and VC-funded companies generally, is on raising the next round of funding. In the case of Y Combinator, months of work and preparation are put into nailing the perfect pitch for a performance in fundraising called demo day. For VC-backed startups, the focus is on what milestones a team must hit in order to be attractive for their next round of funding—ideally at a meaningful multiple of the last funding round price.

With the indie.vc experiment, the sole focus of the investment and support provided is on getting the company to profitability and positive cash flow. "Real businesses bleed black," Bryce likes to say. There is no discussion about what milestones need to be hit to be attractive for the next round of funding. There is no demo day. The investment stake we receive in return is in the form of a convertible note that either can be paid off at a fixed multiple in dividends if and when the company becomes profitable and cash flow positive, or can be converted to equity if the company later decides to take on investors and shoot for an exit.

In using dividends as a form of payout to investors, Bryce is following the same game plan as companies like Basecamp and Kickstarter. Jason Fried, the founder and CEO of Basecamp, notes that Basecamp makes tens of millions of dollars a year in profits and has paid out tens of millions in distributions.

A major selling point for entrepreneurs in taking less cash up front from investors and shooting for a slower growth, cash-flow positive business is not to pay dividends, but to achieve greater independence, freedom, and control. That control allows the startup to keep going as long as customers value the work, independent of the judgment of investors. Marc Hedlund, the founder and CEO of Skyliner, an indie.vc investment, writes: "[W]e and many peers have, in the past, invested huge amounts of time and energy in work we love only to see companies fold and all that effort go to waste. Too many startups fail. Too much of our work as an industry goes into the dustbin if growth is not immediate and meteoric."

That control also allows startups to choose a business model that aligns with their values and purpose. *The Information*, which has become Silicon Valley's go-to source for deep, thoughtful technology reporting, is a great example. Jessica Lessin, the founder and CEO, took no outside investment and made an early commitment to a subscription model because of what she saw as the inevitable corruption brought on by the advertising model, which requires high growth to succeed, and in which the pursuit of clicks and views overrides the pursuit of truth.

The jury is still out on these experiments, but they illustrate the stresses of the current model. Fed up with a system that gives rich and certain returns to venture capitalists but often little or nothing to the entrepreneur, startups are beginning to turn away from the financial market casino and trying to build real businesses again.

DIGITAL PLATFORMS AND THE REAL ECONOMY

In stating the importance of companies rooted in the real market of goods and services, I am not advocating a return to a world of 1950s-era small business, but to a reinvention of small business for the twenty-first century, enabled and empowered by networked platforms. I am also calling for a clearheaded dialogue about the power of those networked platforms to set rules that govern those small businesses—and the government's role in setting rules that govern the platforms.

If the great Silicon Valley platforms are a model for the twenty-first-century company organization, then the people who matter to the platform are not just the employees in the network hub, the corporation proper. The participants in many of those platforms are individuals and businesses operating in the real world of goods and services: the host offering a room on Airbnb, the driver offering a ride on Lyft or Uber, all entrepreneurs of a sort. The iPhone and Android app stores don't just offer products from Apple and Google; they are platforms for independent developers. Facebook and YouTube depend on both their creators and their consumers. Search engines, Yelp, OpenTable, and other similar sites succeed to the extent that they drive traffic to other businesses, not just to themselves.

If they are to break free from the mistakes of the failed philosophy of current financial markets, which too often hollow out the real economy and increase inequality, these platform companies must commit themselves to the health and sustainability of their partner ecosystems. This is not just a matter of idealism. It is a matter of self-interest. When platforms take too much of the value for themselves, they lose their way.

Video-hosting sites like YouTube are a good model for understanding how a network platform can generate new forms of employment while also allowing existing businesses to grow and participate in the platform. Before YouTube, could you imagine the costs of sharing a video with the world? Billions of videos, available to anyone? For free? After ten years, with revenue estimated to be over $9 billion, YouTube is reportedly still not profitable. Its cost structure for hosting and high-speed content distribution is enormous; much of the video has no advertising against it, and when it does monetize video, it shares that monetization with its creators. Fifty-five percent goes out to the video provider; 45% goes to the platform.

There is a thriving small business economy around YouTube. Hank Green, a YouTube star whose various channels with his brother, best-selling young adult author John Green, collectively have nearly 10 million followers, cofounded an organization called the Internet Creators Guild to "support, represent, and connect online creators" on platforms like YouTube and Facebook. Hank estimates that there are more than 37,000 people who make a full-time living posting to YouTube alone (ranging from those who are barely making a living wage to those pulling down seven figures), and almost 300,000 who make supplemental income. And that number is growing. "If 'Internet creator' were a company," Hank says, "it would be hiring faster than any company in Silicon Valley."

YouTube is a testament to the power of supermoney to do good. It is only by borrowing from the future that this infrastructure could have been funded. This is true of all the infrastructure of the Internet, from the providers who bring bandwidth to offices and homes, coffee shops and public spaces, to the countless free services we all enjoy. But not all of those who borrow from the future recognize their obligation to pay back their moral debt by making that a better future. Supermoney is not a gift. It is an obligation.

MEASURING VALUE CREATION

When used properly, the value created in financial betting markets is also realized in the real human economy. Google's founders created enormous wealth for themselves—Larry Page and Sergey Brin are each worth somewhere north of $38 billion—and through stock options distributed to every employee, they have also created wealth for everyone who has worked for Google, as well as for those who invested in the company. But more important, they also have created enormous value for other businesses and for society as a whole.

Company financial statements routinely measure and report on the value captured by the company for its owners. Little is routinely done to measure the value that is created for others. That needs to change.

I was told by James Manyika of the McKinsey Global Institute that at a meeting of global business leaders hosted by *Fortune* at the Vatican in November 2016, CEOs were admitting to each other that perhaps they were measuring the wrong things. "We measure ourselves on shareholder value," one said. "Should we have a metric around job growth, or income growth?"

There are already small steps in this direction.

Every year, Google chief economist Hal Varian and his team publish an economic impact report. In their 2016 report, they estimated that during the prior year, Google increased US economic activity for their customers by $165 billion. They base this figure primarily on a conservative estimate of the expected impact of Google advertising on the increased revenues of their advertisers. If they were to include what must be a far larger economic benefit to businesses found through organic search who are not also advertisers, the total would be far, far larger. That is probably the more important number. After all, Google and other search engines are how people find out about virtually everything. Research conducted by deal site Groupon in 2014 suggested that more than 60% of their traffic came from search.

But even ignoring organic search, and merely using Google's figures for the positive impact of paid advertising, value created for their advertisers in 2015 was almost five times Google's own $34.8 billion in 2015 US revenue. Given that Larry and Sergey founded Google in 1998, you can count the cumulative economic impact in the trillions of dollars. And the consumer surplus provided by free access to vast amounts of online information has to be much, much larger. As users of Google Search, we participate in an exchange of real value, receiving free search services, maps and navigation, office applications, video hosting on YouTube, and much more, in exchange for possibly clicking on some of the advertisements that those paying Google customers placed via the service. Even Thomas Piketty agrees that increased productivity and better diffusion of knowledge create more wealth for society and are among the forces that reduce income inequality.

In short, the trillions of dollars of value created for society as a whole is far larger than the supermoney value created for shareholders of Google (at present, approximately $562 billion). That's what success looks like. It is what happens when a company creates more value than it captures.

Google is not the only company to regularly publish an economic impact report. It is becoming increasingly common for Internet companies to measure their positive economic impact. This is a step in the right direction, but it should ideally be systematized and made part of regular company financial reporting. It would be great to see standardized financial measures of the ratio between the value created for the owners and investors in a company and the value created for other stakeholders. This ratio is particularly important in the winner-takes-all world of online platforms. The value created for the ecosystem should be a paramount concern.

In the summer of 2016, crowdfunding pioneer Kickstarter commissioned a report from a researcher at the University of Pennsylvania, which concluded that since its founding in 2009, Kickstarter had funded a total of $5.3 billion in projects, creating 8,800 new small businesses employing approximately 29,000 people full-time, and working with another 283,000 part-time collaborators. Many of those projects doubtless failed, just like those backed by venture capitalists or started as local businesses, but many have gone on to great success. Some have even joined the supermoney economy. One project, Oculus, was later sold to Facebook for $2 billion, of which Kickstarter received nothing. (Unfortunately, neither did any of the project's backers. It would have set a great precedent if, having won big, the Oculus founders had treated their initial backers as if they had been investors, letting them in on some of the windfall.)

While the absolute numbers are far smaller than those for Google, Kickstarter's ratio of value captured to value created is far better. Since

Kickstarter charges a fee of only 5%, that means the company's total lifetime revenues were roughly $250 million, a tiny fraction of the value created. Because Kickstarter is a private company, and Yancey Strickler, its cofounder and CEO, made clear that he has no plans for the company to sell or go public, it's impossible to estimate what Kickstarter would be worth if it were to do so. But Kickstarter is in the game for the long haul, committed to creating value for its participants rather than extracting it.

Kickstarter has gone so far as to register as a public benefit corporation, a designation that places a legal requirement on the company to consider its impact on society and not just on shareholders. Kickstarter's founders told their venture capital investors from the start that they have no plan to exit, and have instead put in place a mechanism for making regular cash distributions to their shareholders, just like Basecamp and the indie.vc companies.

An aside: I've always had mixed feelings about public benefit corporations and their lighter-weight cousins, benefit corporations, or B corps, which certify to their investors that they do take factors other than shareholder value into account, but are not legally required to do so. I love the idea of public benefit, but I hate to accept the idea that a regular corporation is legally obliged to ignore it. Law professor Lynn Stout's book *The Shareholder Value Myth* makes what appears to be a compelling case that shareholder value primacy has no legal basis, but Leo Strine, the chief justice of the Delaware Supreme Court, argues otherwise. And given that most US corporations are registered under Delaware law, Strine's views carry more legal weight. Frankly, though, if there is legal precedent for the corporate obligation to disregard the interests of all but shareholders, I'd like to see it challenged and overturned.

Etsy, the marketplace for handmade goods, is also a benefit corporation, mindful of the benefits to its sellers. "Etsy sellers personify a new paradigm for business," Etsy's economic impact report announces. "For many years, the conventional and dominant retail model has prioritized delivering goods at the lowest possible price and growth at any cost. . . . In many ways, Etsy sellers represent a new approach to business, where autonomy and independence matter just as much as, if not more than, the bottom line."

Etsy's report is full of softer statistics and personal success stories. On average, sellers report that their creative business contributes 15% of their yearly household income; 17% use their creative business for rent or mortgage payments; 51% "work independently" (that is, their creative business is their sole business, or it is part of a mix of income from a variety of sources); 36% have a full-time job; and 11% identify themselves as unemployed.

Alas, Etsy provides a cautionary tale for those who hope that benefit corporation status will protect them from angry investors. In May 2017, two years after Etsy's IPO, investor anger at the company's lackluster financial results led to the ouster of Chad Dickerson, Etsy's CEO.

Airbnb doesn't do an overall economic impact statement like Google, Kickstarter, or Etsy, but regularly publishes studies of individual cities. For example, in its 2015 study of Airbnb in New York City, the company calculated that visitors staying with Airbnb hosts generated $1.15 billion in economic activity during the prior year and supported more than 10,000 jobs. A 2016 study claimed an economic benefit to the Netherlands of €800 million. There is of course some offsetting loss of income to hotels, so these numbers likely deserve further scrutiny. But it's important to note that Airbnb's benefit is distributed more directly to ordinary people and small businesses than are the profits of large hotel chains. Across all the cities they've studied, 74% of Airbnb properties are outside the main hotel districts. Airbnb guests spend 2.1 times longer than the average hotel stay, and spend 2.1 times more than hotel visitors, with 41% of it spent in local neighborhoods not usually frequented by tourists. While professional Airbnb hosts play a larger role in some markets like Japan, Airbnb is increasingly enforcing a "one host, one home rule" to minimize the conversion of rental housing stock to short-term rentals. Eighty-one percent of hosts share their own home, 52% of hosts have low to moderate income, and 53% say that the income from Airbnb has helped them stay in their home.

Even Uber, the bad boy of the WTF? economy, likes to tout their positive social goals. The origin story on its website concludes: "For the women and men who drive with Uber, our app represents a flexible new way to earn money. For cities, we help strengthen local economies, improve access to transportation, and make streets safer." Consider how much more powerful this statement would be if there were published metrics, backed by reliable data, to support it. There is, at least, some measurement of consumer surplus. A third-party economic study of Uber pricing in North America suggested that during 2015, Uber had actually left $6.8 billion on the table by charging less than they could have.

China's Alibaba, owner of the world's largest e-commerce marketplace, Taobao, doesn't issue an economic impact report, but the numbers speak for themselves: $256 billion in gross merchandise volume from nine million third-party sellers.

Unlike Amazon, which sells both products that it stocks and sells directly and products from third-party sellers, Taobao is like eBay, purely a marketplace for connecting buyers directly with third parties. And unlike eBay, which aggregates all products into a vast catalog, each Taobao merchant has its own storefront. Also unlike eBay, which under then CEO John Donahue was accused of turning away from the small businesses that gave it its start in order to favor more lucrative sales by big brands, Alibaba has segregated global brands into a separate site, Tmall, which has $136 billion in gross merchandise volume. And unlike both Amazon and eBay, Taobao charges no commission for its sales; all of its revenue comes from advertising, which merchants use to increase their visibility on the site. (Taobao's sister site, Tmall, does charge a commission, which ranges from 3 to 6%.)

E-commerce sites like Taobao, eBay, Etsy, and the Amazon marketplace for third-party sellers can play a meaningful role in reinvigorating local economies. They should all measure themselves by the success of their sellers, report on it religiously, and aim to have the metrics for those marketplace participants, not just themselves, go up and to the right. After all, without the sellers, a marketplace is an empty shell.

Small businesses are the bedrock of the economy, providing nearly half of all private-sector employment. Policy makers must understand the role of platforms in bringing small business into the twenty-first century, measure their economic impact, and craft tax policies to encourage the creation of broader economic value, not just the value companies extract for themselves.

THE CLOTHESLINE PARADOX

What we measure matters. I first became fascinated with the curious fact that we often ignore and take for granted many types of economic value when, in 1975, I read an essay by environmentalist Steve Baer published in Stewart Brand's *Co-Evolution Quarterly*, the successor to *The Whole Earth Catalog*. The essay was called "The Clothesline Paradox."

"If you take down your clothesline and buy an electric clothes dryer, the electric consumption of the nation rises slightly," Baer wrote. "If you go in the other direction and remove the electric clothes dryer and install a clothesline, the consumption of electricity drops slightly, but there is no credit given anywhere on the charts and graphs to solar energy, which is now drying the clothes."

> The Clothesline Paradox is a tool for seeing the economy with fresh eyes, essential if we are to correctly rewrite the rules. It is another of those general-purpose concepts that is an aid to seeing what others are blind to.

It is also a good reminder that economic value is realized in different ways at different points in the value chain, and important sources of value are often invisible or taken for granted. For example, Google and Facebook provide free services monetized by advertising, while companies like Comcast charge hefty subscription fees for access to those same services. Meanwhile, Internet users are often accused of being unwilling to pay for content, despite being the source of much of the activity that is monetized by both advertising platforms and Internet service providers.

At least ad-supported media is clear on the nature of the transaction: "We'll give you free services if you'll give us your attention."

Something is clearly wrong, though, with the map being used by the cable companies to frame this discussion. The cable company must pay for professionally produced television content; on the Internet side of its business, the cable company gets much of its content for free, created by the very customers who are paying them for access. Simply by comparing the cost of content for the cable companies and other providers of Internet services versus their content cost for television, you can see that it is the cable company, not the consumer, that is getting the free ride. Debates about net neutrality should be informed by Clothesline Paradox economics, not the extractive economics of financial value capture by big companies!

The Clothesline Paradox is a great way to understand the value of investments in basic research, and in particular, open science, where information is freely shared. Much of the basic research that pays such enormous dividends is funded by taxpayers, yet when government makes a claim on those dividends in the form of corporate or capital gains tax, far too many of the beneficiaries complain or seek to avoid it.

There's an argument to be made that government should get a share at the point of origin, getting a stake in the supermoney outcomes just like investors do. In *The Entrepreneurial State*, Mariana Mazzucato details the role of government in funding the innovations that are embodied in products such as the iPhone, pharmaceutical and agricultural innovation, and the new private space race. She makes the case that startups commercializing government-funded research should pay royalties into a "National Innovation Fund" or issue a "golden share"—an undilutable percentage ownership to the public—precisely in order to capture a portion of the value as and if created.

That being said, it is also true that the value from innovation accrues to society in many unmeasured ways. In a 2004 paper, economist William Nordhaus estimated the amount of "Schumpeterian profits"—"those profits that arise when firms are able to appropriate the returns from innovative activity"—and discovered that from 1948 to 2001, only "a minuscule fraction" (2.2%) of the total value from technological advances was captured by their producers. The increase of human knowledge makes us all richer.

> Sharing rather than hoarding knowledge can also be a powerful lever for competitive advantage. Companies too often assume that the best way to increase their share of the gains from innovation is to keep it proprietary. Yet as the open source pioneers of Linux and the Internet taught us, knowledge compounds when it is shared.

This is also true today in the fierce competition of artificial intelligence research. Yann LeCun, the head of Facebook's AI research group, pointed out to me that most of the cutting-edge AI research today is being done at Google, Facebook, Baidu, and Microsoft. Key to their ability to hire the best people, he said, is these companies' willingness to let their researchers share their work. Apple, which has a culture of secrecy, has been unable to attract top talent, and as a result, has recently had to change its policies.

Understanding where value is created versus where it is captured is equally important when considering the future of work. As we will see in the next chapter, the question of whether the next wave of automation will leave enough jobs for humans is deeply rooted in outdated maps of what counts as paid work, and what we take for granted and expect to be provided for free.

14

WE DON'T HAVE TO RUN OUT OF JOBS

AT THE OUTSET OF THE GREAT DEPRESSION, JOHN MAYNARD Keynes penned a remarkable economic prognostication: that despite the ominous storm that was then enfolding the world, mankind was in fact on the brink of solving "the economic problem"—that is, the quest for daily subsistence.

The world of his grandchildren—*the world of those of us living today*—would, "for the first time . . . be faced with [mankind's] real, his permanent problem—how to use his freedom from pressing economic cares, how to occupy the leisure, which science and compound interest will have won for him, to live wisely and agreeably and well."

It didn't turn out as Keynes imagined. Sure enough, after a punishing depression and a great world war, the economy entered a period of unparalleled prosperity. But in recent decades, despite all the remarkable progress of business and technology, that prosperity has been very unevenly distributed. Around the world, the average standard of living has increased enormously, but in modern developed economies, the middle class has stagnated and for the first time in generations, our children may be worse off than we are. Once again, we face what Keynes called "the enormous anomaly of unemployment in a world full of wants," with consequent political instability and uncertain business prospects.

But Keynes was right. The world he imagined, where "the economic problem" is solved is, in fact, still before us. Global poverty has sunk to all-time lows, and if only we play our cards right, we could still enter the world Keynes envisioned.

Technology and the spread of knowledge have greatly reduced poverty in the world, even as they have created economic challenges for workers in developed countries. As Max Roser, creator of Our World in Data, a remarkable collection of visualizations about how the world has been getting better over the last five hundred years, notes: "Even in 1981 more than 50% of the world population lived in absolute poverty—this is now down to about 14%. This is still a large number of people, but the change is happening incredibly fast. For our present world, the data tells us that poverty is now falling more quickly than ever before in world history."

Much of Keynes's essay, titled "Economic Possibilities for Our Grandchildren," concerns the issue of what people might do with their time when productivity has increased to the point where the machines do all the work.

Is there really not enough work left for humans to do?

Keynes didn't think so in 1930, and I don't think so now. "We are suffering just now from a bad attack of economic pessimism," he wrote. "It is common to hear people say that the epoch of enormous economic progress which characterised the nineteenth century is over; that the rapid improvement in the standard of life is now going to slow down; that a decline in prosperity is more likely than an improvement in the decade which lies ahead of us. I believe that this is a wildly mistaken interpretation of what is happening to us. *We are suffering, not from the rheumatics of old age, but from the growing-pains of over-rapid changes, from the painfulness of readjustment between one economic period and another*" (italics mine).

Sure enough, we are indeed once again hearing the chorus of pessimism and doubt. Automation is going to destroy white-collar jobs in the same way it once destroyed factory jobs. We have an economy that relies on growth, but the age of growth is over. And so on.

Keynes presciently gave a name to the heart of our current angst: *technological unemployment*. He defined it as our inability to find new uses for labor as quickly as we are finding ways to eliminate the need for it. He concluded, "But this is only a temporary phase of maladjustment."

Like Keynes, I remain optimistic. There has already been enormous dislocation, with far more ahead, but if we make the right choices as a society, we will come through it in the end. The short-term pain is very real, and as we've discussed, we must rewrite the rules of our economy and strengthen our safety net to mitigate this pain. If we can manage through the transition without violent revolution, though, history provides plenty of reason for hope.

Back in 1811, weavers in Britain's Nottinghamshire took up the banner of the mythical Ned Ludd (who had supposedly smashed mechanical knitting machines thirty years earlier) and staged a rebellion, wrecking the machine looms that were threatening their livelihood. They were right to be afraid. The decades ahead were grim. Machines did replace human labor, and it took time for society to adjust.

But those weavers couldn't imagine that their descendants would have more clothing than the kings and queens of Europe, that ordinary people would eat the fruits of summer in the depths of winter. They couldn't imagine that we'd tunnel through mountains and under the sea, that we'd fly through the air, crossing continents in hours, that we'd build cities in the desert with buildings a half mile high, that we'd stand on the moon and put spacecraft in orbit around distant planets, that we would eliminate so many scourges of disease. And they couldn't imagine that their children would find meaningful work bringing all of these things to life.

What is possible with the aid of today's technology that we can't yet imagine?

Nick Hanauer once said to me, "Prosperity in human societies is best understood as the accumulation of solutions to human problems. We won't run out of work until we run out of problems."

Are we done yet?

I don't think so. We have yet to deal with the enormous transitions to our energy infrastructure that will be required to respond to climate change; the public health challenges of new infectious diseases; the demographic inversion in which a growing class of elders will be supported by a smaller cohort of workers; rebuilding the physical infrastructure of our cities; providing clean water to the world; feeding, clothing, and entertaining nine billion people. How do we turn millions of displaced people into settlers in the cities of the future rather than refugees in squalid encampments? How do we reinvent education? How do we better care for each other?

History provides another story of jobs being taken by machines, more recent than the Luddites. Thanks to the makers of *Hidden Figures*, a moving film from 2016 about female African American mathematicians who worked at Langley Research Center during the space race of the early 1960s, millions now know how Dorothy Vaughan reacted when she saw what amounted to the Luddites' machine looms. Vaughan supervised a segregated group of "computers," in this case, all women and all African American, who did complex mathematical calculations by hand to power JFK's space program. In the romanticized retelling of her story in the movie, when NACA (the precursor to NASA) bought an IBM 7090 computer (so big they had to break down walls to get it in), Vaughan saw the writing on the wall, and took it upon herself not only to learn FORTRAN, the programming language of this computer, but to teach her staff. Instead of ending up unemployed, they ended up with jobs that had never existed before, making possible something that had never been done before.

Tomorrow, that new work might not come in the form of what we think of as a job. Note that Nick said "we won't run out of work," not "we won't run out of jobs." Part of the problem is that "the job" is an artificial construct, in which work is managed and parceled out by corporations and other institutions, to which individuals must apply to participate in doing the work. Financial markets are supposed to reward people and corporations for accomplishing work that needs doing. But as discussed in Chapter 11, there is a growing divergence today between what financial markets reward and what the economy really needs.

This is what Keynes meant by "the enormous anomaly of unemployment in a world full of wants." Because corporations have different motivations and constraints than individuals, it is possible that a corporation is not able or willing to offer "jobs" even as "work" goes begging. Because of the structure of employment, in uncertain times companies are hesitant to take on workers until they are sure of customer demand. And because of pressure from financial markets, companies often find short-term advantage in cutting employment, since driving up the stock price gives owners a better return than actually employing people to get work done. Eventually "the market" sorts things out (in theory), and corporations are once again able to offer jobs to willing workers. But there is a lot of unnecessary friction and consequential negative side effects—what economists call "externalities."

We've seen how technology platforms are creating new mechanisms that make it easier to connect people and organizations to work that needs doing—a more efficient marketplace for work. You can argue that that is one of the key drivers at the heart of the on-demand revolution that includes companies like Uber and Lyft, DoorDash and Instacart, Upwork, Handy, TaskRabbit, and Thumbtack. The drawbacks of these platforms in providing consistent income and a social safety net shouldn't blind us to what does work about them. We need to improve these platforms so that they truly serve the people who find work through them, not try to turn back the clock to the guaranteed employment structure of jobs in the 1950s.

There is also a leadership challenge: to correctly identify work that needs doing. Think of what Elon Musk has done to catalyze new industries with Tesla, SpaceX, and SolarCity.

Like Elon, I believe that climate change will be for our generation, and the next, what World War II was for our parents and grandparents, a challenge that we must rise to meet or will suffer dire consequences from. But it is in rising to challenges that we can build a better future. It's already clear that transforming our energy infrastructure will provide a great many well-paid human jobs, but it is also clear that technology will play an enormous role. Already in data centers, for example, AI is radically increasing power efficiency. How do we rethink and rebuild our electric grid to be decentralized and adaptive? How do we use autonomous vehicles to rethink the layout of our cities, making them greener, healthier, better places to live? How do we use AI to anticipate ever-more-unpredictable weather, protecting our agriculture, our cities, and our economy?

Mark Zuckerberg and Priscilla Chan's announcement in 2016 that they are funding an initiative that aims to cure all disease within their children's lifetimes is another example of a bold dream that leaps over the feeble imagination of the current market. It's hard to imagine that

AI and machine learning won't play a major role in striving toward that ambitious goal, along with our growing control over human genetics and biology. Already AI is being used to analyze millions of radiology scans at a level of resolution and precision impossible for humans, as well as helping doctors to keep up with the flood of medical research at a level that can't be accomplished by a human practitioner. It's also hard to imagine that there isn't plenty of work for humans in eliminating disease and disability for everyone.

Markets are not infallible. Government can play a role, as it did with the Internet, GPS, and the Human Genome Project. That role is not limited just to investments in basic research or to projects that require coordinated effort beyond the capability of even the largest commercial actors. Government must also deal with market failure. This can be the failure of the commons, outright malfeasance by commercial actors, or the misdirected fitness function of financial markets and the bad maps of economists, which are strangling the economy today.

But the change can and must begin with corporate "self-interest, properly regarded." Jeff Immelt, Jack Welch's successor as CEO at GE, has rejected the purely financial calculus of the old GE, and has recommitted the company to "solving the world's hardest problems," as he told me at my 2015 Next:Economy Summit. Jeff believes that it should be a paramount concern for all of us that there is a shortage of good jobs around the world. "We need to be investing in this next generation of who's employable and what skills they need. And that's the purpose of companies just like it is of schools." That is, good jobs, not just profits, or even great products, are one of the key outputs of a great company. Executives can't just complain about not being able to hire the right people. They have to take responsibility for training the people they need for the jobs of the future. "If there's going to be a competitive workforce," he continued, "we need to be at the leading edge of who is going to create that."

The question is not whether there will be enough work to go around, but the best means by which to fairly distribute the proceeds of the productivity made possible by the WTF? technologies of what Erik Brynjolfsson and Andy McAfee call "the second machine age."

Reducing working hours for the same amount of pay is one of the most fundamental ways that the benefits of rising productivity have traditionally been distributed more widely. In 1870, the average American (male) worked 62 hours per week; by 1960, that number was down to just over 40 hours, where it has roughly hovered since. Yet our material standard of living is far higher. Unpaid work in the home (mostly done by women) has declined even more sharply, from 58 hours in 1900 to 14 in 2011. One key question is why external paid labor hours have not fallen further in the past fifty years, matching the increase in productivity for domestic labor. The case can be made that the entry of women into the paid external workforce, then global access to workers in low-wage countries, and direct legislative action have reduced the bargaining power of labor, allowing companies to allocate the surplus to corporate profits rather than reducing working hours and paying higher hourly wages, as happened in the past.

Education is another way that we effectively reduced working hours. Young children once went to work; in the nineteenth century, we sent them instead to school. In the first half of twentieth century, the high school movement extended schooling by another six years; in the second half, college added two to four more. As we will discuss in Chapter 15, education will need to be extended again to meet the changing needs of the twenty-first century.

Something must be done to end this "temporary phase of maladjustment," which has gone on far too long and created so much economic pain for too many!

It is deeply unfortunate how difficult it is for humans to practice foresight. In his wise and insightful book, *The Wealth of Humans*, senior editor for the *Economist* Ryan Avent traces the lessons that we could and should take from the centuries of economic and political struggle that led from the innovations of the industrial revolution to the successful economies of the second half of the twentieth century. Prosperity came when the fruits of productivity were widely shared; enmity, political turmoil, and even outright warfare were the harvest of rampant inequality. It is obvious that generosity is the robust strategy.

MACHINE MONEY AND PEOPLE MONEY

Universal basic income (UBI) is one proposed mechanism for achieving the transition between today's system and a more human-centered future. This proposal, that every human being should be given an income sufficient to meet the basic needs of life, appeals to progressives as a basic human right, and to conservatives as a way of radically simplifying the complex rules of the present welfare state.

Fabled labor leader Andy Stern left his job as the head of the Service Employees International Union (SEIU) to write a book making the case for UBI; Y Combinator Research has begun a pilot program in Oakland, California; and peer-to-peer charity GiveDirectly is asking its users to fund a pilot in Kenya. The GiveDirectly experiment is fascinating on two fronts: It is crowdfunded by ordinary people, who already use the platform to provide aid in the form of direct cash transfers to the needy; and in a developing country, the costs are lower so the program can be more extensive, and thus allows for a true randomized control trial.

These experiments tell us how far the idea has come since it was proposed by Thomas Paine in 1795, and more recently by Milton Friedman in 1962 (and Paul Ryan in 2014). There are many arguments against UBI, most notably the cost of making it truly universal, and that providing the income to people whether they need it or not will starve existing programs that provided targeted aid to those who actually need it. At the very least, though, UBI provides a compelling exercise in imagining a radically different way of building a social safety net, and, in thinking through how we might pay for it, a radically different way of dividing up the economic pie.

I asked MIT labor economist David Autor whether there were any natural experiments in universal basic income, and what they teach us. He cited the contrast between Saudi Arabia and Norway. Both countries have enormous oil wealth, he noted, but in Saudi Arabia, the bulk of the wealth goes to a small percent of the population. Much of the work of everyday society is looked down on and is done by an underclass of low-paid "guest workers" while an elite works at sinecure jobs or enjoys idle pursuits. In Norway, by contrast, Autor said, "All kinds of work are valued. Everybody works, they just work a little less." The generous redistribution of oil profits and a strong social safety net funded by the wealth that is understood to belong to all makes Norway one of the happiest and wealthiest countries in the world.

For a technology perspective, I turned to Paul Buchheit, creator of Gmail and now a partner at Y Combinator, and Sam Altman, the head of Y Combinator. In a 2016 conversation, Paul said to me: "There may need to be two kinds of money: machine money, and human money. Machine money is what you use to buy things that are produced by machines. These things are always getting cheaper. Human money is what you use to buy things that only humans can produce."

The idea that there should be different kinds of "money" is a provocative metaphor rather than a concrete proposal. Money is already a method for agreeing on the exchange rate between radically different kinds of goods and services. Why should we need different kinds of money? I'm not sure that Paul meant this literally. What he was pointing to is that at different times in history, the primary lever for the creation of money has changed. Ownership of land was once the key to great wealth. During the industrial era, we built mechanisms that were optimized for converting a regimented combination of human and machine labor into money. In the twenty-first century, we need to recognize and optimize for a different kind of value.

Paul's argument is that the key thing that humans offer that machines do not is "authenticity." You can buy a cheap table made by a machine, he said, or a handcrafted table made by a person. In the long term, the price of the former (in machine money) should decline, but the latter will always cost about the same in human money (some quantity roughly proportional to the number of hours required to make it).

Paul believes that the right name for what many are calling "universal basic income" should be "the citizen's dividend," the name given to it in Thomas Paine's *Agrarian Justice*. Paine made the appeal for sharing the value of unimproved land with every citizen of the new United States; Buchheit suggests that all of mankind should have some claim on the fruits of technological progress. That is, we should use tax policy to capture some amount of the bounty from machine productivity, and provide that to all people as a stipend with which they can meet the needs of everyday existence. Similarly, in 2017, Bill Gates proposed a "robot tax," with the proceeds being used to fund caring for children or the elderly, or for education.

Paul believes that the bounty from the next generation of machine productivity should be distributed sufficiently, so that everyone can have enough "machine money" to meet their basic needs. Meanwhile, that productivity should also provide goods at ever-lower costs, increasing the value of the citizen's dividend. This is the world of prosperity that Keynes envisioned for his grandchildren.

How might we pay for a universal basic income? The entire amount the United States federal government spends on social welfare programs—$668 billion in 2014—would amount to only $2,400 per person. Rutger Bregman, the author of *Utopia for Realists*, a book about basic income, divides the pie differently, pointing out that rather than providing an income to those who don't need it, we could use a negative income tax to give cash only to those who actually need it. Writers Matt Bruenig and Elizabeth Stoker calculated that in 2013, the amount needed to bring all of the Americans living below the poverty line up to at least its level would cost only $175 billion.

Sam Altman explained that those who argue about how we would pay for a universal basic income today miss the point. "I am confident that if we need it, we will be able to afford it," he said in a 2016 discussion of UBI at venture capital firm Bloomberg Beta with Andy Stern and the Aspen Institute's Natalie Foster. One major factor that isn't being considered, as Sam expanded on it in our subsequent conversation, is that the possible productivity gains from technology are enormous, and these gains can be used to reduce the price of any goods produced by machines—a basket of goods and services sufficient to support basic needs that costs $35,000 today might cost $3,500 in a future where the machines have put so many people out of work that a universal basic income is required.

Hal Varian agrees. "In fact, it has to work that way," he told me. "If people adopt a technology because it produces more output at a lower cost, then the size of the pie gets bigger. The real question is how that additional value is divided."

Neither Paul nor Sam addressed the point that not all goods become evenly cheaper—in many cities, for instance, the price of housing has gone up far faster than the price of consumables has gone down. Nor do they address the political obstacles to dividing that bounty. Nonetheless, there is enough truth in this idea to support Paul's metaphor that machine money could operate by different rules from human money. In a profound way, the value of machine money inflates not as a currency normally inflates, but because the lower costs provided by machine productivity constantly increase its purchasing power. Meanwhile, the declining cost of anything made by machines would argue that the work that humans alone can do should become more rather than less valuable.

The remainder of this chapter will discuss some ways in which that future is and is not unfolding.

> The chorus of doubt about the jobless future sounds remarkably similar to the one that warned of the death of the software industry due to open source software. Clayton Christensen's Law of Conservation of Attractive Profits holds true here too. When one thing becomes commoditized, something else becomes valuable. We must ask ourselves what will become valuable as today's tasks become commoditized.

CARING AND SHARING

What might we do with our time, if there were a universal basic income sufficient to meet the necessities of life, or if paid working hours were reduced by the same amount as domestic labor, and wages increased? Keynes was right. The key question for mankind *should be* how to use our freedom from pressing economic cares, how to occupy our leisure, and how "to live wisely and agreeably and well."

What might we do with our time, if we didn't have to work for a living? The things that require a human touch, for starters. Caring for our parents and our friends. Reading aloud to a child. And things we do for love. Enjoying a meal with a loved one is not something that machines can make more efficient.

I love Paul's distinction between two types of money, but I do wonder whether it is complete. His notion of human money encompasses two very different classes of goods and services: those that involve a human-to-human touch—parenting, teaching, caregiving of all kinds—and those that involve creativity.

Perhaps "human money" should be further subdivided into "caring money" and "creativity money." Caring is a necessity of life, just as is food and shelter, and should not be denied to anyone in a just society. In an ideal world, caring is a natural outgrowth of family and community, as we care for those we love.

Time is a key currency of caring. And that brings us full circle, back to the on-demand economy as an alternative to traditional employment. For many people, an on-demand platform that allows a better blend of personal *human* time and machine money time may be a real step forward into a far better labor economy than an attempt to fit everyone back into the regimented industrial age world of jobs with regular forty-hour workweeks.

Anne-Marie Slaughter, the president of New America and author of *Unfinished Business: Women Men Work Family*, notes that the on-demand economy "will reshape not only ways of working but also patterns of consumption." She hopes for a future where the choice to take off time to raise children or take care of parents will not be a career-ending move. "Care is unpredictable and work traditionally has been fixed. And that doesn't work," she told me in an onstage interview at my 2015 Next:Economy Summit in San Francisco. "So when you are able to schedule your own work, that is the solution to the care problem. But it is only a solution to the care problem if we can let people make a living *and* support the families they are caring for."

However, economies thrive on exchange, and even in the world of caring, money is a substitute for time. And so there is also a caring economy of paid professionals, including teachers, doctors, nurses, eldercare assistants, babysitters, hairdressers, and massage therapists. Back in 1950, who would have guessed that in 2014 there would be nearly 300,000 "fitness trainers" in the United States?

If you look at the current shape of the economy, there are huge and growing numbers of service jobs of this kind. A study of UK census data by consulting firm Deloitte found that in 1871, caring economy jobs represented 1.1% of the total labor economy. By 2011, they were 12.2%. The report also noted that between 1992 and 2014, the number of nursing auxiliaries and assistants rose tenfold, and the number of teaching assistants rose nearly sevenfold.

In a society with an inverted demographic pyramid, in which there are far more of the elderly than young people to support them, as we will see in many developed countries by 2050, there may not be enough people to do the work of caring and machines may even be called on to fill the gap. This problem is not limited to developed countries; China's rapidly growing middle class is an eager consumer of caring services.

On-demand technology has promise for growing the market even further. Seth Sternberg, founder of Honor, a service making it easier for older adults to remain in their own homes, pays its caregivers as full-time employees with benefits, but uses on-demand technology to make care more flexible and affordable for consumers. Seth told me that being able to purchase just the amount of care you need, when you need it, means that people who could never afford the service find themselves able to do so, and the market expands.

The economic problem is that caregiving is insufficiently valued in our society. If there were ever a case to be made for the Clothesline Paradox, this is it. Why is it that work that is so valuable to society is expected to be provided for free, or when paid, is paid so poorly?

> If we are working from a new map, in which our objective is to value human effort, not to dispense with it, we surely must start by assigning an economic value to caregiving.

If you think about it, this is in fact what most countries (and progressive employers in the United States) are doing with extended paid parental leave for both women and men, or when countries provide public financing for eldercare services. (The United States is one of only two countries in the world that don't mandate any paid maternity or paternity leave; the other is Papua New Guinea.)

Parental leave is just the beginning. Early childhood education could be revolutionized by an economic system that provided basic income and the flexibility for parents to spend time with their children. Hiring more teachers at better salaries and reducing class size in public schools to the level of the best private schools would be another pragmatic way to transition to the caring economy. It is slowly being recognized that the cost of insufficient care for children gets paid one way or another, if not up front, then in healthcare or prison costs later in life.

Even in the absence of changes in childcare, eldercare, or education spending, I suspect that if we successfully tackle the problem of creating a better income distribution across all levels of society by some other means, people will naturally allocate more of that income to caring, education, and similar activities. After all, we already know that given sufficient income, people routinely pay more for better, more personal service. The rich still live in a world where doctors make house calls and personal tutoring is the norm.

> Might it not be the case that in a world where routine cognitive tasks are commoditized by artificial intelligence, it is the human touch that will become more valuable, the source of competitive advantage?

The issue remains whether a combination of market forces and political action can increase the earnings of those who do the work that is not going to be automated away. Even if we never run out of jobs, we must still ask what sort of lives those jobs will pay for. A world in which a small number of people enjoy productive, highly paid work and can indulge in expensive leisure activities and superb personal care while others are ground underfoot is not a world that any of us should aspire to.

THE BIG, BEAUTIFUL ART MARKET

As suggested in the first part of this chapter, the principal work of the twenty-first century should be to harness the power of today's digital and cognitive technologies to achieve leaps of presently inconceivable progress analogous to those that our nineteenth- and twentieth-century forebears accomplished with their industrial tools. It may be that we will need fewer human hours to do that work, just as over the past centuries we have vastly reduced the amount of labor required to feed ever-larger numbers of people.

But the work of the nineteenth and twentieth centuries included innovations not just in food production, commerce, transportation, energy, sanitation, and public health but also in new ways for far more people to consume the vast variety of goods and services made possible by those innovations. So too, the cognitive era will bring forth new types of consumption. This is the realm of creativity money. Creativity is an indomitable wellspring within all of us. It is part of what makes us human, and in many ways, it is entirely independent of the monetary economy.

It is a mistake to think that "the creative economy" is limited to entertainment and the arts. Creativity is the focus of a competition for accumulation as intense as any that characterizes Paul Buchheit's machine money. It is at the heart of industries like fashion, real estate, and luxury goods, all of which depend on the competition among people who are already rich to own more, to enjoy, or sometimes just to show off their wealth.

Creativity money is another way of saying that we pay a premium for the good things of life beyond the basics. Sports, music, art, storytelling, and poetry. The glass of wine with friends. The night out at the movies or the local music venue. The beautiful dress and the sharp suit. The combination of design, manufacturing, and marketing that goes into the latest LeBron James basketball shoe.

People at all levels of society pay that premium as a way of expressing and experiencing beauty, status, belonging, and identity. Creativity money is what someone pays for the difference between a Mercedes

C-Class and a Ford Taurus, for a meal at a world-famous restaurant like the French Laundry rather than the local French bistro, or at that same bistro rather than at a McDonald's. It is why those who can afford it pay three dollars for an individually crafted cappuccino rather than drinking Folger's coffee from a five-pound can, as our parents did. It is why people pay huge prices or wait years to see *Hamilton*, while tickets for the local dinner theater are available right now.

Dave Hickey, an art critic and MacArthur "genius" grant winner, describes how Harley Earl of General Motors, the first-ever head of design for a major American company, turned the post–World War II automobile into "an art market." Hickey defines that as a market in which products are sold on the basis of what they *mean*, not just what they *do*. The annual turnover of new models was one way that Detroit soaked up the enormous postwar productive capacity of America's factories.

Turning the computer into an "art market" is also a perfect explanation of what Steve Jobs accomplished when he returned to Apple in 1997. "Think Different" was a powerful statement that buying from Apple was a statement about *who you are*. Yes, the products were beautiful and useful, but just like the automobile when it was the ultimate object of consumer desire, the Mac, and later the iPhone, became a statement of identity. Design was not just a functional improvement, but a way of making a statement. In a world where personal computers had become a commodity, design became a unique source of added value. Once again, attractive profits were conserved.

In the late eighteenth century, in his short novel *Rasselas*, Samuel Johnson wrote that the Great Pyramid "seems to have been erected only in compliance with that hunger of imagination which preys incessantly upon life, and must be always appeased by some employment. Those who have already all that they can enjoy must enlarge their desires. He that has built for use till use is supplied must begin to build for vanity, and extend his plan to the utmost power of human performance that he may not be soon reduced to form another wish." That is, even in a world where every need is met, there will still be "a world full of wants."

Given an income sufficient to the necessities of life, some people will choose to step off the wheel—to spend more time with family and friends, in creative pursuits, or whatever they damn well please. But even if the machines do most of the essential work and everybody gets a stipend to cover the cost of basic living expenses, competition for additional creativity money will likely drive an economy in which some people just get by while others develop solid middle-class incomes and still others amass vast fortunes.

I'm fascinated by a comment that Hal Varian, Google's chief economist, made to me over dinner one night: "If you want to understand the future, just look at what rich people do today." It's easy to think of this as a heartless libertarian sentiment. Our dinner companion, Hal's former student and coauthor Carl Shapiro, fresh from his stint at Obama's Council of Economic Advisers, seemed horrified. But when you think about it for a moment, it makes sense.

Dining out was once the province of the wealthy. Now far more people do it. In our most vibrant cities, a privileged class experiences a taste of a future that could be the future for everyone. Restaurants compete on the basis of creativity and service, "everyone's private driver" whisks people around in comfort from experience to experience, and one-of-a-kind boutiques provide unique consumer goods. Rich people once took the European grand tour; now soccer hooligans do it. Cell phones, designer fashion, and entertainment have all been democratized. Mozart had the Holy Roman Emperor as his patron; Kickstarter, GoFundMe, and Patreon extend that opportunity to millions of ordinary people.

This rings of bubble talk from the privileged coasts. Yet it is true far more broadly. Cell phones are found even in the poorest parts of the world. The variety of clothing, food, and consumer goods available at a Walmart would astonish even wealthy people from fifty years ago.

Restaurants—and food in general—teach us something profound about the future of the economy. Everywhere, food is blended with *ideas* to make it more valuable. As Korzybski said, "People don't just eat food, but also words." This isn't just ordinary coffee. It's fair-trade, single-origin coffee. And look, we have six different kinds. You must try them all. These aren't ordinary fruits and vegetables. They are organic, farm-to-table. This bread is gluten-free. Is that North Carolina barbecue or Texas barbecue? KFC or Church's Fried Chicken?

At every price level, there is competition to provide a unique experience. Food is a commodity, yet, just as Christensen pointed out, when one thing becomes a commodity, something adjacent becomes valuable. In a flourishing city, there is a dizzying array of creative, multicultural dining options.

In 2016, I met with a staffer from the White House who wanted my advice on which Silicon Valley entrepreneur President Obama should sit down with onstage at the Global Entrepreneurship Summit. "We're here in a wonderful restaurant in Oakland," I said. "Boot and Shoe Service is one of three restaurants created by a man named Charlie Hallowell. They are part of why people now say Oakland is a great place to live. We need more Charlie Hallowells more than we need another Mark Zuckerberg." After all, a great platform like Facebook is a rare thing, not easily duplicated. You can count the people who succeed like Zuck did on your fingers; not so the tens of thousands of Charlie Hallowells who characterize a truly rich and diverse economy.

New industries driven by the human touch are everywhere. In the United States, more than 4,200 craft breweries now make up more than 10% of the market, and command a price double that of a mass-produced beer. In the first quarter of 2016, 25 million customers purchased handcrafted and artisan goods on Etsy. These are small green shoots in an economy dominated by mass-produced products, but they teach us something important.

What is happening in entertainment may be another interesting harbinger of the future. While blockbusters still dominate in Hollywood and New York publishing, a larger and larger proportion of people's entertainment time is spent on social media, consuming content created by their friends and peers. Anne-Marie Slaughter notes that "millennials' definition of quality of life now involves more time and less stuff." They want to spend their money on experiences, not things.

That profound shift in media consumption has most visibly enriched Facebook, Google, and the current generation of media platforms, but it has also created new opportunities for professional media creators. The *New York Times* or Fox News article shared on Facebook has something added to it that a copy picked up from a newsstand never did: the endorsement of someone you know. The art of sharing things that will spread virally often now involves remixing them in some way—combining a quote with an image, or framing the subject with a pithy observation of your own.

Social media is also increasingly creating paying jobs for a growing number of individual media creators. YouTube star and VidCon impresario Hank Green wrote, "I started paying my bills with YouTube money around the time I hit a million views a month." Millions of teens use "Hank and John EXPLAIN!" videos to learn about current events, and they get a deeper dive in a five-minute video than they would in hours of mass-produced "news." Millions more learn math, science, music, and philosophy from other YouTube channels like Khan Academy, or One-Minute Physics, or Hank's own Crash Course. When my young niece learned that I knew Larry Page and Mark Zuckerberg and Bill Gates, she said, "Meh!" But when she heard I knew Hank and John Green, she was really impressed.

Keep in mind that "YouTube money," as Hank names it, is only one of many new forms of creative money that are available via online platforms. There's Facebook money, Etsy money, Kickstarter money, App Store money, and more. Who would have thought ten years ago that people could make six-figure earnings playing video games while millions of others follow along on YouTube or Twitch?

To those who worry that these small signs of a new economy could not possibly replace the jobs of today, I would once again cite Gibson's observation that "the future is here. It just isn't evenly distributed yet." Every flourishing harvest begins with the smallest of shoots poking their heads through the soil.

Some of these marketplaces are further along than others in creating opportunities for individuals and small companies to convert attention (the raw material of creativity money) into cash. The next few years will see an explosion of startups that find new ways to convert more and more of the attention that is spent online into traditional money.

Jack Conte, half of the musical duo Pomplamoose and founder and CEO of crowdfunding patronage site Patreon, told me that he founded

Patreon after "Nataly and I got 17 million views of our music videos, and it turned into $3,500 in ad revenue. Our fans value us more than that." Tens of thousands of artists now receive enough patronage via the platform that they can now concentrate on their work. As crowdfunding sites like Patreon (and, of course, Kickstarter, Indiegogo, and Go-FundMe) show, there are increasingly new opportunities for ordinary people to compete for real currency, not just attention. These sites are still a relatively small part of the overall economy, but they have a lot to teach us about its possible future direction.

Perhaps the right answer, though, is not to monetize creativity in the old way, by converting it to machine money, but to build an entirely new kind of economy. In his 2003 novel, *Down and Out in the Magic Kingdom*, science fiction writer and activist for a better future Cory Doctorow wrote of a future economy where advanced technology has made it essentially free to meet any physical need. The economy instead is based on a reputation currency called "whuffie." The economic competition is to get other people to approve of and support your creative projects. Kickstarter campaigns and Facebook likes may be early prototypes of that future currency.

Creativity can be the focus of an intense competition for status, so that "he who has built for use till use is supplied must begin to build for vanity," but it can also be the key to a future human economy that would let all enjoy the fruits of leisure that are brought to us by machine productivity while also encouraging entirely new kinds of creative work and social consumption.

Work gives a sense of purpose, and it's also worth considering how many things people work at that are currently unpaid, or low paid, that are actually far more valuable to them than things we have been mistakenly trained to pay for. Aspiring actors and musicians working as baristas to pay the rent consider the constant training and auditions in hope of future success to be their real work. It is not at all inconceivable to add "I'm working on my YouTube channel" or "I'm building my Facebook following" to the list of things that give a higher proportion of purpose than of remuneration. Dave Hickey writes that his dad "thought money was something you turned into music, and music, ideally, was something you turned into money." It was music, not money, that gave him purpose and made him happy.

Purpose and meaning are also essential to the caring economy. Jen Pahlka told me the story of a Lyft driver she met in Indianapolis, who leaves a couple of hours early every morning to pick up strangers because he doesn't get enough human contact in his job as a highly paid engineer. He donates his earnings from Lyft to charity.

The volunteer at a homeless shelter may derive far deeper meaning from that unpaid care for other human beings than from the rushed busywork of even a fulfilling career. The amateur athlete may consider her or his training and competitions more important to happiness than earning big bucks at the investment bank. A father or mother who stays home to raise children is not "opting out." He or she is opting *in* to something potentially far more meaningful and important.

This is the possibility that Keynes foresaw when he wrote: "The strenuous purposeful money-makers may carry all of us along with them into the lap of economic abundance. But it will be those peoples, who can keep alive, and cultivate into a fuller perfection, the art of life itself and do not sell themselves for the means of life, who will be able to enjoy the abundance when it comes."

Research on what demographers Gianni Pes and Michel Poulain called "blue zones"—areas with the highest percentage of centenarians, so-called because they originally marked them with blue circles drawn on a map—identified the key characteristics that lead to longer, happier lives. There were a number of dietary factors (an approach that author Michael Pollan summarized as "Eat food. Not too much. Mostly plants"), moderate, regular alcohol intake, especially wine, and moderate, regular physical activity. But even more important were a sense of purpose, engagement in spirituality or religion, and engagement in family and social life.

We know what the good life looks like. We have the resources to provide it to everyone. Why have we constructed an economy that makes it so difficult to achieve?

When faced with questions of how we adapt society and the economy to the current wave of technological change, our goal should not be to have the future look like the past. We must make it new. Writing about the political challenges we face today, Jen Pahlka put her finger on what must always be a key principle for thinking about the future:

> The status quo isn't worth protecting. It's so easy to be in reaction, on the defensive, fighting for the world we had yesterday. Fight for something better, something we haven't seen yet, something we have to invent.

15

DON'T REPLACE PEOPLE, AUGMENT THEM

THESE ARE TWO SEPARATE QUESTIONS: WHETHER THE KIND of cognitive work described in the previous chapter can ever replace the mass employment in the factories of the twentieth century; and whether it can be well paid enough for the flywheel of prosperity to continue.

In answer to the first question, let me simply say that it was inconceivable during the agricultural era that so many people could find employment in factories and in cities. Yet automation and far lower cost of production led to huge increases in demand for previously unavailable products and services. It is up to us once again to put people to work in fulfilling ways, creating new kinds of prosperity. The lessons of technology innovation remind us that progress always entails thinking the unthinkable, and then doing things that were previously impossible.

As to the second question, it is up to us to ensure that the fruits of productivity are shared. The first step is to prepare people for the future that awaits them.

From 2013 through 2015, I was part of the Markle Foundation Rework America task force, exploring the future of the US economy. The question before the task force was how to provide opportunities for Americans in the digital age. One of the moments that stuck with me was a remark from political scientist and author Robert Putnam, who said, "All of the great advances in our society have come when we have made investments in other people's children."

He's right. Universal grade school education was one of the best investments of the nineteenth century, universal high school education of the twentieth. We forget that in 1910, only 9% of US children graduated from high school. By 1935, that number was up to 60%, and by 1970 nearing 80%. The GI Bill sent returning World War II veterans to college, enabling a smooth transition from wartime to peaceful employment.

In the face of today's economic shifts, there were proposals in the 2016 presidential election for universal free community college. In January 2017, the city of San Francisco went beyond proposals and agreed to make the City College of San Francisco, its community college, free for all residents. This is a great step.

But we don't just need "more" education, or free education. We need a radically different kind of education. "If the students we are training today are going to live to be 120 years old, and their careers are likely to span 90 years, but their training will only make them competitive for 10 years, then we have a problem," notes Jeffrey Bleich, former US ambassador to Australia and now chair of the Fulbright scholarship board. Advances in healthcare and technology, and the changing nature of employment, are compounding to obsolete our current educational model, which viewed schooling as preparation for a lifetime of work at a single employer.

We need new mechanisms to support education and retraining throughout life, not just in its early stages. This is already true for professionals in every field, whether athletes or doctors, computer programmers or skilled manufacturing workers. For them, ongoing learning is an essential part of the job; access to training and educational resources is one of the most prized perks, used to attract top employees. And as "the job" is deconstructed, the need for education doesn't go away. If anything, it is increased. But the nature of that education also needs to change. In a connected world where knowledge is available on demand, we need to rethink what people need to know and how they come to know it.

THE AUGMENTED WORKER

If you squint a little, you can see the Apple Store clerk as a cyborg, a hybrid of human and machine. Each store is flooded with smartphone-wielding salespeople who are able to help customers with everything from technical questions and support to purchase and checkout. There are no cash registers with lines of customers waiting with products pulled from the piles on the shelves. The store is a showroom of products to explore. When you know what you want, a salesperson fetches it from the back room. If you're already an Apple customer with a credit card on file (and as of 2014, there were 800 million of us), all you need to provide is your email address to walk out the door with your chosen product. Rather than using technology to eliminate workers and cut costs, Apple has equipped them with new powers in order to create an amazing user experience. By so doing, they created the most productive retail stores in the world.

As a design pattern, this is remarkably similar to one of the key business model elements of Lyft and Uber, discussed in Chapter 3. The Apple Store has nothing to do with on-demand, the map that most people use to understand these new platforms, yet it has a great deal in common with them as a lesson plan for constructing a magical user experience made possible by networked, cognitively augmented workers connected to a data-rich platform that recognizes its customers and tailors its services to them.

The Apple Stores are also a testament to the truth that it is not technology itself that is transformative. It is its application to rethinking the way the world works, not inventing something new but applying newly latent capabilities to do an old thing so much better as to change it utterly.

Even the very first advances in civilization had this cyborg quality. The marriage of humans with technology is what made us the masters of other species, giving us weapons and tools harder and sharper than the claws of any animal, projecting our strength at greater and greater distance until we could bring down even the greatest of beasts in the hunt, not to mention engineer new crops that produce far more food than their wild forebears, and domesticate animals to make us stronger and faster.

I remember once reading an account of the crossing of the land bridge between Siberia and Alaska that used a curious fact as part of its analysis of the possible date. It couldn't have happened *before the invention of sewing*, the authors noted, which made possible the piecing together of close-fitting garments that allowed humans to live in cold climes. Sewing! Sewing with bone needles was once a WTF?

technology, making possible something that had previously been un-thinkable.

Every advance in our productivity, getting more output from an equivalent amount of labor, energy, and materials, has come from the pairing of human and machine. It is the acceleration and compounding of that productivity that has produced the riches of the modern world. For example, agricultural production doubled over the hundred years from 1820 to 1920, but it took only thirty years for the next doubling, fifteen for the doubling after that, and ten for the doubling after that.

The ultimate source of productivity increases is innovation. Abra-ham Lincoln, no economist, but an acute judge of the forces of human history, wrote:

> Beavers build houses; but they build them in nowise differently, or better, now than they did five thousand years ago. . . . Man is not the only animal who labors; but he is the only one who improves his workmanship. These improvements he effects by Discoveries and Inventions.

A discovery or invention only improves the livelihood of all, though, when it is shared. Consider one of the world's most heralded inven-tions. Can you imagine the first woman (I like to imagine that it was a woman) who built a controlled fire? How amazed her companions were. Perhaps afraid at first. But soon warmed and fed by her boldness. Even more important than fire itself, though, was her ability to tell others about it.

It was language that was our greatest invention, the ability to pass fire from mind to mind. In periods where knowledge is embraced and widely shared, society advances and becomes richer. When knowledge is hoarded or disregarded, society becomes poorer.

The adoption of movable type and the printed book in fifteenth-century Europe led to our modern economy, a remarkable flowering of both knowledge and of freedom, as the discoverers of the new could pass the fire of knowledge to people not yet born and to those living thousands of miles away. Those inventions and discoveries took centuries to reach their full potential, as the value of literacy fed on itself, and a better-educated population further increased the rate of invention and the spread of new ideas, creating demand for even more learning, discovery, and consumption. The Internet was another great leap. But the web browser—words and pictures online—was only a halfway house. It was an increase in accessibility and the speed of dissemination of knowledge, but not a change in kind from the physical forms that preceded it.

The final step by which knowledge is shared is via *embedding in tools*. Consider maps and directions. The path from physical maps through GPS and Google Maps to self-driving cars illustrates what I call "the arc of knowledge." Knowledge sharing goes from the spoken to the written word, to mass production, to electronic dissemination, to embedding knowledge into tools, services, and devices.

In the past, I could ask someone for directions. Or I could consult the stored knowledge in a paper map. The first online maps were merely facsimiles of paper maps. Now I can see exactly where I am and how to get where I want to go in real time. The next step is for me to forget about all that and just let the car take me to my destination. The step after that is to imagine what we might do differently when transportation is as reliable as running water.

This embedding of knowledge into tools isn't something new. It has always been a critical enabler of the productivity gains that come from mastery over the physical world. And it inevitably leads to massive changes in society.

When Henry Maudslay built the first screw-cutting lathe in 1800, creating a machine that could reproduce *exactly* the same pattern every time—something impossible for even the most skilled human craftsman equipped only with hand tools—he made possible a world of mass production. From the first nuts and bolts with threads identical to within thousandths of an inch, first hundreds then thousands then millions of products descended, the children, grandchildren, and great-grandchildren of Maudslay's mind.

So too, when Henry Bessemer invented the first process for cheaply mass-producing steel in 1856, he didn't just remove carbon and impurities from iron: He added knowledge. Knowing how to make vast quantities of cheap steel made entirely different futures possible. Andrew Carnegie made his fortune and took over leadership of the worldwide steel industry from Britain by manufacturing the rails that tied together a country far vaster. Steel girders enabled skyscrapers; steel cables enabled elevators and vast suspension bridges. Each of these nineteenth-century WTF? technologies built on the others, much as today's advances do.

The three-part process of creating new knowledge, sharing it, and then embedding it into tools so that it can be used by less skilled workers is illustrated neatly by the rise of big data technologies. Google had to develop entirely new techniques in order to deal with the growing scale of the web. One of the most important of these was called MapReduce, which splits massive amounts of data and computation into multiple chunks that could be farmed out to hundreds or thousands of computers working in parallel. MapReduce turned out to be relevant to a large class of problems, not just search.

Google published papers about MapReduce in 2003 and 2004, laying bare its secrets, but it didn't take off more widely until Doug Cutting created an open source implementation of MapReduce called Hadoop in 2006. This allowed many other companies, at that time facing problems similar to those that Google had encountered years earlier, to more easily adopt the technique.

This process is key to the progress of software engineering. New problems beget new solutions, which are essentially handcrafted. Only later, when they are embodied into tools that make them more accessible, do these remarkable innovations become the workaday life of the next generation of developers. We are currently at the beginning of the transition from handcrafted machine learning models to tools that will make it possible for workaday developers to produce them. Once that happens, AI will infuse and change our entire society in the same way that mass manufacturing transformed the nineteenth and twentieth centuries.

The vastly improved productivity of agriculture provides a bit more nuance in understanding the mix of mind and matter in new tools. Agricultural productivity has come not just from the use of machines to do much of the work of planting and harvesting and from energy-intensive fertilizers (another industrial product), but through the development of more productive cultivars of the foods themselves. When Luther Burbank bred the Russet Burbank potato, now the most widely grown potato, he enhanced productivity with a very different balance of knowledge and material inputs than Hiram Moore did with the invention of the combine harvester.

In short, the two types of augmentation, physical and mental, are in a complex dance. One frontier of augmentation is the addition of sensors to the physical world, allowing data to be collected and analyzed at a previously unthinkable scale. That is the real key to understanding what is often called the "Internet of Things." Things that once required guesswork are now knowable. (Insurance may well be the native business model of the "Internet of Things" in the same way that advertising became the native business model of the Internet, because of the data-driven elimination of uncertainty.) It isn't simply a matter of smart, connected devices like the Nest thermostat or the Amazon Echo, the Fitbit and the Apple Watch, or even self-driving cars. It's about the data these devices provide. The possibilities of the future cascade in unexpected ways.

When Monsanto bought Climate Corporation, the big data weather insurance company founded by former Google employees David Friedberg and Siraj Khaliq, and paired it with Precision Planting, the data-driven control system for seed placement and depth based on soil composition, they showcased that the new focus of productivity in agriculture *is in data and control*. Less seed, less fertilizer, and less water are needed when an eye in the sky can tell the farmer with precision the state of his land and the progress of his crop, and automatically guide his equipment to act on that knowledge.

This is true in engineering and materials science as well. Remember Saul Griffith's comment: "We replace materials with math." One of Saul's companies, Sunfolding, sells a sun-tracking system for large-scale solar farms that replaces steel, motors, and gears with a simple pneumatic system made from an industrial-grade version of the same material used for soft drink bottles, at a tiny fraction of the weight and cost. Another project replaces the giant carbon containment vessels for natural gas storage with an intestine of tiny plastic tubules, allowing natural gas tanks to fit any arbitrary shape as well as reducing the risk of catastrophic rupture. It turns out that when you properly understand the physics, you can indeed replace materials with math.

"In 1660, Robert Hooke described what is now known as Hooke's Law," Saul told me. (Hooke's Law states that the force needed to compress or extend a spring, or to deform a material, is proportional to the distance times the stiffness of the material.) "This meant that we could model all materials as linear springs," Saul continued. "This was important in pre-computer days because it made the math simple when designing trusses and structures to take loads. In the real world, no materials are perfectly linear, and especially not plastics and rubbers. Now we have so much computation available we can design entirely new types of machines and structures that we simply couldn't do the math on before."

The new design capabilities go hand in hand with new manufacturing techniques like 3-D printing. 3-D printing doesn't just provide low-cost prototyping and local manufacturing. It makes possible different kinds of geometries than traditional manufacturing. That requires software that encourages human designers to explore possibilities far afield from the familiar. The future is not just one of "smart stuff," tools and devices infused with sensors and intelligence, but of new kinds of "dumb stuff" made with smart tools and better processes for making that stuff.

Autodesk, the design software firm, is all over that concept. Their next-generation tool set supports what is called "generative design." The engineer, architect, or product designer enters a set of design constraints—functionality, cost, materials; a cloud-based genetic algorithm (a primitive form of AI) returns hundreds or even thousands of possible options for achieving those goals. In an iterative process, man and machine together design new forms that humans have never seen and might not otherwise conceive.

Most intriguing is the use of computation to help design radically new kinds of shapes and materials and processes. For example, Arup, the global architecture and engineering firm, showcases a structural part designed using the latest methods that is half the size and uses half the material, but can carry the same load. The ultimate machine design does not look like something that would be designed by a human.

The convergence of new design approaches, new materials, and new kinds of manufacturing will ultimately allow for the creation of new products as astonishing as the Eiffel Tower was to the world of 1889. Might we one day be able to build the fabled space elevator of science fiction, or Elon Musk's Hyperloop transportation system?

The fusion of human with the latest technology doesn't stop there. Already there are people trying to embed new senses—and make no mistake of it, GPS is already an addition to the human sensorium, albeit still in an external device—directly into our minds and bodies. Might we one day be able to fill the blood with nanobots—tiny machines—that repair our cells, relegating the organ and hip replacements of today, marvelous as they are, to a museum of antiquated technology? Or will we achieve that not through a perfection of the machinist's art but through the next steps in the path trod by Luther Burbank? Amazing work is happening today in synthetic biology and gene engineering.

George Church and his colleagues at Harvard are beginning a controversial ten-year project to create from scratch a complete human genome. Ryan Phelan and Stewart Brand's Revive and Restore project is working to use gene engineering to restore genetic diversity to endangered species, and perhaps one day to bring extinct species back to life. Technologies like CRISPR-Cas9 allow researchers to rewrite the DNA inside living organisms.

Neurotech—direct interfaces between machines and the brain and nervous system—is another frontier. There has been great progress in creating prosthetic limbs that provide sensory feedback and respond directly to the mind. On the further edges of innovation, Bryan Johnson, the founder of Braintree, an online payments company sold to PayPal for $800 million, has used the proceeds to found a company whose goal is to build a neural memory implant as a cure for Alzheimer's disease.

Bryan is convinced that it's time for neuroscience to come out of the labs and fuel an entrepreneurial revolution, not merely repairing damaged brains but enhancing human intelligence.

Bryan is not the only high-profile neurotech entrepreneur. Thomas Reardon, the creator of Microsoft's Internet Explorer web browser, retired from Microsoft to pursue a PhD in neuroscience and in 2016 cofounded a company called CTRL-Labs to produce the first consumer brain-machine interface. As Reardon noted in an email to me, "Every digital experience can and should be controlled by the neurons which deliver the output of your thoughts, those neurons which directly innervate your muscles." This is a brilliant combination of neuroscience and computer science. "The kernel of our work is held in the Machine Learning models which translate biophysical signals—yes, even at the level of individual neurons—to give you control over digital experiences."

Elon Musk joined the parade in 2017 with a company called Neuralink, that is, according to Elon, "aiming to bring something to market that helps with certain severe brain injuries (stroke, cancer lesion, congenital) in about four years." But as Tim Urban, the author of the "Wait But Why" blog, who was given extensive access to the Neuralink team, explains, "[W]hen Elon builds a company, its core initial strategy is usually to create the match that will ignite the industry and get the Human Colossus working on the cause." Proving that a profitable, self-sustaining business can be created in an untried area is a way to get everyone else piling on to the new opportunity. That is, like Bryan Johnson, Elon's vision is not just to build a company, but to build a new industry.

In the case of Neuralink, that new industry is a generalized Brain-Machine interface that would allow humans and computers to interoperate far more efficiently. "You're already digitally superhuman," Elon notes, referring to the augmentation that our digital devices already give to us. But, he notes, our interfaces to those devices are painfully slow—typing on keyboards or even speaking aloud. "We should be able to improve that by many orders of magnitude with a direct neural interface."

These technologies raise questions and fears as profound as any in the world of artificial intelligence. Like other tools of enormous power, they may come into common use through a tumultuous, violent adolescence. Yet I suspect that in the end, we will find ways to use them to make ourselves live longer, happier, more fulfilled lives.

When I was a kid, I read science fiction, a novel a day for years. And for so long, the future was a disappointment to me. We achieved so much less than I had hoped. Yet today, I am seeing progress toward many of my youthful dreams.

And that brings me back to AI. AI is not some kind of radical discontinuity. AI is not the machine from the future that is hostile to human values and will put us all out of work. AI is the next step in the spread and usefulness of knowledge, which is the true source of the wealth of nations. We should not fear it. We should put it to work, intentionally and thoughtfully, in ways that create more value for society than they disrupt. It is already being used to enhance, not replace, human intelligence.

"We've already seen chess evolve to a new kind of game where young champions like Magnus Carlsen have adopted styles of play that take advantage of AI chess engines," notes Bryan Johnson. "With early examples of unenhanced humans and drones dancing together, it is already obvious that humans and AIs will be able to form a dizzying variety of combinations to create new kinds of art, science, wealth and meaning." Like Elon Musk, Bryan Johnson is convinced that we must use neurotech to directly enhance human intelligence (HI) to make even more effective use of AI. "To truly realize the potential of HI+AI," he says, "we need to increase the capacity of people to take in, process, and use information, by orders of magnitude." But even without direct enhancement of human intelligence in the way that Bryan envisions, entrepreneurs are already building on the power of humans augmented by AI.

Paul English, the cofounder of Kayak, the travel search site that helped put many travel agents out of work, has a new startup called Lola, which pairs travel agents with an AI chatbot and a back-end machine learning environment, working to get the best out of both human and machine. Paul describes his goal with Lola by saying, "I want to make humans cool again." He is betting that just as a human chess master paired with a chess computer can beat the smartest chess computer *or* the smartest human grandmaster, an AI-augmented travel consultant can handle more customers and make better recommendations than unaugmented travel agents—or travelers searching for deals and advice on their own using traditional search engines.

The arc between travel agents and Kayak and Lola, the embedding of what was once the specialized knowledge of a travel agent into ever-more-sophisticated tools, teaches us something important. Kayak used automation to replace travel agents with search-enabled self-service. Lola puts humans back into the loop for *better service*. And *when we say "better service," we usually mean "more human, less machinelike service."*

Sam Lessin, the founder and CEO of Fin, an AI-based personal assistant startup, makes the same point: "People in the technology community frequently ask me 'how long will it take to replace the Fin operations team with pure AI?'" he wrote in an email. "At Fin, however, our mission is not automation for its own sake. Our guiding principle is providing the best experience for users of Fin. . . . Technology is clearly part of the equation. But people are also a critical part of the system that results in the best possible customer experience. And the role of technology at Fin is largely to empower our operations team to focus their time and effort on the work that requires decidedly human intelligence, creativity, and empathy."

We are back to Clayton Christensen's Law of Conservation of Attractive Profits. When something becomes a commodity, something else becomes valuable. As machines commodify certain types of human mental labor—the routine, mechanical parts—the truly human contributions will become more valuable.

> Searching out the frontier for enhancing human value is the great challenge for the next generation of entrepreneurs, and for all of society.

In addition to enabling better, more human service, automation can expand access by making other jobs cheap enough to be worth doing.

After receiving what he believed was an unfair parking ticket, Josh Browder, a young British programmer, took a few hours to write a program to protest the ticket. When the ticket was cleared, he realized he could turn this into a service. Since then, DoNotPay, which Josh calls "the Robot Lawyer," has cleared more than 160,000 parking tickets. Josh has since moved on to building a chatbot in Facebook Messenger to automate the application for asylum in the United States, Canada, and the United Kingdom on behalf of refugees.

There are many jobs—like protesting unfair parking tickets—that don't get done because they are too expensive, and making the job cheaper conflicts with the business model of existing companies. Tim Hwang, a programmer who is also trained as a lawyer, told me that when he worked at a law firm, he set out to make himself obsolete. "Every day, I'd get a set of tasks to do, and each night I'd go home and write programs to do them for me the next time I got asked to do them," he said. "I got more and more efficient at doing the work more quickly, and this started to become a problem for the law firm because their business model depends on billable hours. I quit just ahead of getting fired."

ACCESS TO OPPORTUNITY

An Uber or Lyft driver demonstrates two different kinds of augmentation. The first is provided by Google Maps and similar services, which embed knowledge of the layout of city into a tool, so that drivers no longer need to know the city like the back of their hand. Google does that for them. The other augmentation is provided by the Uber or Lyft app itself. This app provides *access to opportunity*, alerting the driver that there are passengers to be picked up, and just where to find them. A real innovation in on-demand applications is the lighter-weight, more flexible methods they provide for matching workers with those who need their services.

Seth Sternberg, the founder of Honor, which matches home care workers with patients, describes better matching as central to what his company does. Unlike Uber, Honor's caregivers are employees of the company, but the need for their services comes and goes. Some caregivers settle into an ongoing relationship with a patient, while others are called on demand for short-term needs. Getting the right match of caregiver and patient is important, Seth told me. It isn't just location that matters but also skills. Some patients might need someone strong enough to lift them; others might need specialized nursing. A platform that helps the workers know in advance what they are getting into creates better, longer-lasting relationships, happier customers, and a more efficient system.

More effective matching is also an essential part of Upwork, the platform for connecting companies with freelancers in categories such as programming, graphic design, writing, translation, search engine optimization, accounting, and customer service. Stephane Kasriel, the CEO of Upwork, pointed out that if you want to understand the dynamics of job marketplaces, there is no better place to do it than on Upwork, because the "velocity of jobs" is so high. A typical job lasts days or weeks rather than years. Stephane told me that there are three kinds of workers on Upwork, and how the job of the platform is different for each of them.

First, Stephane said, are those who already have marketable skills, and good reputations on the platform, and are getting all the work they need because they are "in the flow." The platform doesn't need to do much to help these people.

Second, there are workers who have marketable skills but have not yet built a reputation and are not getting enough work. A lot of the focus of Upwork's internal data science team is to find these people and to point them to the right open jobs. The challenge is not just helping them find a perfect match with the work they have the skills for; often it is pointing them to new areas where there is not enough supply, where some study or retraining will let them get a foothold in the virtuous circle of reputation and recommendation. For example, Stephane pointed out that a few years ago, there were plenty of Java developers, but not enough Android developers, and the best way for people in this second group to get traction in the system (and better pay, since Android was paying more than Java) was to gain new skills.

Today there aren't enough workers with data science skills, and there's a pay premium to be had there.

The third group consists of workers who don't have the right skills for the jobs that they are applying for. Here the right thing to do is to discourage people from applying to the wrong jobs. "The time they spend applying for the wrong jobs is time they could spend working," Stephane told me.

Upwork has developed its own skills assessment system; the company performs 100,000 hours of assessment a month. What's so fascinating about Upwork's assessments is that they are immediately verifiable, because someone either is able to do a job to the satisfaction of the customer, or they aren't. This is in stark contrast to many of the assessment tools sold by education companies, which provide paper certifications but little evidence that workers with those certifications can actually do the job.

All of these points suggest that we may be reaching a tipping point where we escape the shackles of the current labor mindset, and instead rediscover how to use technology to empower and augment workers, finding their strengths and matching them with opportunity, building tools that make it easier and more effective to work together, and creating dynamic labor marketplaces in which on-demand, "high freedom," and the "high velocity" of work go hand in hand.

LEARNING: THE MASTER AUGMENTATION

> One key to understanding the future is to realize that as prior knowledge is embedded into tools, a different kind of knowledge is required to use it, and yet another to take it further. Learning is an essential next step with each leap forward in augmentation.

I've observed this throughout my career educating programmers about the next stages in technology. When I wrote my first computer manual in 1978, for Digital Equipment Corporation's "LPA 11K Laboratory Peripheral Accelerator," it described how to transfer data from high-speed laboratory data acquisition devices using assembly language, the low-level language corresponding closely to the actual machine code that is still hidden deep inside our computers. The directions to the computer had to be very specific: move the data from this device port to that hardware memory register; perform this calculation on it; put the result into another memory register; write it out to permanent storage.

While some programmers still need to delve into assembly language, machine code is usually produced by compilers and interpreters as output from higher-level languages like C, C++, Java, C#, Python, JavaScript, Go, and Swift, which make it easier for programmers to issue broader, high-level instructions. Meanwhile, those programmers in turn are creating user interfaces that allow people who don't even know how to program to invoke powerful capabilities that a few decades earlier were impossible without knowing the exact memory layout and instruction set of the computer.

But even "modern" languages and interfaces are only an intermediate stage. Google, which employs tens of thousands of the most-sought-after software engineers on the planet, is now realizing that they need to retrain those people in the new disciplines of machine learning, which use a completely different approach to programming, training AI models rather than explicit coding. They are doing it not by sending them back to school, but with an apprenticeship.

This highlights a point that I've observed again and again through my career: Technology moves far faster than the education system. When BASIC was the programming language of the early personal computer, programmers learned it from each other, from books, and by looking at the source code of programs shared via user groups. By the time the first classes teaching BASIC appeared in schools, the industry had moved far beyond it. By the time schools were teaching how to build websites with PHP, the bigger opportunity was in building smartphone apps or in mastering statistics and big data.

That lag was key to O'Reilly's success over the past few decades as the publisher of record on emerging technologies. No one was teaching what people needed to know. We had to learn it from each other. All of our bestselling books were created by finding people who were at the edges of innovation and either getting them to write down what they knew, or pairing experts with writers who could extract their knowledge. This led us to document the cutting edge of Linux; the Internet; new programming languages like Java, Perl, Python, and JavaScript; the best practices of the world's leading programmers; and more recently, big data, DevOps, and AI. When, in 2000, our ad on the cover of *Publishers Weekly* baldly stated, "The Internet Was Built on O'Reilly Books," everyone accepted it as the simple truth.

As the pace of technology has increased, bringing people together at live events became a more important part of our work. We also built a knowledge-sharing platform that allows anyone with unique technology or business skills to teach them to our customers. The platform, which we called Safari as a homage to the nineteenth-century woodcuts of animals that graced the covers of our books, now includes tens of thousands of ebooks from hundreds of different publishers, not just our own, plus thousands of hours of video training, learning paths, learning environments with integrated text, video, and executable code, and live online events with leading experts teaching cutting-edge techniques.

One of the big changes in our business is that technologies that were once the realm of adventurers at the edges of innovation have moved into the mainstream. Fortune 500 companies, not just individual programmers or small startups, have to learn at the pace at which technology itself evolves. What we do is in a period of profound transformation, but I know that whatever techniques and delivery methods we use for new knowledge, some things will remain constant:

People need a base—knowing enough to ask the right questions and to take in new knowledge.

People learn from each other.

People learn best by doing, solving real problems and pulling in the knowledge they need on demand.

People learn best when what they are doing is so compelling that they want to do it on their own time, not just because the job asks them to do it.

TECHNOLOGY ON YOUR OWN TIME

When we launched the first issue of *Make:* magazine in January 2005, the cover story featured Charles Benton, who'd built a rig so he could take aerial photos from a kite, before GoPro had shipped its first action camera and long before drone video was a gleam in anyone's eye; another story described how to make a homebrew videocam stabilizer. Yet another explained how Natalie Jeremijenko had added sensors to Sony's AIBO robotic dog so it could be used to sniff out toxic waste. A fourth story included plans for building a device that would let you see just what information was stored on the magnetic stripe of a credit card or hotel room key.

Dale Dougherty, who conceived *Make:*, had been struck by the fact that early issues of magazines such as *Popular Mechanics* were very different from their modern equivalents. The modern versions were whiz-bang tours of technology products you could *buy*. Forty years ago they were full of projects that you could *do*.

Go back to the days of the Wright Brothers and you'll find how-to books like *The Boy Mechanic*. You couldn't buy an airplane, but you could dream of building one.

> This design pattern, that the future is built before it can be bought, is an important one to recognize. The future is created by people who can make and invent things and those who can tinker and improve and put inventions into practice. These are people who learn by doing.

In a later issue of *Make:*, Dale published an "Owner's Manifesto," which opened with the words, "If you can't open it, you don't own it." The truth of that statement has been proven many times since, as companies have increasingly used "Digital Rights Management" software to drive up profit by locking in customers, denying them the right to repair or even resupply the devices they nominally own. Printers, coffee-makers, and most recently high-tech tractors and other farm equipment have been the locus of battles between companies and their customers over who controls products that the consumers nominally own.

But it wasn't just the power grab represented by DRM and sealed hardware that you can't open without special tools, or are forbidden to service under the terms of a shrink-wrapped license agreement, that bothered Dale and the makers he represents. The idea was that if we really want to have mastery over our tools, we have to be able to get inside them, understand how they work, and modify them.

When you get a smartphone, a tablet, or a computer today, you get a slick computer product that's been designed to be easy to use, but is difficult to modify or repair. It wasn't like that for those of us who began working with computing in the 1970s and 1980s (or even earlier). We started with something relatively primitive, a blank slate that we had to teach to do anything useful. That teaching is called programming. Only a small number of the billions of people who own a smartphone today know how to program; back then, with a few limited exceptions, a computer wasn't very useful at all unless you learned to program it yourself.

We taught ourselves how to program by solving problems. Not random, artificial exercises to teach us programming. Real problems that we needed to solve. Since Dale and I were writers, this meant creating programs to help us write and publish—editing scripts to enforce consistency of terminology across a documentation set, to correct common grammar mistakes, to build an index, or to format and typeset a manuscript. We got so good at doing this that we wrote a book together called *Unix Text Processing*. And we put our newfound skills to work building a publishing company that could send a book to the printer within days after the author and editor finished working on it, rather than waiting months, as is still common with most traditional publishers.

Unix was a creature of the odd transition between the proprietary hardware systems of the first age of computing and the commodity PC architectures of the second. It was designed to be a portable software layer across a wide range of computers with different hardware designs. And so, whenever we heard about an interesting new program, we couldn't just download it and run it, we often had to "port it" (modify it so it would run on the type of computer we were using). And because every computer had a programming environment, we could easily add our own custom software. When we started publishing and selling books via mail order in 1985, I didn't buy an order-entry and accounting system; I wrote my own.

When we discovered the web, building things got even more fun. Because the web had been designed to format online pages using a text markup language—HTML, the Hypertext Markup Language—it played right to our strengths. HTML meant that whenever you saw a neat new feature on a web page, you could pull down a menu, select View Source, and see how the trick was done.

The early web was very simple. Clever new "hacks" were introduced all the time, and we gleefully copied them from each other with abandon. Someone came up with a clever solution; it quickly became the common property of everyone else who had the same problem.

In our early days at O'Reilly, we'd written documentation for hire. But we soon realized that there was an enormous opportunity in following the explosive wavefront of innovation, documenting technologies that were just being invented, capturing the knowledge of the people who were learning by doing, because they were doing things that had never been done before.

Emulation is key to learning. In our early years, we used to describe our books as trying to re-create the experience of looking over the shoulder of someone who knew more than you did, and watching how they worked. This was an important attraction of open source software. Back in 2000, when the software industry was trying to come to grips with this new idea, Karim Lakhani, then at MIT's Sloan School of Management, and Robert Wolf of the Boston Consulting Group did a study of motivations of people working on open source software projects. What they found was that along with adapting software to meet their own specialized needs, learning and the sheer joy of intellectual exploration were more important than traditional motivators like higher salaries or career success.

Dale recognized this pattern playing out again in the world of new kinds of hardware. Cheap sensors, 3-D printers, and lots of old, disposable hardware waiting for creative reuse meant that the physical world was starting to experience the same kind of malleability that we'd long associated with software. But in order to take advantage of that opportunity, people had to be able to take things apart and put them back together in new ways.

That is the essence of the Maker movement. Making for the joy of exploration. *Making to learn.*

There's no joy in our current education system. It is full of canned solutions to be memorized when it needs to be a vast collection of problems to be solved. When you start with what you want to accomplish, knowledge becomes a tool. You seek it out, and when you get it, it is truly yours.

Stuart Firestein, in his book *Ignorance*, makes the case that science is not the collection of what we know. It is the practice of investigating what we don't know. Ignorance, not knowledge, drives science.

There's also an essential element of play in both science and learning. In his autobiography, physicist Richard Feynman described the origin of the breakthrough that led to his Nobel Prize. He was burned-out and found himself unable to concentrate on work. Physics was no longer fun. But he remembered how it used to be. "When I was in high school, I'd see water running out of a faucet growing narrower, and wonder if I could figure out what determines that curve," he wrote. "I didn't have to do it; it wasn't important for the future of science; somebody else had already done it. That didn't make any difference. I'd invent things and play with things for my own entertainment."

So Feynman resolved to go back to having fun, and stop being so goal-driven in his research. Within a couple of days, while watching someone in the Cornell University cafeteria spinning a plate in the air, he noticed that the wobbling rim of the plate was rotating faster than the university logo in its center. Just for fun, he began to calculate the equations for the rate of spin. Bit by bit he realized that there were lessons for the spin of electrons, and before long, he was deep into the work that eventually became known as quantum electrodynamics.

This is true in corporate learning as well. I remember a powerful conversation with David McLaughlin, director of developer relations at

Google. We had both agreed to speak at a technology advisory meeting for a huge software firm. The company wanted to know how to get more developers for its platform. David asked a key question: "Do any of them play with it after work, on their own time?" The answer was no. David told them that until they fixed that problem, reaching out to external developers was wasted effort.

The importance of fun to learning was the source of Dale's original subtitle for *Make:* magazine, "Technology on your own time." That is, this is stuff you want to do even though no one is asking you to. In 2006, we followed up the magazine with Maker Faire, a vast "county fair with robots" that now draws hundreds of thousands of people each year. It is packed with kids eager to learn about the future, and parents rediscovering the wonder of learning.

We have far too little fun in most formal learning, and people are hungry for it. If you can't inspire curiosity, chances are you are on the wrong path.

THE POWER OF PULL

Once you have curiosity, the Internet has provided powerful new ways to feed it. In their book, *The Power of Pull*, John Hagel III, John Seely Brown, and Lang Davison outline a fundamental change in the nature of twenty-first century learning. The book opens with the story of a group of young surfers, on the brink of becoming professional competitors, who improved their surfing skills by creating, watching, and analyzing videos of themselves surfing, and by comparing themselves to surf footage of experts available online. They posted their own footage to YouTube, and as their skills grew, they were discovered by sponsors and invited to competitions.

This combination of learning by doing, social sharing, and on-demand expertise is central to how people—especially young people—learn today. Brit Morin, the founder and CEO of millennial lifestyle site *Brit + Co*, explains that "I'm beginning to feel like I'm no longer part of the popular crowd at school." Marketers, she says, are now obsessed with what they call "Generation Z"—fourteen- to twenty-four-year-olds. This is a generation that doesn't remember a time when you couldn't use the Internet to look up anything you want. She notes that "sixty-nine percent of them say they go to YouTube to learn 'just about everything' and prefer it as a learning mechanism far beyond teachers or textbooks."

That's certainly my experience with my thirteen-year-old step-daughter. Recently we were having some guests over to the house for a business dinner. "Can I make dessert?" she asked. We agreed, not quite sure what to expect. What we got was astonishing, worthy of a high-end restaurant. Ice cream with a scattering of berries in perfect, eggshell-thin dark chocolate cups.

"How did you do that?" I asked.

"I melted the chocolate, then formed it on balloons." She'd learned how to do it on YouTube. She is not someone who has spent years learning to cook either. She got interested when a friend of hers competed on the Food Network kids-baking reality TV show. She started watching food videos and duplicating them in the kitchen. This was one of her first efforts.

The power of on-demand access to information is the key to the next generation of learning. Those concerned about technology and the future of work should take note. So too the switch to short snippets of video as a preferred learning mechanism. More than 100 million hours of how-to videos were watched on YouTube in North America during the first four months of 2015.

"Employers must recognize this change and begin valuing skills/competencies that are learned in nontraditional ways," says Zoë Baird of the Markle Foundation, who led the Rework America initiative with Starbucks CEO Howard Schultz. "The key is embracing skills-based hiring and employment practices. Too many employers use a four-year degree as a proxy for hiring, even for jobs that don't require one." She pointed out to me that a majority of the jobs projected to have the largest growth rates in America through 2024 don't require a bachelor's degree. If that is the case, we surely need to transform our outdated labor market into one that values skills. The Markle Foundation, LinkedIn, the state of Colorado, Arizona State University, and others have been working to address this over the past year through Skillful, an effort to transform America's outdated labor market to reflect the needs of the digital economy.

There's another important point to add. Access to an unlimited world of information is a powerful augmentation of human capability, but it still has prerequisites. Before she could learn how to make an exquisite dessert by watching a YouTube video, my stepdaughter had to know how to use an iPad. She had to know how to search on YouTube. She had to know that a world of content was there for the taking. At O'Reilly, we call this *structural literacy*.

Users without structural literacy about how computers work struggle to use them. They learn by rote. Going from an iPhone to Android, or the reverse, or from PC to Mac, or even from one version of software to another, is difficult for them. They aren't stupid. These same people have no trouble getting into a strange car and orienting themselves. "Where is that darned lever to open the gas cap?" they ask. They know it's got to be there somewhere. Someone with structural literacy knows what to look for. They have a functional map of how things ought to work. Those lacking that map are helpless.

When I used to personally write and edit computer books, the first chapter was always designed to provide a kind of structural literacy about the topic. My goal was for readers of that first chapter to understand the topic well enough that they could drop in anywhere in the book, looking for a specific piece of information, and have enough context to find their way around and understand what they come across.

The level and type of structural literacy required differs with the type of work you do. Today's startups, increasingly embedding software and services into devices, require foundational skills in electrical and mechanical engineering, and even "trade" skills such as soldering. An experienced software developer today probably needs to up his or her game with regards to tensor calculus in order to work with machine learning algorithms. Teachers are far more effective if they are broadly familiar with the culture and context of their students.

One of the problems with many online learning platforms for teaching new technology is that structural literacy is all they provide.

They are good for teaching beginners who know nothing about a topic—getting them to structural literacy about programming by teaching them JavaScript, for example, or providing a course on digital marketing—when what people need next is just-in-time learning about very specific topics.

We had a telling experience with one of our Safari customers, a large international bank, when it came time to renew their annual subscription. "No need to pitch us," they said. "We had a failure in one of our systems, and we found the documentation we needed in Safari, averting millions of dollars in losses." Pat McGovern, the founder of technology media giant IDG, once told me that his working principle was that as technology advances, "the specific drives out the general."

In the end, on-demand education is not that dissimilar from on-demand transportation. You need a rich marketplace of people who know things, and others who need to know them. The way that knowledge is delivered—book, video, face-to-face teaching—gets a lot of attention, but the bigger question is how to bootstrap a rich *knowledge network*.

AUGMENTED REALITY AND THE FUTURE OF ON-DEMAND LEARNING

If being able to search for instructions on YouTube or on a specialized platform like Safari is the heart of today's on-demand learning, augmented reality is surely tomorrow's. Aircraft mechanics at Boeing are engaged in a pilot project using Microsoft HoloLens to give them schematics and diagrams overlaid on the work they are doing, guiding them through complex tasks that otherwise would take years of experience to master. At various architectural firms, architects and their clients equipped with augmented or virtual reality are stepping into their own models, modifying them, and seeing what they wish to build before they actually create anything in the physical world.

Despite the much-publicized failure of Google Glass and the premature hype around virtual reality platforms such as Oculus Rift, there is plenty of evidence that augmented reality and virtual reality will have a powerful impact in on-demand learning. Smartphones and tablets alone are already being used effectively in areas like telehealth and shop floor communications and on-the-job training, and with Microsoft's investment in HoloLens, continued experiments like Snap's Spectacles, and rumored new products from Apple, not to mention that a next generation of Google Glass is likely still under development, I'm confident that there will be plenty of news on this front.

> Once you understand that a trend is happening, you can watch it unfold. Your mental map cues you to be alert to signs that it is gaining steam, and to explore ways that it can be applied.

You can start looking for and tracking interesting news, like the $200 head-mounted augmented reality display for infantry soldiers demonstrated at a DARPA event in 2015, or the deep commitment Microsoft has made to human augmentation of all kinds as a key part of its corporate strategy.

LEARNING AND SOCIAL CAPITAL

There's a deeper economic story too, one that has been explored by James Bessen in his book *Learning by Doing*. He attempts to answer the question "Why does it take so long for the productivity advances from new technology to show up in people's wages?" Looking at the history of the nineteenth-century cotton mills in Lowell, Massachusetts, as well as the introduction of modern digital technology, he comes to the conclusion that our traditional narrative about innovation is wrong. The bulk of the gains in productivity come over time, as innovations are implemented and put into practice.

Bessen describes how major innovations, such as the introduction of the steam mill, involve both de-skilling and up-skilling, the replacement of one set of skills with another. It is mythology, he notes, that automation replaced skilled crafters with unskilled workers. In fact, by measuring the productivity difference between beginners and fully competent crafters and doing the same for workers in the new factories,

it is possible to determine that in the 1840s, it took a full-year investment in training for either to reach full productivity. Using training
time as a proxy for skill, it is clear that they were equally but differently
skilled.

The new skills, Bessen notes, were not the result of schooling.
"They were mostly learned on the factory floor," he notes. This continues today. "Economists' common practice of defining 'skilled workers'
as those with four years of college is particularly misleading," he writes.
"The skill needed to work with a new technology often has little to do
with the knowledge acquired in college."

That was certainly true of me. I studied Greek and Latin in college. Everything I learned about computers, I learned on the job. The
knowledge I learned in college was useless to me. The habits of mind
that I formed were what mattered, the foundational skills of study, and
particularly the ability to recognize patterns. The struggle to parse
complex Greek texts that were, quite frankly, beyond my skill in the
language was great preparation when I took on the challenge of documenting programs written in programming languages that at first I
barely understood. It is not just knowledge that we have to teach, it is
the ability to learn. To learn constantly. Over the course of my career,
learning itself has been the most important part of my ongoing work.

The struggle to find work, which affects far too many people in our
economy, has many causes, but if there is one solution that anyone can
take into his or her own hands, it is the power to learn. It is the one
essential skill we must teach our children if they are to adapt to a constantly changing world. A broad general education and love of learning
may be more important than specific skills that will soon be out of date.

During the industrial revolution, the new generation of workers was
surprisingly well educated. Bessen notes that when Charles Dickens
visited the mills in Lowell in 1842, he "reported several 'surprising
facts' back to his English readers: the factory girls played pianos, they
nearly all used circulating libraries, and they published quality periodicals."

People typically entering the new workforce were less productive
at first, and there was no pool of experienced workers to draw from.

Turnover was high as people tried out the new style of work, and not all of them succeeded. The machine mills and looms didn't become truly productive for decades after their introduction. Bessen explains that "what matters to a mill, an industry, and to society generally is not how long it takes to train an individual worker, but what it takes to create a stable, trained labor force." This is also exactly what I've observed in my own career.

> The skills needed to take advantage of new technology proliferate and are developed over time through communities of practice that share expertise with each other. Over time, the new skills are routinized and it becomes easier to train lots of people to exercise them. It is at that point that they begin to affect productivity and improve the wages and incomes of large numbers of people.

Part of the secret of Silicon Valley's success, so difficult to replicate elsewhere, has been that there is a large pool of people who have the necessary skills to go to work at virtually any high-tech company and get productive fairly quickly. This concentrated labor force is not yet available everywhere. As the necessary knowledge penetrates society, though, we can expect the achievements of Silicon Valley to be both more replicable and less remarkable. The unicorn will be fading into the ordinary.

Writing in *Wired*, Clive Thompson asks a provocative question: Is coding becoming a blue-collar job? "These sorts of coders won't have the deep knowledge to craft wild new algorithms for flash trading or neural networks," he writes. "But any blue-collar coder will be plenty qualified to sling JavaScript for their local bank." As coding becomes routinized, the educational needs of those practicing it become less demanding. For many types of programming, people need the equivalent of vocational training rather than an advanced software engineering or math degree. And that's exactly what we see with the rise of coding academies and boot camps.

But there's more to it than that. The rise of the web didn't just require (and reward) people with the skills of programming. As the technology matured, it also called into being entirely new jobs. An early "webmaster" was the jack-of-all-trades, from programming and system administration to web design. But before long, a successful website needed specialized designers, front-end developers whose skills combined programming and design, back-end developers with deeper experience in databases, experts in search engine optimization and social media, and much, much more. The expertise embodied in a successful media website of 2016, like BuzzFeed, is radically different from the expertise at Yahoo! in 1995. As technology penetrates every sector of our society, it will create many more specialized jobs.

Ryan Avent, the author of *The Wealth of Humans*, has a further insight, that the success of new technology depends on social capital, which he describes as "contextually dependent know-how, which is valuable when shared by a critical mass of people." He distinguishes this from the concept of human capital, which includes skills and knowledge that are not especially context dependent, and can belong to a single person. (It is also distinct from the notion of social capital as originally defined by Glenn Loury and James Coleman or popularized by Robert Putnam. For Loury and Coleman, it is the networks, whom we know, and how we can use our networks as resources, rather than the "know-how," and for Putnam, how these networks are strengthened by civic engagement. But Avent's usage overlaps profoundly: It is only when there is a substantial network of people with shared knowledge that a technology can really take hold in the economy.)

Anyone who has visited Google and seen the flyers in the bathroom stalls with names like "Testing on the Toilet" and "Learning on the Loo," with many of the weekly updates focused on how to use Google's internal systems, understands how even at a firm so rich in expertise there is a constant need to educate people about the specialized, context-specific knowledge of how Google itself operates.

This kind of social capital is key to the shared expertise that differentiates firms. Describing his work as a senior editor at the *Economist*, Avent notes: "The general sense of how things work lives in the heads of long-time employees. That knowledge is absorbed by newer employees over time, through long exposure to the old habits. What our firm is, is not so much a business that produces a weekly magazine, but a way of doing things consisting of an enormous set of processes. You run that programme, and you get a weekly magazine at the end of it."

But, Avent continues, "[T]he same internal structures that make production of the print edition so magically efficient hinder our digital efforts." And, he notes that "simply bringing in tech-savvy millennials isn't enough to kick an organization into the digital present; the code must be rewritten." (That is, of course, the central lesson of Amazon's platform transformation as well.) One of the critical roles of the entrepreneur, Avent adds, is to create space for new ways of doing things. This is true within existing firms as well as at startups.

The process of integrating new technology into business and society is far from over. New skills are proliferating faster than they can be learned in any school. Meanwhile, the advantages accruing to firms from new technology are deeply wrapped up in their ability to train their workforce and change their workflows to accommodate it.

This retraining was central, for example, to former IBM CIO Jeff Smith's attempt to transform IBM's internal software development culture to one that mirrors the agile, user-centered, data-driven, and cross-functional approach that characterizes today's Silicon Valley startups. Except that instead of doing it at a startup, he was doing it for a software development team of 20,000 people, in support of a company with more than 400,000 employees.

Laura Baldwin, president and COO at O'Reilly Media, tells our customers, "You have to go to war with the army you have." Yes, it's essential to bring in new talent with the latest skills, but retraining your existing team and building new ways for people to work together is also essential.

The presence of a stable, trained workforce is not something to be achieved and then taken for granted. The mill owners of Lowell invested in their workforce; the decisions in America over the past decades to ship manufacturing jobs overseas have effectively been a commitment to de-skilling without re-skilling. As new small-batch techniques now make manufacturing cost-effective in America again, the necessary skilled labor force is missing. According to a 2015 study by Deloitte and the Manufacturing Institute, more than two million manufacturing jobs will go unfilled over the next decade. Even if China's costs rise to match those of the United States, the United States would not be competitive without a major investment in manufacturing skills development.

A lot of companies complain that they can't hire enough people with the skills they need. This is lazy thinking. Graham Weston, the cofounder and chairman of managed hosting and cloud computing company Rackspace, based in San Antonio, Texas, proudly showed me Open Cloud Academy, the vocational school his company founded to create the workforce he needs to hire. He told me that Rackspace hires about half of the graduates; the rest go to work in other Internet businesses.

At the speed with which technology changes today, we can expect the traditional education establishment to provide a foundation, but it will be the job of every company that wants to succeed to invest in the unique and ever-changing skills of its workforce. Our education system must be rethought for a world of lifelong learning. If Bessen is right, it is not just technology innovation, but the diffusion of knowledge about how to use that technology through society that makes a difference in making us all richer. Accelerating that diffusion is one of the most important ways we can work to create a better future.

16

WORK ON STUFF THAT MATTERS

WHEN CLAYTON CHRISTENSEN INTRODUCED THE TERM *DISruptive technology* in his 1997 business classic, *The Innovator's Dilemma*, he asked a very different question than "How can I get funded by convincing VCs that there's a huge market I can blow up?" He wanted to know why existing companies fail to take advantage of new opportunities. He discovered that breakthrough technologies that are not yet mature first succeed *by finding radically new markets*, and only later disrupt existing markets.

When I first met Clay in person, at Matt Asay and Bryce Roberts's Open Source Business Conference in 2004, he retold the story of how RCA had spent billions of dollars in current value trying without success to make the radio and television sound quality of transistors as good as that of vacuum tubes. Sony's brilliant business innovation wasn't to improve the transistor—that came later. It was to find a market—portable radios, initially only for young people—where the quality didn't matter as much as low price and the previously unattainable possibility of a radio that you could carry around with you.

The point of a disruptive technology is not the market or the competitors that it destroys. It is the new markets and the new possibilities that it creates. Just like transistor radios or the early World Wide Web, these new markets are often too small for established companies to consider them worth pursuing. By the time they wake up, an upstart has taken a leadership position in the emerging segment.

This was true of Microsoft, of Google, Facebook, and Amazon, and it is also true of current disruptors like Uber, Lyft, and Airbnb or the researchers who are taking us pell-mell into a future of self-driving cars and other applications of artificial intelligence. They started out trying to solve a problem.

I spend a lot of time urging Silicon Valley entrepreneurs to forget about disruption, and instead to work on stuff that matters. What do I mean by that? There are a number of litmus tests that I've learned by watching innovators in science and in open source software and the Internet, and that I try to pass on to young entrepreneurs. Here's what I like to tell them.

1. WORK ON SOMETHING THAT MATTERS TO YOU MORE THAN MONEY.

Remember that financial success is not the only goal or the only measure of achievement. It's easy to get caught up in the heady buzz of making money. You should regard money as fuel for what you really want to do, not as a goal in and of itself.

> Money is like gas in the car—you need to pay attention or you'll end up on the side of the road—but a successful business or a well-lived life is not a tour of gas stations.

Whatever you do, think about what you really value. If you're an entrepreneur, the time you spend thinking about your values will help you build a better company. If you're going to work for someone else, the time you spend understanding your values will help you find the right kind of company or institution to work for, and when you find it, to do a better job.

Don't be afraid to think big. Business author Jim Collins said that great companies have "big hairy audacious goals." Google's motto, "access to all the world's information," is an example of such a goal. I like to think that my own company's mission, "changing the world by spreading the knowledge of innovators," is also such a goal. Nick Hanauer likes to say, "Solve the biggest problem you can."

> Pursue something so important that even if you fail, the world is better off for you having tried.

There's a wonderful poem by Rainer Maria Rilke that retells the biblical story of Jacob wrestling with an angel, being defeated, but coming away stronger from the fight. It ends with an exhortation that goes something like this: "What we fight with is so small, and when we win, it makes us small. What we want is to be defeated, decisively, by successively greater beings."

One test of a bubble is how many entrepreneurs are focused on their upcoming payday rather than on the big things they hope to accomplish. Me-too products are almost always payday-focused; the entrepreneurs who first made the market often had much less expectation of easy success, and were instead wrestling, like Jacob with the angel, with a hard problem that they weren't even sure that they could solve, but that they believed they could at the very least make a dent in. Those who follow are too often just trying to cash in.

The most successful companies treat success as a by-product of achieving their real goal, which is always something bigger and more important than they are. Satya Nadella, the CEO of Microsoft, makes the same point when talking about the opportunity for AI. "The challenge will be to define the grand, inspiring social purpose for which AI is destined," he writes. "In 1961, when President Kennedy committed America to landing on the moon before the end of the decade, the goal was chosen in large part due to the immense technical challenges it posed and the global collaboration it demanded. In similar fashion, we need to set a goal for AI that is sufficiently bold and ambitious, one that goes beyond anything that can be achieved through incremental improvements to current technology."

When I asked Satya for an example of what he meant, he spoke movingly of his disabled son. "I have a special needs kid, and he's locked in, and so I always think, 'Wow, if only he could speak.' And I think about what a brain-machine connection could do. Someone who's got visual impairment could see or someone who's got dyslexia could read. This is finally that technology that truly brings inclusiveness."

Former Google executive Jeff Huber is also chasing this kind of bold dream of using technology to make transformative advances in healthcare. Jeff's wife died unexpectedly of an aggressive, undetected cancer. After doing everything possible to save her and failing, he committed himself to making sure that no one else has that same experience. He has raised more than $100 million from investors in the quest to develop an early-detection blood test for cancer. That is the right way to use capital markets. Enriching investors, if it happens, will be a by-product of what he does, not his goal. He is harnessing all the power of money and technology to do something that today is impossible. The name of his company—Grail—is a conscious testament to the difficulty of the task. Jeff is wrestling with the angel.

2. CREATE MORE VALUE THAN YOU CAPTURE.

It's pretty easy to see that a financial fraud like Bernie Madoff wasn't following this rule, and neither were the titans of Wall Street who ended up giving out billions of dollars in bonuses to themselves while wrecking the world economy. But most businesses that prosper do create value for their communities and their customers as well as themselves, and the most successful businesses do so in part by creating a self-reinforcing value loop with and for others. They build or are part of a platform on which people who don't work directly for them can build their own dreams.

Investors as well as entrepreneurs must be focused on creating more value than they capture. A bank that loans money to a small business sees that business grow, perhaps borrow more money, hire employees who make deposits and take out loans, and so on. An investor who bets on the future of an unproven technology can do the same. The power of this cycle to lift people out of poverty has been demonstrated for centuries.

If you're succeeding at the goal of creating more value than you capture, you may sometimes find that others have made more of your ideas than you have yourself. *It's okay.* I've had more than one billionaire (and an awful lot of startups who hope to follow in their footsteps) tell me how they got their start with a couple of O'Reilly books. I've had entrepreneurs tell me that they got the idea for their company from something I've said or written. *That's a good thing.* I remember back in the early days of the Internet, when Carla Bayha, the computer book buyer at Borders, told me after one of my talks, "Well, you've just given your competitors their publishing program for the year."

> If my goal is really "changing the world by spreading the knowledge of innovators," I'm thrilled when my competitors jump on the bandwagon and help me spread the word.

Look around you: How many people do you employ in fulfilling jobs? How many customers use your products to make their own living? How many competitors have you enabled? How many people have you touched who gave you nothing back?

There's a wonderful section in Victor Hugo's brilliant, humane novel *Les Misérables* about the good that his protagonist Jean Valjean does as a businessman (operating under the pseudonym of Father Madeleine since he is an escaped convict). Through his industry and vision, he makes an entire region prosperous, so that "there was no pocket so obscure that it had not a little money in it; no dwelling so lowly that there was not some little joy within it." And the key point: "Father Madeleine made his fortune; but a singular thing in a simple man of business, it did not seem as though that were his chief care. He appeared to be thinking much of others, and little of himself."

Focusing on solving big problems rather than on making money, and focusing on creating more value than you capture, are closely related principles. The first one is a test that applies to those starting something new; the second is the harder test that you must pass in order to create something enduring.

3. TAKE THE LONG VIEW.

The musician Brian Eno tells a story about the experience that led him to conceive of the ideas that evolved into the Long Now Foundation, a group that works to encourage long-term thinking. In 1978, Brian was invited to a rich acquaintance's housewarming party, and as the neighborhood his cab drove through became dingier and dingier, he began to wonder if he was in the right place. "Finally [the cab driver] stopped at the doorway of a gloomy, unwelcoming industrial building," he writes. "Two winos were crumpled on the steps, oblivious. There was no other sign of life in the whole street."

But he was at the right address, and when he stepped out on the top floor, he discovered a multimillion-dollar palace.

"I just didn't understand," he explains. "Why would anyone spend so much money building a place like that in a neighbourhood like this? Later I got into conversation with the hostess. 'Do you like it here?' I asked. 'It's the best place I've ever lived,' she replied. 'But I mean, you know, is it an interesting neighbourhood?' 'Oh—the neighbourhood? Well . . . that's outside!' she laughed."

In the talk many years ago where I first heard him tell this story, Brian went on to describe the friend's apartment, the space she controlled, as "the small here," and the space outside, full of winos and derelicts, as "the big here." He went on from there, along with others, to come up with the analogous concept of the Long Now. We need to think about the long now and the big here, or one day our society will enjoy neither.

It's very easy to make local optimizations, but they eventually catch up with you. Our economy has many elements of a Ponzi scheme. We borrow from other countries to finance our consumption, and we borrow from our children by saddling them with debt, using up nonrenewable resources, and failing to confront great challenges in income inequality, climate change, and global health.

Every new company trying to invent the future has to think long-term. What happens to the suppliers whose profit margins are squeezed by Walmart or Amazon? Are the lower margins offset by higher sales or do the suppliers faced with lower margins eventually go out of business or lack the resources to come up with innovative new products? What happens to driver income when Uber or Lyft cut prices for consumers in an attempt to displace competitors? Who will buy the products of companies that no longer pay workers to create them?

Walter Reuther, the pioneer UAW organizer, told the story of a conversation with a Ford executive who was showing Reuther his new factory robots. "How are you going to collect union dues from all these machines?" he asked. Reuther said he replied, "You know, that is not what's bothering me. I'm troubled by the problem of how to sell automobiles to them." The question of who will have the money to buy tomorrow's products in an increasingly automated world should be central to every entrepreneur's thinking.

It's essential to get beyond the idea that the only goal of business is to make money for its shareholders. I'm a strong believer in the social value of business done right. We should aim to build an economy in which the important things are a natural outcome of the way we do business, paid for in self-sustaining ways rather than as charities to be funded out of the goodness of our hearts. Pierre Omidyar, the founder of eBay who went on to become a pioneer in what is now sometimes called "West Coast philanthropy," which uses both traditional charitable giving and strategic startup investing as tools toward the same social goals, once told me, "I invest in businesses that can only do well by doing good."

Whether we work explicitly on causes and the public good, or work to improve our society by building a business, it's important to think about the big picture, and what matters not just to us, but to building a sustainable economy in a sustainable world.

4. ASPIRE TO BE BETTER TOMORROW THAN YOU ARE TODAY.

I've always loved the judgment of Kurt Vonnegut's novel *Mother Night*: "We are what we pretend to be, so we must be careful about what we pretend to be." This novel about the postwar trial of a Nazi propaganda minister who was secretly a double agent for the Allies should serve as a warning to those (politicians, pundits, and business leaders alike) who appeal to people's worst instincts but console themselves with the thought that the manipulation is for a good cause.

But I've always thought that the converse of Vonnegut's admonition is also true: Pretending to be better than we are can be a way of setting the bar higher, not just for ourselves but for those around us.

People have a deep hunger for idealism. The best entrepreneurs have the courage that comes from aspiration, and everyone around them responds to it. Idealism doesn't mean following unrealistic dreams. It means appealing to what Abraham Lincoln so famously called "the better angels of our nature."

That has always been a key component of the American dream: We are living up to an ideal. The world has looked to us for leadership not just because of our material wealth and technological prowess, but because we have painted a picture of what we are striving to become.

If we are to lead the world into a better future, we must first dream of it.

DEVELOPING A ROBUST STRATEGY

The future is fundamentally uncertain. No matter how hard we try to map the future, we will be surprised. As Hamlet said, "The readiness is all."

Fortunately, there's actually a management discipline designed specifically to address this issue. It's called scenario planning. Scenario planning takes for granted that the future is uncertain. But it also notes that there are deep trends shaping the future that we can observe and take into account. Some of them are fairly certain—population growth and demographics, for instance, or for many years, technological trends like Moore's Law—while others, such as political elections, technology innovation, and terrorist attacks, constantly surprise us.

Even in the areas where we are surprised, in retrospect we often realize we could have seen changes coming. World War I followed what was widely regarded as "the perfect summer" at the height of the British Empire's success. A mad assassin lit the fuse but the kegs of powder had been set in place by decades of bad decisions by the great powers.

The near collapse of the world economy in 2008 as a result of financial industry excesses happened while Ben Bernanke, an expert on the 1929 stock market crash and its aftermath, who should have known better, was chairman of the Federal Reserve.

Scenario planning takes for granted that it is hard for human beings to imagine the future as being radically different from the present. As a result, its practitioners don't try to predict the future; they work to prepare companies and countries to develop "robust strategies" that work in the face of radically different futures.

The goal is not to identify what *will* happen, but to stretch the mind to think about what *might* happen. A scenario-planning exercise therefore asks participants to imagine four radically different futures that could come about as a result of current trends. As Peter Schwartz, one of the originators of the technique, wrote in the introduction to his book about it, *The Art of the Long View*, the scenario is "a vehicle . . . for an imaginative leap into the future."

The first step is to identify some key vectors that may influence the future. Remember, a vector is defined in mathematics as a quantity that can only be fully described by both a magnitude and a direction.

It's worth noting that both velocity and acceleration are vectors. But velocity is the speed that something is going in a particular direction, while acceleration is the rate of increase in speed. Trends that are accelerating are especially worth taking note of. One mistake many entrepreneurs and investors make is to look at the size of something, decide that it's "big" or inevitable, and go all in on it. But of course, it's often much more useful to recognize something when it's small, and growing fast.

There are many trends that are big and inevitable, but growing more slowly than the entrepreneurial time horizon. Others are growing too fast. That's why one of the measures that we've tried to use at O'Reilly Media in looking at emerging technologies or other trends is the rate of change. A robust strategy has to take your own resources and time horizon into account. Sunil Paul was a victim of this problem. He correctly identified a huge opportunity, but at first it wasn't happening fast enough, and then later, it was moving so fast he couldn't catch up to it.

Bill Gates once wrote, "We always overestimate the change that will occur in the next two years and underestimate the change that will occur in the next ten. Don't let yourself be lulled into inaction." That was true of Microsoft itself. Despite Gates's warning (in the 1996 afterword to the revised edition of his book *The Road Ahead*, which had been updated to repair the omission of the Internet from the first edition a year earlier) and despite massive efforts to catch up, Microsoft missed the Internet wave and was surpassed by companies with radically new technology and business models.

In a scenario-planning exercise, the vectors are drawn in such a way that they cross each other and divide the possibility space into quadrants. Those quadrants are the basis for four scenarios, typically developed over a period of days by a small group of company executives, military planners, or government policy makers together with a set of invited experts.

Let me illustrate the technique by imagining what such an exercise might look like at an energy industry company faced with the possibility of human-caused climate change.

There has been for many decades fairly incontrovertible scientific evidence that anthropogenic climate change is a reality. But for the purposes of this example, let's assume that instead, it remains one of the critical uncertainties. After all, one major political party in the United States has staked its policies on the notion that anthropogenic climate change is a hoax. And even if it is not a hoax, the magnitude and speed of the change remains unclear even in our best climate models.

So let the first uncertainty vector be whether or not potentially catastrophic anthropogenic climate change is happening, how fast it goes, and how bad it gets.

Let the second vector be the magnitude and urgency of humanity's response to the problem, and our ability to come up with ingenious solutions to it in time to make a difference.

You might end up with a quadrant map that looks like the one shown in the figure below:

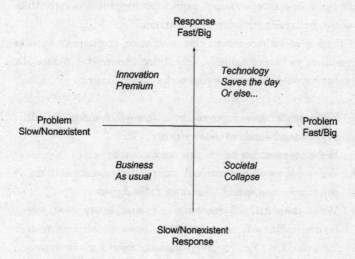

It seems apparent that if you're a businessperson thinking through these scenarios, the "robust strategy" is to assume that climate change is happening, and to respond. In the lower half of the scenario quadrant, we see no opportunity—it's either business as usual, or societal collapse. In the upper half, there is business opportunity whether the climate scientists or the skeptics are correct.

What makes the strategy robust is that we don't need to be sure that the worst fears of climate scientists are correct in order to act. The strategy is a good one even if they are wrong.

Climate change provides us with a modern version of Pascal's Wager (the argument of the seventeenth-century philosopher and mathematician for acting as though you believe in God even if you don't). If catastrophic global warming turns out not to happen, the steps we'd take to address it are still worthwhile. Given that there's even a reasonable risk of disruptive climate change, any sensible person should decide to act. It's insurance. The risk of your house burning down is small, yet you carry homeowner's insurance; you don't expect to total your car, but you know that the risk is there, and again, most people carry insurance; you don't expect catastrophic illness to strike you down, but again, you invest in insurance.

If there is no human-caused climate change, or the consequences are not dire, and we've made big investments to avert it, what's the worst that happens? In order to deal with climate change:

- We've made major investments in renewable energy, which pay off handsomely to those who make them.
- We've invested in a potent new source of jobs.
- We've improved our national security by reducing our dependence on oil from hostile or unstable regions.
- We've mitigated the enormous economic losses from pollution. (China has estimated these losses to its economy as 10% of GDP.) We currently subsidize fossil fuels in dozens of ways, by allowing power companies, auto companies, and others to keep environmental costs "off the books," by funding the infrastructure for autos via fuel taxes while demanding that railroads and other forms of public transportation pay for their own infrastructure, and so on.
- We've renewed our industrial base, investing in new industries rather than propping up old ones. Climate critics like to cite the cost of dealing with global warming. But the costs are similar to the "costs" incurred by record companies in the switch to digital music distribution, or the costs to newspapers implicit in the rise of the web. That is, they are costs to existing industries, but ignore the opportunities for new industries that exploit the new technology. I have yet to see a convincing case made that the costs of dealing with climate change aren't principally the costs of protecting incumbent industries.

By contrast, let's assume that the climate skeptics are wrong. We face the displacement of hundreds of millions of people, droughts, floods and other extreme weather, species loss, and economic harm that will make us long for the good old days of the 2008 financial industry meltdown.

It really is like Pascal's Wager. On one side, the worst outcome is that we've built a more robust innovation economy. On the other side, the worst outcome truly is hell. In short, we do better if we believe in climate change and act on that belief, even if we turn out to be wrong.

That's what scenario planners mean by a "robust strategy."

I doubt that Elon Musk did a conscious scenario-planning exercise, but all of his business decisions are consistent with the model above. Tesla, SolarCity, and SpaceX have all turned out to be robust business opportunities even though the worst ravages of climate change have not yet hit us. Musk's leadership in electric vehicles, rooftop solar, and human space exploration have all been bets worth placing. Similarly, countries such as China that invested heavily in solar energy have built huge new industries. Germany and Scandinavia are far ahead in decoupling their economies from fossil fuels. The United States, which largely chose the "business as usual" scenario, has lagged behind.

This may be about to change, if we can move beyond the old left–right divide on the issue. The Climate Leadership Council, an organization led by a Who's Who of conservative economists and former government and business leaders, recently came out with a report titled "The Conservative Case for Carbon Dividends," calling for a carbon tax whose proceeds would be rebated directly to all Americans, as a kind of citizen's dividend similar to those discussed in the previous chapter. So many of our problems come from being stuck in a bad map that we are unwilling to rewrite, even though it's clear that it no longer matches reality.

There is enormous opportunity in transforming our energy economy. My son-in-law, energy researcher and inventor Saul Griffith, has drawn a massive Sankey diagram—a map of all the energy sources and uses in the US economy to a resolution better than 1%. Standing next to the map, he explains to a visitor that any pathway on this map that is as big as his little finger (about 1% of energy flow) represents a $30–100 billion annual opportunity.

Saul has used that analysis to help guide the choice of projects that his company, Otherlab, works on. Natural gas storage. Much cheaper ways of building large-scale solar arrays that more efficiently track the sun. Air-conditioning that uses half as much energy to heat or cool a room. Soft robots able to tackle the trillion-dollar market for combating corrosion in infrastructure and building, sanding and repainting airplanes and bridges at a fraction of today's cost. Otherlab's expertise is in cutting-edge materials science and the mathematics of structure and manufacturing. Where they apply that expertise is based on an analysis of the big problems that need solving this century in energy and climate change.

"If anyone thinks that nine billion people are going to live as well as two billion people do today without radical changes to the economy," Saul said to me, "they must be crazy."

Thinking about the conjoined long-term trends of global population growth, worldwide rise in living standards, and the energy intensity of modern civilization, it's pretty clear that a huge part of our future is going to require a radical shift in the amount of energy we use per unit of consumption.

It is possible to construct a similar scenario grid for the questions of technology and the future of the economy that we have been exploring in this book.

Such a scenario grid might look something like this:

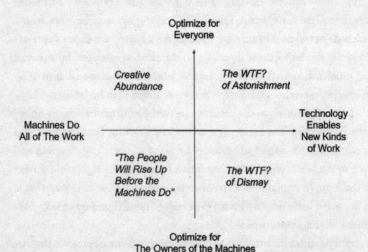

Let the first vector be the speed with which technology destroys jobs versus the speed with which it enables new kinds of work.

Let the second vector be the extent to which we use technology solely to optimize for the wealth of the owners of the machines, or to optimize for the wealth of all participants in the global economy.

Even if the machines do all of the work, and technology destroys jobs outright, we can build an economy of creative abundance if we use the fruits of machine productivity for the benefit of everyone. The challenges in the top left quadrant will be to knit a new social fabric where learning, creativity, and the human touch are valued very differently than they are today. We must craft policies that support, encourage, and reward the kinds of work that only humans can do for each other. Networked marketplace platforms can be a powerful tool in shaping this next economy.

On the top right, human beings are augmented to do things that were previously impossible. This is the WTF? of astonishment and delight, the future of bringing extinct species back to life or creating whole new ones, of extending human life spans and traveling to other planets, of eliminating disease, of engaging all of humanity in great challenges and fairly distributing the rewards of mastering those challenges.

In my optimistic moments, I think we can build a robust future across both of these quadrants.

In the lower two quadrants, though, we have the world we are heading pell-mell toward, at worst a world of revolution, social upheaval, and perhaps even warfare like we saw in the early days of the first industrial revolution, and at best, the WTF? of dismay, as technology births new wonders whose benefits are reserved for privileged elites while most of humanity barely gets by.

It doesn't have to be that way.

Even when a dark future seems to be staring us in the face, though, we lack the courage to do what must be done. Despite our best efforts, most of the time we fail to respond to potentially catastrophic consequences of changes that are already well under way. And despite the lessons of history, we haven't yet made the hard choices to fundamentally restructure our economy.

Instead, we argue over which of the failed recipes of the past we will try again. Political leaders and policy makers could learn a lot from Jeff Bezos.

In the employee Q&A at a March 2017 all-hands meeting at Amazon, where Jeff continues to remind his employees that "it's still Day 1," someone asked him, "What does Day 2 look like?" Jeff gave a passionate response, which he recounted in his annual shareholder letter a few weeks later: "Day 2 is stasis. Followed by irrelevance. Followed by excruciating, painful decline. Followed by death." That is a dire prognosis for a company or a society, yet it is what we face if we accept the status quo or the WTF? of dismay.

Jeff continued with four tips for staving off Day 2: "customer obsession, a skeptical view of proxies, the eager adoption of external trends, and high-velocity decision making." Customer obsession is the key to the WTF? of delight: "Even when they don't yet know it," Jeff wrote, "customers want something better, and your desire to delight customers will drive you to invent on their behalf." Whether you're in business or public policy, don't settle for rehashing tired solutions. Keep looking for that positive astonishment that means you've accomplished something wonderful for the people you serve. Jeff continued: "Staying in Day 1 requires you to experiment patiently, accept failures, plant seeds, protect saplings, and double down when you see customer delight."

Regarding "resisting proxies," Jeff noted that one of the traps that leads to Day 2 is that "you stop looking at outcomes and just make sure you're doing the process right." We can't just accept whatever results we get from following old rules; we must constantly measure our actions against their results. And when we see that the results don't measure up to our dreams, we must rewrite the rules.

Jeff also urged his employees to embrace powerful trends in technology and the economy: "If you fight them, you're probably fighting the future. Embrace them and you have a tailwind." Artificial intelligence isn't just for companies like Amazon, Google, and Facebook; like the Internet, open source software, and data science, it will transform every business and ultimately our entire society. Genetic engineering and neurotech are not far behind.

Jeff's last point, about the speed of decision making, is the last ingredient for dealing effectively with the task of creating not just a better company but a better future. Jeff's advice is priceless:

> First, never use a one-size-fits-all decision-making process. Many decisions are reversible, two-way doors. Those decisions can use a light-weight process. For those, so what if you're wrong? . . . Second, most decisions should probably be made with somewhere around 70% of the information you wish you had. If you wait for 90%, in most cases, you're probably being slow. Plus, either way, you need to be good at quickly recognizing and correcting bad decisions. If you're good at course correcting, being wrong may be less costly than you think, whereas being slow is going to be expensive for sure. Third, . . . If you have conviction on a particular direction even though there's no consensus, it's helpful to say, "Look, I know we disagree on this but will you gamble with me on it? Disagree and commit?"

The future is full of uncertainty. But our society is deep into Day 2, and the path we are on does indeed lead to stasis, irrelevance, and decline. Bold decision making; reversal of course when we find we are wrong; understanding of technological, demographic, and economic trends; and a relentless focus on making a better world for everyone can bring us renewal and the opportunity to rediscover Day 1 for our economy.

WORK, NOT JOBS

Even without doing a scenario-planning exercise, asking yourself "What happens if this goes on?" is a great way to prepare for the future—and to spot entrepreneurial opportunities.

Whether through observing positive trends like Moore's Law or the decreasing cost of gene sequencing (which is accelerating even faster than Moore's Law did), you can often anticipate the direction of new breakthroughs. You can also anticipate the negative disruptions that can result from failing to come to grips with a problem like income inequality or algorithms being perfected to faithfully fulfill the wrong fitness functions.

Entrepreneurship and invention require a kind of intellectual arbitrage, understanding the gap between what is possible and what has been accomplished so far.

It isn't just in technology that you can apply this kind of thinking. One of my favorite moments during the Markle Foundation Rework America initiative was the talk by fellow task force member Mike McCloskey, the founder and CEO of Select Milk Producers, the sixth-largest dairy cooperative in the country, and of Indiana's Fair Oaks Farms, his own dairy.

Mike looks a bit like the character Ron Swanson in the TV show *Parks and Recreation*, except bigger, and he talks like him too, slowly and with impact. "Some people would say we're an agribusiness," he said, "but I like to think we're still a family farm. I work on the farm. My wife and kids work on the farm. And ten thousand other families live and work on our farms."

Mike had been asked to talk about the importance of the agricultural sector to the economy, but he had much more to say than that. What he said was, to me, the most important statement of the multiyear exercise. "The way I see it, we have a job to do. There are going to be nine billion people in the world, and they are going to need protein. There are going to be three billion people in the middle class, and they are going to want better protein."

Mike took a hard look at the world and the way things are going, and made a determination of what work needs doing. That should be the goal of every entrepreneur.

Mike's comments struck me as so much more actionable than the hand-wringing we'd all been seeing about the decline of good middle-class jobs. While I agree with the urgency of that problem, Mike had put his finger on the answer. Not hoping that "the market" could somehow be incented to produce those good middle-class jobs again. "*We have a job to do*," he said.

Not "we need jobs." As Nick Hanauer also pointed out, there are two very different concepts tied up in that same word. The first, the one that Mike was using, is about the work that needs doing. The second, which pervades too many discussions of the economy, is a pallid, passive echo of the first, a job as something that you acquire from someone else, like you might find a product on the grocery shelf. If they are all gone, you're out of luck. "Work," not "jobs," should be the organizing principle for our map of the future labor economy. There is plenty of work to be done.

IT'S UP TO US

At my 2015 Next:Economy Summit, an event I'd organized to explore the impact of technology on the future of work, Limor Fried, the founder and CEO of Adafruit, appeared via Skype, giving us a virtual tour of her factory and warehouse in New York City. She showed us the design workstation where she creates innovative electronic devices and kits; a few steps away were the pick-and-place machines for placing chips on the circuit boards she develops herself, as well as other small-scale manufacturing equipment. Forty feet away she showed us the video studio where she records her popular "Ask an Engineer" show as well as free online tutorials for how to do everything from circuit design to 3-D printing. Then we took a walk around her warehouse, taking a look at the more than $30 million worth of products and parts she sells each year to her enthusiastic online audience, and meeting some of her 100+ employees.

I remember when Limor, an MIT-trained engineer, and her husband, Phil Torrone, a creative genius who formerly worked in the ad industry but now helps to build Limor's online presence, had their living quarters behind a curtain in her first small office. Limor built the business without venture capital, using credit cards to finance her original investments in office and inventory, and then bootstrapped her way up to success by creating products people really want, using the tools of modern media— YouTube, Twitch, email, and the web—to promote them. Championing open source hardware and engineering education, Limor had become a media star, gracing the cover of *Wired* magazine and being named a White House "Champion of Change" by President Obama. But perhaps the thing that makes her proudest is the mom who wrote in after watching "Ask an Engineer" with her daughter to tell Limor that the seven-year-old had asked, "Mom, can boys be engineers too?"

A year later, at the second Next:Economy Summit, we had another livestreamed presentation, this time from a cavernous hangar that had just been built in an open field in Rwanda. Keller Rinaudo, Zipline's cofounder and CEO, had just concluded an event with the president of the country to celebrate the formal launch of the California-based company's on-demand blood-delivery drones. Rwanda is a country with underdeveloped hospital infrastructure and often-impassable roads. Postpartum hemorrhage is one of the leading causes of death among women. It has never been possible to stock enough of the various blood types needed at remote clinics, but Keller and his cofounders have found a way to leapfrog over the lack of twentieth-century infrastructure and instead use the WTF? technologies of the twenty-first to solve a seemingly intractable problem. From only three drone airfields paired with blood storage facilities, the company can get blood to clinics anywhere in the country within fifteen minutes.

The company has raised $43 million in venture capital, with the latest round of $25 million intended to build distribution in other markets, including Vietnam, Indonesia, and, if regulatory barriers can be overcome, the United States. In the United States, services could mean delivering blood or medicine in rural areas, but it could also mean medical supplies for urgent needs, such as an Epi-Pen or snakebite antivenin on demand for an unexpected life-threatening emergency.

In the months in between the two events, I talked to hundreds of innovators, including many people whom you might not think of as entrepreneurs inventing the future. One of the most vivid and important meetings in shaping my thinking about a possible future for the economy was my nighttime walk around Central Park with social media sensation Brandon Stanton, creator of the Facebook feed *Humans of New York*. It was the only time he had to meet, while walking the dog, he said. He is too busy during the day.

Brandon is a photographer and storyteller. He searches, he told me, for people who look like they might have time to talk. His photographs, each accompanied by a paragraph with a key quote that captures the essence of his long conversation with the subject, have garnered his feed more than 25 million followers on Facebook and other social media platforms.

He originally started publishing his photos online in hopes that he could just make a living from doing what he loves. Unlike most people who have large social media followings, though, Brandon didn't try to cash in via advertising. He has created two bestselling books from his photos and stories, and is a frequent speaker to businesses and at college commencements. But he reserves the direct power of his social media following to raise money for causes inspired by the people whose stories he tells.

Brandon didn't set out to be an online fundraiser. His instinct for the importance of human connection led him to it. Vidal Chastanet, a thirteen-year-old from Brownsville, an area of Brooklyn with one of the highest crime rates in New York City, told Brandon that Nadia Lopez, the principal of his school, was the most inspiring person in his life. That led Brandon to a photo series on Mott Hall Bridges Academy. "Up until this moment, I didn't know I mattered," said Lopez. "I didn't know that anybody cared what I was doing." In her interview, Nadia confessed that one of her dreams was to take her students on a trip to Harvard, to remind them that anything could be possible for them. Brandon asked his followers (then 12 million) to chip in. He thought he might raise $30,000. Social media fans contributed $1.2 million.

A sad-looking woman sitting on a bench led him into the world of childhood cancer and a series on the families and the healthcare professionals who battle it. He ended up raising $3.8 million for research into the condition that took the life of that mother's young son. And so on. Refugees. Veterans. Inmates. The homeless. Ordinary people of every race, religion, and age, no longer just in New York but around the world. Brandon plumbs their souls, tells their stories, and shows us their faces. And millions of us respond.

Limor, Keller, and Brandon illustrate why, despite the fears of those who say that the next wave of automation will put everyone out of work, we don't have to run out of jobs. *It isn't technology that puts people out of work; it's the decisions we make about how to apply it.*

Limor has applied technology as a tool of creativity and teaching, bootstrapping her business by finding customers willing to pay for what she does; she has committed much of her time and effort to educating others about how she does her work both as an engineer and as an entrepreneur, so they can do it too.

Keller has used technology as a tool to solve a previously insoluble problem, using venture capital to build out the infrastructure of the future. If Zipline invents the new model for on-demand healthcare delivery, it won't be because they set out to disrupt healthcare. It will be because they first solved the problem for people half the world away, people whom the last wave of prosperity had passed by.

Brandon has used technology to create and distribute humane works of enormous beauty and insight, and has wielded the power of his social media following to shed light on and to support causes that matter.

Almost everything we need to know about the future of work in a world where machines take away many of today's jobs can be found in these three stories. Given a fair distribution of the fruits of machine productivity, people will entertain, educate, care for, and enrich each other's lives. And given a focus on solving real human problems, people can invent amazing futures.

Entrepreneurs like Limor and Keller and Brandon give me hope because they would do what they do even in a world where machines had made all of the necessities of life so cheap that no one *needed* to work. There are millions—nay, billions—more people who can follow in their footsteps.

The great political eruptions of 2016 also give me hope because they signal the beginning of the end of a failed economic theory. In the cracks in our society that they have exploited, and so unmasked, we can see that it is time to renew ourselves.

This is my faith in humanity: that we can rise to great challenges. Moral choice, not intelligence or creativity, is our greatest asset. Things may get much worse before they get better. But we can choose instead to lift each other up, to build an economy where people matter, not just profit. We can dream big dreams and solve big problems. Instead of using technology to replace people, we can use it to augment them so they can do things that were previously impossible.

ACKNOWLEDGMENTS

FOR STARTERS, I'D LIKE TO THANK HOLLIS HEIMBOUCH, MY EDitor and publisher, for taking a chance on an unusual combination of memoir, business book, and polemic. Your enthusiasm for it has driven me to express ideas that have sometimes surprised even me. Most of all, I thank you for giving me the chance to reach an audience very different than the one I usually touch. A book is a dialogue with its readers, and finding the right readers is as important as finding the right author. As Michael Lewis once said, "You never know what book you wrote till you find out what book people read." I am eager to hear from the readers you find for me. Thanks also to your marvelous team, including Stephanie Hitchcock, Cindy Achar, Nikki Baldauf, Thomas Pitoniak, Rachel Elinsky, and Penny Makras.

John Brockman, thanks for pushing me since 1993 to write a book that I didn't publish myself, and to you and Max Brockman for finding such a great home for the project.

Nick Hanauer, thanks for that 2012 TED University talk that got me thinking more deeply about the problems of technology and the economy. Thanks also to Zoë Baird, Howard Schultz, and my colleagues in the Markle Rework America initiative, through which I developed much of my exposure to others coming to grips with those problems. James Manyika, you in particular have been a guiding spirit. I'd also like to thank all of the speakers and participants in my Next:Economy Summit, through which I have explored not just the problems but solutions to the problems we face.

Bill Janeway, Hal Varian, and Peter Norvig—your willingness to read multiple drafts and to take the time to educate me about areas where my knowledge wasn't up to par has made this book so much stronger than it would have been otherwise. Hal and Bill, you've given me a master class in economics. If the student was not up to the level of his teachers, that is no fault of yours. Benedict Evans, Margaret

Levi, Laura Tyson, James Manyika, and Kevin Kelly—you also saved me from some egregious errors and omissions, and your challenges to my thinking clarified it. Jay Schaefer, Mike Loukides, and Laurent Haug, your close reading and comments strengthened both my ideas and my writing. Sunil Paul, Logan Green, Kim Rachmeler, Matt Cutts, Danny Sullivan, and Dave Guarino, you filled in critical details and context for key moments in this history. Satya Nadella, Reid Hoffman, Jeff Immelt, Peter Schwartz, Peter Bloom, Andy McAfee, Erik Brynjolfsson, David Autor, Larry Katz, Anne-Marie Slaughter, Sebastian Thrun, Yann LeCun, Joaquin Quiñonero Candela, Mike George, Rana Foroohar, Robin Chase, David Rolf, Andy Stern, Natalie Foster, Betsy Masiello, Jonathan Hall, Lior Ron, Paul Buchheit, Sam Altman, Esther Kaplan, Carrie Gleason, Zeynep Ton, Mikey Dickerson, Wael Ghonim, Tim Hwang, Henry Farrell, Amy Sellars, Mike McCloskey, Hank Green, Brandon Stanton, Jack Conte, Limor Fried, Phil Torrone, Seth Sternberg, Palak Shah, Keller Rinaudo, Stephane Kasriel, Bryan Johnson, Patrick Collison, Roy Bahat, Paddy Cosgrave, Steven Levy, Lauren Smiley, Bess Hochstein, Nat Torkington, Clay Shirky, Lawrence Wilkinson, Jessi Hempel, Mark Burgess, Carl Page, Maggie Shiels, Adam Davidson, and Winnie King, you also gave me the gift of your time and insight during the research and writing that led up to this book.

I'd also like to thank the people who taught me much of what I've shared in this book. As the poet Elizabeth Barrett Browning wrote, "What I do and what I dream include thee, as the wine must taste of its own grapes."

From my father and mother, Sean and Anne O'Reilly, I learned to think of good fortune as something to be shared. My father used to borrow so he could meet his "charitable obligations"; after his death, my mother demonstrated that a small amount of money well circulated could go far in building shared prosperity among our family. She loaned me money at a critical time in my company's history, with the only requirement that I pass it on to others when the crisis had passed.

From my former father-in-law, Jack Feldmann, I learned to love business, and to see it as an opportunity for creativity as great as any in art or literature. My ex-wife, Christina Isobel, taught me that business must always be infused with the values we want in the world; it must not operate by rules of its own. What I made of O'Reilly Media was deeply shaped by your rootedness in the human rather than the machine. My daughters, Arwen and Meara, my stepdaughter, Clementine, and my grandchildren, Huxley and Bronte, remind me daily why it matters that we hand off a better world to those who follow us.

Jen Pahlka, you are my partner in life and thought. This book is the culmination of a journey that began in that moment after I ended my 2008 talk "Why I Love Hackers" by reciting Rilke's poem about struggling with angels greater than ourselves, and you came up to me with shining eyes and told me, "I need a talk like that for my conference, except for entrepreneurs." Since then, you have taken things that for me were only ideas, and made them real in the world. You are a perfect exemplar and inspiration for the advice that formed the core of the talk I developed for you: "Work on stuff that matters." Your reading and comments on the book made it better in the same way that your thoughtful nudges make me better and our life together a constant exploration of what is possible when people work together in a perfect team.

My colleagues at O'Reilly Media, Maker Media, and O'Reilly AlphaTech Ventures, most notably Dale Dougherty, Laura Baldwin, Brian Erwin, Mike Loukides, Edie Freedman, Sara Winge, Gina Blaber, Roger Magoulas, Mark Jacobsen, and Bryce Roberts, but really, all of you who've been part of it over the years, helped to build something remarkable, with an impact far larger than I ever dreamed when I started in 1978. You are my second family. You inspire me and are a testament to the fact that a corporation too is a human augmentation, enabling us to do things that we could never accomplish on our own.

Over my years in the technology industry, I'd like to single out as mentors and sources of inspiration, directly or indirectly, Stewart

Brand, Dennis Ritchie, Ken Thompson, Brian Kernighan, Bill Joy, Bob Scheifler, Larry Wall, Vint Cerf, Jon Postel, Tim Berners-Lee, Linus Torvalds, Brian Behlendorf, Jeff Bezos, Larry Page, Sergey Brin, Eric Schmidt, Pierre Omidyar, Ev Williams, Mark Zuckerberg, Saul Griffith, and Bill Janeway. I have drawn my map by studying the world you have helped to create.

NOTES

INTRODUCTION

x beat the world's best human Go player: Cade Metz, "In Two Moves, AlphaGo and Lee Sedol Redefined the Future," *Wired*, March 16, 2016, https://www.wired.com/2016/03/two-moves-alphago-lee-sedol-redefined-future/.

x An AI running on a $35 Raspberry Pi computer: Cecille de Jesus, "an AI Just Defeated Human Fighter Pilots in an Air Combat Simulator," *futurism.com*, June 28, 2016, http://futurism.com/an-ai-just-defeated-human-fighter-pilots-in-an-air-combat-simulator/.

x it wants an AI to make three-fourths of management decisions: Olivia Solon, "World's Largest Hedge Fund to Replace Managers with Artificial Intelligence," *Guardian*, December 22, 2016, https://www.theguardian.com/technology/2016/dec/22/bridgewater-associates-ai-artificial-intelligence-management.

x Oxford University researchers estimate: Carl Benedikt Frey and Michael A. Osborne, "The Future of Employment: How Susceptible Are Jobs to Computerisation?," Oxford Martin School, September 17, 2013, http://www.oxfordmartin.ox.ac.uk/downloads/academic/The_Future_of_Employment.pdf.

x Airbnb has more rooms on offer: Andrew Cave, "Airbnb Is on Track to Be the World's Largest Hotelier," *Business Insider*, November 26, 2013, http://www.businessinsider.com/airbnb-largest-hotelier-2013-11.

xi Uber is still losing $2 billion every year: Eric Newcomer, "Uber Loses at Least $1.2 Billion in First Half of 2016," *Bloomberg Technology*, August 25, 2016, http://www.bloomberg.com/news/articles/2016-08-25/uber-loses-at-least-1-2-billion-in-first-half-of-2016.

xi *Fortune* magazine started keeping a list: "The Unicorn List," *Fortune*, retrieved March 29, 2017, http://fortune.com/unicorns/.

xi "Unicorn Leaderboard": "Crunchbase Unicorn Leaderboards," *TechCrunch*, retrieved March 29, 2017, http://techcrunch.com/unicorn-leaderboard/.

xii "the name we give to the common experience": Tom Stoppard, *Rosencrantz & Guildenstern Are Dead* (New York: Grove Press, 1967), 21.

xiii complaining about it: If you haven't seen the late-night TV rant by comedian Louis CK, "Everything Is Amazing and Nobody's Happy," watch it now! Retrieved March 29, 2017, https://www.youtube.com/watch?v=q8LaT5Iiwo4.

xiv "They transform their customers": Michael Schrage, *Who Do You Want Your Customers to Become?* (Boston: Harvard Business Review Press, 2012), ebook retrieved March 29, 2017, https://www.safaribooksonline.com/library/view/who-do-you/9781422187852/chapter001.html#a002.

xvi Sixty-three percent of Americans: Pew Research Center in Association with the Markle Foundation, *The State of American Jobs*, retrieved March 29, 2017, https://www.markle.org/sites/default/files/State-of-American-Jobs.pdf.

xvi life expectancy is actually declining in America: Olga Khazan, "Why Are So Many Americans Dying Young?," *Atlantic*, December 13, 2016, https://www.theatlantic.com/health/archive/2016/12/why-are-so-many-americans-dying-young/510455/.

xix health monitoring by wearable sensors: Darrell Etherington, "Google's New Health Wearable Delivers Constant Patient Monitoring," *TechCrunch*, June 23, 2015, https://techcrunch.com/2015/06/23/googles-new-health-wearable-delivers-constant-patient-monitoring/.

xxi what happens to the companies that depend on consumer purchasing power: Nicholas J. Hanauer, "The Capitalist's Case for a $15 Minimum Wage," *Bloomberg View*, June 19, 2013, https://www.bloomberg.com/view/articles

/2013-06-19/the-capitalist-s-case-for
-a-15-minimum-wage.

xxi had the same experience: Richard Dobbs, Anu Madgavkar, James Manyika, Jonathan Woetzel, Jacques Bughin, Eric Labaye, and Pranav Kashyap, "Poorer than Their Parents? A New Perspective on income inequality," McKinsey Global Institute, July 2016, http://www.mckinsey.com/global-themes/employment-and-growth/poorer-than-their-parents-a-new-perspective-on-income-inequality.

xxi Top US CEOs now earn 373x the income of the average worker: Melanie Trottman, "Top CEOs Make 373 Times the Average U.S. Worker," *Wall Street Journal*, May 13, 2015, http://blogs.wsj.com/economics/2015/05/13/top-ceos-now-make-373-times-the-average-rank-and-file-worker/.

xxi that chance has fallen to 50%: David Leonhardt, "The American Dream, Quantified at Last," *New York Times*, December 8, 2016, https://mobile.nytimes.com/2016/12/08/opinion/the-american-dream-quantified-at-last.html.

xxi household debt is over $12 trillion: "Quarterly Report on Household Debt and Credit," Federal Reserve Bank of New York, August 2016, https://www.newyorkfed.org/medialibrary/interactives/householdcredit/data/pdf/HHDC_2016Q2.pdf.

xxi 80% of gross domestic product, or GDP: St. Louis Fed, "Household Debt to GDP for United States," https://fred.stlouisfed.org/series/HDTGPDUSQ163N.

xxi Seven million borrowers in default: "The Digital Degree," *Economist*, June 27, 2014, http://www.economist.com/news/briefing/21605899-staid-higher-education-business-about-experience-welcome-earthquake-digital.

xxii $30 trillion of cash is sitting on the sidelines: "Cash on the Sidelines: How to Unleash $30 Trillion," panel discussion at the Milken Institute Global Conference, April 20, 2013, http://www.milkeninstitute.org/events/conferences/global-conference/2013/panel-detail/4062.

xxiii four major disruptive forces: Richard Dobbs, James Manyika, and Jonathan

Woetzel, *No Ordinary Disruption* (Philadelphia: PublicAffairs, 2015), 4–7.

CHAPTER 1: SEEING THE FUTURE IN THE PRESENT

3 "The skill of writing": I don't remember where I first heard this quote. It may have been in a public radio interview around 1980 or so. I once asked Ed Schlossberg, and he didn't remember either.

5 Mark Twain is reputed to have said: "History Does Not Repeat Itself, but It Rhymes," *Quote Investigator*, retrieved March 27, 2017, http://quoteinvestigator.com/2014/01/12/history-rhymes/.

6 free as in freedom: Sam Williams, *Free as in Freedom: Richard Stallman's Crusade for Free Software* (Sebastopol, CA: O'Reilly, 2002). See also Richard Stallman, "The GNU Manifesto," retrieved March 29, 2017, http://www.gnu.org/gnu/manifesto.en.html.

8 "The Cathedral and the Bazaar": Originally published at http://www.unterstein.net/su/docs/CathBaz.pdf. Book version: Eric S. Raymond, *The Cathedral & the Bazaar* (Sebastopol, CA: O'Reilly, 2001).

9 "Hardware, Software, and Infoware": Tim O'Reilly, "Hardware, Software, and Infoware," in *Open Sources: Voices from the Open Source Revolution* (Sebastopol, CA: O'Reilly, 1999), available online at http://www.oreilly.com/openbook/opensources/book/tim.html.

12 sales of 250,000 units in the first five years: Edwin D. Reilly, *Milestones in Computer Science and Information Technology* (Westport, CT: Greenwood, 2003), 131.

12 rumored to have sold 40,000 on the first day: Sol Libes, "Bytelines," *Byte* 6, no. 12, retrieved March 29, 2017, https://archive.org/stream/byte-magazine-1981-12/1981_12_BYTE_06-12_Computer_Games#page/n315/mode/2up.

14 *Just for Fun:* Linus Torvalds and David Diamond, *Just for Fun* (New York: Harper Business, 2001).

15 "cut off Netscape's air supply": Joel Klein, "Complaint: United States v. Microsoft in the United States District for the District of Columbia, Civil Action No. 98–1232 (Antitrust), Filed: May 18, 1998," retrieved March 30, 2017,

https://www.justice.gov/atr/complaint-us-v-microsoft-corp.

19   "It's just not evenly distributed yet": I believe I first heard Gibson say this in a 1999 NPR interview. I like to think that my frequent use of the quote in my talks from the time I first heard it has been the source of much of its usage, because it usually appears in the slightly misquoted form in which I remembered it. For Gibson's account of its origin, see "The future has arrived," *Quote Investigator*, retrieved March 30, 2017, http://quoteinvestigator.com/2012/01/24/future-has-arrived/.

21   a tin of biscuits wrapped in brown paper: I believe I first heard this story from George Simon. It is also recounted in "Alfred Korzybski," *Wikipedia*, retrieved March 30, 2017, https://en.wikipedia.org/wiki/Alfred_Korzybski#cite_note-4.

22   sufficiently to use them in real life: Richard Feynman, *Surely You're Joking, Mr. Feynman* (New York: Norton, 1984), 212.

22   "Their knowledge is so fragile!": Ibid., 36.

CHAPTER 2: TOWARD A GLOBAL BRAIN

23   "The Open Source Paradigm Shift": Tim O'Reilly, "The Open Source Paradigm Shift," in *Perspectives on Free and Open Source Software*, ed. J. Feller, B. Fitzgerald, S. Hissam, and K. R. Lakhani (Cambridge, MA: MIT Press, 2005). Also available at http://archive.oreilly.com/pub/a/oreilly/tim/articles/paradigmshift_0504.html.

24   "an intellectual property destroyer": Jim Allchin, quoted in Tim O'Reilly, "My Response to Jim Allchin," *oreilly.com*, February 18, 2001, http://archive.oreilly.com/pub/wlg/104.

24   emerge at an adjacent stage: Clay Christensen, "The Law of Conservation of Attractive Profits," *Harvard Business Review* 82, no. 2 (February 2004): 17–18.

25   I told free software advocates: There is a transcript of our exchange during the Q&A after my 1999 talk at the "Wizards of OS" conference in Berlin. Richard actually said it didn't matter that Amazon's software wasn't free: "[T]he issue of free software versus proprietary arises for software that we're going to have on our computers and run on our computers. We're gonna have copies and the question is, what are we allowed to do with those copies? Are we just allowed to run them or are we allowed to do the other useful things that you can do with a program? If the program is running on somebody else's computer, the issue doesn't arise. Am I allowed to copy the program that Amazon has on it's computer? Well, I can't, I don't have that program at all, so it doesn't put me in a morally compromised position. . . ." See http://www.oreilly.com/tim/archives/mikro_discussion.pdf.

27   Google is now running on well over a million servers: Google does not actually disclose this information, but in July 2013, Microsoft then-CEO Steve Ballmer noted that Microsoft Bing was running on almost that number; Google serves many more users, and the number has only grown since then. See Sebastian Anthony, "Microsoft Now Has One Million Servers—Less than Google, but More than Amazon, Says Ballmer," *Extremetech*, retrieved March 29, 2017, https://www.extremetech.com/extreme/161772-microsoft-now-has-one-million-servers-less-than-google-but-more-than-amazon-says-ballmer.

28   one of the most significant books of the twentieth century: Elizabeth Diefendorf, ed., *The New York Public Library's Books of the Century* (New York: Oxford University Press, 1996), 149.

29   "What Is Web 2.0?": Tim O'Reilly, "What Is Web 2.0?," *oreilly.com*, September 30, 2005, http://www.oreilly.com/pub/a/web2/archive/what-is-web-20.html.

31   open letter to the company: David Stutz, "On Leaving Microsoft," *synthesist.net*, retrieved March 29, 2017, http://www.synthesist.net/writing/onleavingms.html.

32   global collaboration around open source projects: Tim O'Reilly, "Open Source: The Model for Collaboration in the Age of the Internet," keynote at the Computers, Freedom, and Privacy Conference, Toronto, April 6, 2000, http://www.oreillynet.com/pub/a/network/2000/04/13/CFPkeynote.html.

40 "increased by orders of magnitude": Tim O'Reilly and John Battelle, "Web Squared: Web 2.0 Five Years On," *oreilly .com*, retrieved March 30, 2017, https:// conferences.oreilly.com/web2summit /web2009/public/schedule/detail /10194.

41 "search a possible for its possibleness": Wallace Stevens, "An Ordinary Evening in New Haven," in *The Palm at the End of the Mind*, ed. Holly Stevens (New York: Vintage, 1972), 345.

41 "a raid on the inarticulate": T. S. Eliot, "East Coker," *The Four Quartets*, New York, Houghton Mifflin Harcourt, 1943, renewed 1971.

42 The @ symbol to reply to another user: "The First Ever Hashtag Reply and Retweet as Twitter Users Invented Them," retrieved March 29, 2017, http:// qz.com/135149/the-first-ever-hashtag -reply-and-retweet-as-twitter-users -invented-them/. Interestingly, even participants don't get their history right. In a 2016 conversation, Jack Dorsey, the co-founder of Twitter, told me definitively that I had invented the retweet, and would not be deterred by my demurrals. I certainly was one of its most prolific early adopters, but I remember picking up the trick from someone else. I was particularly inspired by Leisa Reichelt's lovely term for this new practice of sharing what you were reading rather than what you were doing: mindcasting.

42 proposed the use of the # symbol: Chris Messina, Twitter update, retrieved March 29, 2017, https://twitter.com/chrismessina /status/223115412. Note that Joshua Schachter had earlier used # as a symbol for tags in his link-saving site del. icio.us.

42 during the San Diego wildfires: Chris Messina, "Twitter Hashtags for Emergency Coordination and Disaster Relief," retrieved March 29, 2017, https:// factoryjoe.com/2007/10/22/twitter -hashtags-for-emergency-coordination -and-disaster-relief/.

42 The app had already begun showing "trending topics": "To Trend or Not to Trend," *Twitter Blog*, retrieved March 29, 2017, https://blog.twitter.com/2010 /to-trend-or-not-to-trend.

43 features that the platform developer itself hadn't imagined: "Twitpic," retrieved March 29, 2017, https://en.wikipedia .org/wiki/TwitPic.

43 Jim Hanrahan posted the first tweet: Jim Hanrahan, Twitter update, retrieved March 29, 2017, https://twitter.com /highfours/status/1121908186.

43 passengers standing on the wing of the downed plane: "There's a plane in the Hudson. I'm on the ferry going to pick up the people. Crazy." Twitter update, retrieved March 29, 2017, https://twitter .com/jkrums/status/1121915133.

43 "We Are All Khaled Said": Facebook page, retrieved March 29, 2017, https:// www.facebook.com/ElShaheeed.

43 "an internal life of its own": Michael Nielsen, *Reinventing Discovery* (Princeton, NJ: Princeton University Press, 2011), 53.

44 "A theory is a species of thinking": Thomas Henry Huxley, "The Coming of Age of 'The Origin of Species,'" *Collected Essays*, vol. 2, as reprinted at http://aleph0.clarku.edu/huxley/CE2 /CaOS.html.

45 "the way a genome runs on a multitude of cells": George Dyson, *Turing's Cathedral* (New York: Pantheon, 2012), 238–39.

46 income that it can't deliver: Sami Jarbawi, "Uber to Pay $20 Million to Settle FTC Case," Berkeley Center for Law, Business and the Economy, January 31, 2017, http://sites.law.berkeley .edu/thenetwork/wp-content/uploads /sites/2/2017/01/Uber-to-Pay-20-Million -to-Settle-FTC-Case.pdf.

46 technology to deflect their investigations: Mike Isaac, "How Uber Deceives the Authorities Worldwide," *New York Times*, March 3, 2017, https://www .nytimes.com/2017/03/03/technology /uber-greyball-program-evade-authorities .html.

47 Rivals sue over claims of stolen technology: Alex Davies, "Google's Lawsuit Against Uber Revolves Around Frickin' Lasers," *Wired*, February 5, 2017, https://www. wired.com/2017/02/googles-lawsuit -uber-revolves-around-frickin-lasers/.

47 tolerates sexual harassment: Susan J. Fowler, "Reflecting on One Very, Very

Strange Year at Uber," Susan J. Fowler's blog, February 19, 2017, https://www.susanjfowler.com/blog/2017/2/19/reflecting-on-one-very-strange-year-at-uber.

CHAPTER 3: LEARNING FROM LYFT AND UBER

51 I called these meme maps: Tim O'Reilly, "Remaking the Peer-to-Peer Meme," in *Peer to Peer*, ed. Andy Oram (Sebastopol, CA: O'Reilly, 2001). The essay is also available online at http://archive.oreilly.com/pub/a/495.

54 "a remote control for real life": Kara Swisher, "Man and Uber Man," *Vanity Fair*, December 2014, retrieved March 30, 2017, http://www.vanityfair.com/news/2014/12/uber-travis-kalanick-controversy.

54 "That's what it's all about": Brad Stone, *The Upstarts* (New York: Little, Brown, 2017), 52.

55 they observed in Zimbabwe: As told to me by Logan Green in 2015.

55 the incentives: Stone, *The Upstarts*, 71.

57 "everyone benefits": "The Uber Story," *uber.com*, retrieved March 30, 2017, https://www.uber.com/our-story/.

57 one customer in Los Angeles: Priya Anand, "People in Los Angeles Are Getting Rid of Their Cars," *BuzzFeed*, September 2, 2016, https://www.buzzfeed.com/priya/people-in-los-angeles-are-getting-rid-of-their-cars.

59 one of the most difficult exams in the world: Jody Rosen, "The Knowledge, London's Legendary Taxi-Driver Test, Puts Up a Fight in the Age of GPS," *New York Times Magazine*, November 24, 2014, https://www.nytimes.com/2014/11/10/t-magazine/london-taxi-test-knowledge.html.

59 it does save them money: "Workforce of the Future: Final Report (Slide 12)," *Markle*, retrieved March 30, 2017, https://www.markle.org/workforce-future-final-report.

63 Tesla seems to have other plans: Dan Gillmor, "Tesla Says Customers Can't Use Its Self-Driving Cars for Uber," *Slate*, October 21, 2016, http://www.slate.com/blogs/future_tense/2016/10/21/tesla_says_customers_can_t_use_its_self_driving_cars_for_uber.html.

64 "digital sharecroppers": Nicholas Carr, "The Economics of Digital Sharecropping," *Rough Type*, May 4, 2012, http://www.roughtype.com/?p=1600.

67 "proximity to the market": From an unpublished preprint sent to me by Laura Tyson of Laura Tyson and Michael Spence, "Exploring the Effects of Technology on Income and Wealth Inequality," in *After Piketty*, ed. Heather Boushey, J. Bradford DeLong, and Marshall Steinbaum (Cambridge, MA: Harvard University Press, 2017).

68 on-demand drivers in their own cars: "Amazon Flex: Be Your Own Boss. Great Earnings. Flexible Hours," *Amazon*, retrieved March 30, 2017, https://flex.amazon.com.

69 Kalanick burst out: Eric Newcomer, "In Video, Uber CEO Argues with Driver over Falling Fares," *Bloomberg Technology*, February 28, 2017, https://www.bloomberg.com/news/articles/2017-02-28/in-video-uber-ceo-argues-with-driver-over-falling-fares.

70 "you'll never be the same again": PBS, *One Last Thing*, 2011, video clip of Steve Jobs 1994 comment republished April 24, 2013, http://mathiasmikkelsen.com/2013/04/everything-around-you-that-you-call-life-was-made-up-by-people-that-were-no-smarter-than-you/.

CHAPTER 4: THERE ISN'T JUST ONE FUTURE

71 an appeal from Richard Stallman: Richard published his email to me as an open letter. https://www.gnu.org/philosophy/amazon-rms-tim.en.html.

71 an email to Jeff Bezos: Tim O'Reilly, "Ask Tim," *oreilly.com*, February 28, 2000, http://archive.oreilly.com/pub/a/oreilly/ask_tim/2000/amazon_patent.html.

74 an open letter: Tim O'Reilly, "An Open Letter to Jeff Bezos," *oreilly.com*, February 28, 2000, http://www.oreilly.com/amazon_patent/amazon_patent.comments.html.

74 I had 10,000 signatures: Tim O'Reilly, "An Open Letter to Jeff Bezos: Your Responses," *oreilly.com*, February 28, 2000, http://www.oreilly.com/amazon_patent/amazon_patent_0228.html.

74 wasn't introduced till six years later: Jeff Howe, "The Rise of Crowdsourcing," *Wired*, June 1, 2006, https://www.wired.com/2006/06/crowds/.

74 We did award the bounty: Tim O'Reilly, "O'Reilly Awards $10,000 1-Click Bounty to Three 'Runners Up,'" *oreilly.com*, March 14, 2001, http://archive.oreilly.com/pub/a/policy/2001/03/14/bounty.html.

74 after I published my open letter: Tim O'Reilly, "My Conversation with Jeff Bezos," *oreilly.com*, March 2, 2000, http://archive.oreilly.com/pub/a/oreilly/ask_tim/2000/bezos_0300.html.

76 Determining an Efficient Transportation Route: Sunil Paul, System and Method for Determining an Efficient Transportation Route, US Patent 6,356,838, filed July 25, 2000, and issued March 12, 2002.

77 Uber matched as UberPool: As told to me by Logan Green, CEO of Lyft, his company was about to roll out the feature when Uber pre-empted them with an announcement the day before. However, Uber was simply announcing that they were going to introduce the feature, while Lyft was actually launching it, supporting Logan's belief that Uber had heard about and was copying their idea.

78 "with a single touch": This was the Apple Pay page as of September 30, 2014, when I wrote "What Amazon, iTunes, and Uber Teach Us About Apple Pay," *oreilly.com*, September 30, 2014. This language is no longer present on the Apple site as of March 30, 2017, https://www.apple.com/apple-pay/.

78 automatically debit your account: "Introducing Amazon Go," *Amazon*, retrieved March 30, 2017, https://www.amazon.com/b?ie=UTF8&node=16008589011.

81 income and demographics: *Sizing the Internet Opportunity* (Sebastopol, CA: O'Reilly, 2004).

81 till the end of 1993: "Robert McCool," *Wikipedia*, retrieved March 30, 2017, https://en.wikipedia.org/wiki/Robert_McCool.

81 opposed to the idea of third-party apps on the iPhone: Killian Bell, "Steve Jobs Was Originally Dead Set Against Third-Party Apps for the iPhone," *Cult of Mac*, October 21, 2011, http://www.cultofmac.com/125180/steve-jobs-was-originally-dead-set-against-third-party-apps-for-the-iphone/.

81 skeptical of the peer-to-peer model: Stone, *The Upstarts*, 199–200.

86 "how the world *does* work": Aaron Levie, Twitter update, August 22, 2013, https://twitter.com/levie/status/370776444013510656.

CHAPTER 5: NETWORKS AND THE NATURE OF THE FIRM

90 "Apps can do now what managers used to do": Esko Kilpi, "The Future of Firms," *Medium*, February 6, 2015, https://medium.com/@EskoKilpi/movement-of-thought-that-led-to-airbnb-and-uber-9d4da5e3da3a.

90 "support megacorporations": Hal Varian, "If There Was a New Economy, Why Wasn't There a New Economics?," *New York Times*, January 17, 2002, http://www.nytimes.com/2002/01/17/business/economic-scene-if-there-was-a-new-economy-why-wasn-t-there-a-new-economics.html.

90 largest media company in the world: "Google Strengthens Its Position as World's Largest Media Owner," Zenith Optimedia, retrieved March 30, 2017, https://www.zenithmedia.com/google-strengthens-position-worlds-largest-media-owner-2/.

90 surpassed those of the largest traditional media companies: Tom Dotan, "Facebook Ad Revenue (Finally) Tops Media Giants," *The Information*, November 22, 2016, https://www.theinformation.com/facebook-ad-revenue-finally-tops-media-giants?shared=Xmjr9tlVIXs.

90 watch more video on YouTube: Andy Smith, "13–24 Year Olds Are Watching More YouTube than TV," *Tubular Insights*, March 11, 2015, http://tubularinsights.com/13-24-watching-more-youtube-than-tv/.

90 world's most valuable retailer: Shannon Pettypiece, "Amazon Passes Wal-Mart as Biggest Retailer by Market Value," *Bloomberg Technology*, July 24, 2015, https://www.bloomberg.com/news/articles/2015-07-23/amazon-surpasses-wal-mart-as-biggest-retailer-by-market-value.

91 "publish, then filter": Clay Shirky, *Here Comes Everybody* (New York: Penguin, 2008), 98.

93 6,300 companies operating 171,000 taxicabs: *2014 TLPA Taxicab Fact Book*, available from https://www.tlpa.org/TLPA-Bookstore.

94 total number of tellers: James Pethokoukis, "What the Story of ATMs and Bank Tellers Reveals About the 'Rise of the Robots' and Jobs," *AEI Ideas*, June 6, 2016, http://www.aei.org/publication/what-atms-bank-tellers-rise-robots-and-jobs/.

95 making house calls to deliver flu shots: "Uber Health," Uber, November 21, 2015, https://newsroom.uber.com/uberhealth/.

95 bringing elderly patients to doctors' appointments: Zhai Yun Tan, "Hospitals Are Partnering with Uber to Get Patients to Checkups," *Atlantic*, August 21, 2015, https://www.theatlantic.com/health/archive/2016/08/hospitals-are-partnering-with-uber-to-get-people-to-checkups/495476/.

95 from having 1,400 robots in its warehouses to 45,000: Sara Kessler, "The Optimist's Guide to the Robot Apocalypse," *Quartz*, March 19, 2017, https://qz.com/904285/the-optimists-guide-to-the-robot-apocalypse/.

95 It added 110,000: Todd Bishop, "Amazon Soars to More than 341K Employees—Adding More than 110K People in a Single Year," *Geekwire*, February 2, 2017, http://www.geekwire.com/2017/amazon-soars-340k-employees-adding-110k-people-single-year/.

96 the comparison between Kodak: Scott Timberg, "Jaron Lanier: The Internet Destroyed the Middle Class," *Salon*, May 12, 2013, http://www.salon.com/2013/05/12/jaron_lanier_the_internet_destroyed_the_middle_class/.

97 5% of GDP in developed countries: "The Internet Economy in the G20," *BCG Perspectives*, retrieved March 30, 2017, https://www.bcgperspectives.com/content/articles/media_entertainment_strategic_planning_4_2_trillion_opportunity_internet_economy_g20/?chapter=2.

97 80 billion in the days of Kodak: Benedict Evans, "How Many Pictures?," *ven-evans.com*, August 19, 2015, http://ben-evans.com/benedictevans/2015/8/19/how-many-pictures. These figures are from 2015, and are likely even larger today.

97 "we decided to rent them out": AllEntrepreneur, "Travel Like a Human with Joe Gebbia, Co-founder of AirBnB!," *AllEntrepreneur*, August 26, 2009, https://allentrepreneur.wordpress.com/2009/08/26/travel-like-a-human-with-joe-gebbia-co-founder-of-airbnb/.

98 "thick marketplace": Alvin E. Roth, *Who Gets What—and Why?* (Boston: Houghton Mifflin, 2015), 8–9.

101 hundreds of millions of websites: "Netcraft Web Surver Survey, March 2017," Netcraft, https://news.netcraft.com/archives/category/web-server-survey/.

101 trillions of web pages: "How Search Works," Google, retrieved March 30, 2017, https://www.google.com/insidesearch/howsearchworks/thestory/. The actual number cited is 130 trillion!

101 Craig Newmark recalled the process: Dylan Tweney, "How Craig Newmark Built Craigslist with 'No Vision Whatsoever,'" *Wired*, June 5, 2007, https://www.wired.com/2007/06/no_vision_whats/.

102 seventh-most-trafficked site on the web: Tim O'Reilly, "When Markets Collide," in *Release 2.0: Issue 2, April 2007*, ed. Jimmy Guterman (Sebastopol, CA: O'Reilly, 2007), 1, available online at http://www.oreilly.com/data/free/files/release2-issue2.pdf.

102 Today it is still No. 49: Or #116, depending on whose panel you follow, Alexa or SimilarWeb. "List of Most Popular Websites," *Wikipedia*, retrieved March 30, 2017, https://en.wikipedia.org/wiki/List_of_most_popular_websites.

103 now enrolled in Amazon Prime: Alison Griswold, "Jeff Bezos' Master Plan to Make Everyone an Amazon Prime Subscriber Is Working," *Quartz*, July 11, 2016, https://qz.com/728683/jeff-bezos-master-plan-to-make-everyone-an-amazon-prime-subscriber-is-working/.

103 200 million active credit card accounts: Horace Dediu, Twitter update, April 28, 2014, https://twitter.com/asymco/status/460724885120380929.

103 begin their search at Amazon: "State of Amazon 2016," Bloomreach, retrieved March 30, 2017, http://go.bloomreach.com/rs/243-XLW-551/images/state-of-amazon-2016-report.pdf.

103 46% of all online shopping: Olivia LaVec-
chia and Stacy Mitchell, "Amazon's
Stranglehold," Institute for Local Self
Reliance, November 2016, 10, https://
ilsr.org/wp-content/uploads/2016/11
/ILSR_AmazonReport_final.pdf.

103 remove the "Buy" button: Doreen Car-
vajal, "Small Publishers Feel Power
of Amazon's 'Buy' Button," New York
Times, June 16, 2008, http://www
.nytimes.com/2008/06/16/business
/media/16amazon.html. O'Reilly Media
distributes books for a number of smaller
publishers, and several of them faced these
kinds of tactics as a way to extract "co-op
marketing" payments. And of course,
Amazon notably used this technique
during its 2014 dispute with Hachette.
Amazon never used this technique with
us, perhaps because they knew we had a
strong direct online presence of our own.
Shortly before the Hachette-Amazon
dispute became public, Hachette
approached me to inquire whether, and
how quickly, we could license to them our
platform for selling ebooks. Given that
we had made a strong policy decision that
using our platform required all ebooks
to be DRM-free, and Hachette wasn't
willing to take that step, I declined.

103 "a privately controlled one": LaVecchia and
Mitchell, "Amazon's Stranglehold," 13.

104 "dwarf their activities on behalf of outside customers": O'Reilly, "When Markets Collide," 9.

105 "an architecture of participation": Tim
O'Reilly, "The Architecture of Participation, oreilly.com, June 2004, http://archive
.oreilly.com/pub/a/oreilly/tim/articles
/architecture_of_participation.html.

105 "than from the programs themselves":
Brian W. Kernighan and Rob Pike, The
Unix Programming Environment (Engle-
wood Cliffs, NJ: Prentice-Hall, 1984), viii.

105 "small pieces loosely joined": David
Weinberger, Small Pieces Loosely Joined
(New York: Perseus, 2002). See also
http://www.smallpieces.com.

106 "a working simple system": John Gall,
Systemantics: How Systems Work and
Especially How They Fail (New York:
Quadrangle, 1977), 52.

107 "rough consensus and running code":
Paulina Borsook, "How Anarchy Works,"
Wired, October 1, 1995, https://www
.wired.com/1995/10/ietf/.

108 "Be liberal in what you accept from others": Jon Postel, "RFC 761: Transmission
Control Protocol, January 1980," IETF,
https://tools.ietf.org/html/rfc761.

108 "TCP/IP promises to allow for easy mi-
gration to OSI": Robert A. Moskowitz,
"TCP/IP: Stairway to OSI," Computer
Decisions, April 22, 1986.

CHAPTER 6: THINKING IN PROMISES

110 to pitch Jeff on the idea: Tim O'Reilly,
"Amazon.com's Web Services Opportunity," PowerPoint deck March 8, 2001,
uploaded to SlideShare March 30, 2017,
https://www.slideshare.net/timoreilly
/amazoncoms-web-services-opportunity.

110 Amazon all-hands meeting in May 2003:
Tim O'Reilly, "Amazon.com and the Next
Generation of Computing," PowerPoint
deck May 20, 2003, uploaded to
SlideShare March 30, 2017, https://www
.slideshare.net/timoreilly/amazoncom
-and-the-next-generation-of-computing.

111 "coordination between those two groups":
Om Malik, "Interview: Amazon CEO Jeff
Bezos," GigaOm, June 17, 2008, http://
www.i3businesssolutions.com/2008/06
/interview-amazon-ceo-jeff-bezos-gigaom/.

112 "Anyone who doesn't do this will be
fired": Steve Yegge, "Stevey's Google
Platform Rant." The original October
12, 2011, post on Google Plus was deleted, but it was preserved in a number of
places, notably on Github: https://gist
.github.com/chitchcock/1281611.
Yegge explains why he deleted the post in
a follow-up on Google Plus: https://plus
.google.com/110981030061712822816
/posts/bwJ7kAELRnf. However, he (and
Google) did allow others to preserve the
post. Note: Referring to point 6, "Anyone
who doesn't do this will be fired," Kim
Rachmeler told me, "I know Stevey said
this but I don't think Jeff ever did." That
being said, it does get across the commitment that is required to make this kind of
digital transformation.

113 "externally or internally": Werner Vogels, "Working Backwards," All Things
Distributed, November 1, 2006, http://
www.allthingsdistributed.com/2006
/11/working_backwards.html.

114 "we are not always so helpful": Mark Burgess, *Thinking in Promises* (Sebastopol, CA: O'Reilly, 2015), 6.

114 "communication is terrible!": Janet Choi, "The Science Behind Why Jeff Bezos's Two-Pizza Team Rule Works," *I Done This Blog*, September 24, 2014, http://blog.idonethis.com/two-pizza-team/.

115 "both to technology and to the workplace": Burgess, *Thinking in Promises*, 1.

116 animated explainer videos: Henrik Kniberg, "Spotify Engineering Culture (Part 1)," Spotify, March 27, 2014, https://labs.spotify.com/2014/03/27/spotify-engineering-culture-part-1/, and "Spotify Engineering Culture (Part 2)," September 20, 2014, https://labs.spotify.com/2014/09/20/spotify-engineering-culture-part-2/.

116 low-alignment/low-autonomy organization: A still from Spotify's animated video illustrates this nicely: https://spotifylabs.files.wordpress.com/2014/03/spotify-engineering-culture-part1.jpeg.

116 "Follow the orders I would have given you": This is not a verbatim quote, but my recollection from the interview with General Stanley McChrystal and Chris Fussell by Charles Duhigg on March 1, 2016, http://nytconferences.com/NWS_Agenda_2016.pdf. It is also possible that the statement was made in my conversation with General McChrystal after his talk.

117 "'or you can't sell this product'": John Rossman, *The Amazon Way* (Seattle: Amazon Createspace, 2014), Kindle ed., loc. 250.

118 rescue the failed healthcare.gov website: Steven Brill, "Obama's Trauma Team," *Time*, February 27, 2014, http://time.com/10228/obamas-trauma-team/.

118 working under sixty different contracts: John Tozzi and Chloe Whiteaker, "All the Companies Making Money from Healthcare.gov in One Chart," *Bloomberg Businessweek*, August 28, 2014, https://www.bloomberg.com/news/articles/2014-08-28/all-the-companies-making-money-from-healthcare-dot-gov-in-one-chart.

120 "Junta": Venky Harinarayan, Anand Rajaraman, and Anand Ranganathan, Hybrid Machine/Human Computing Arrangement, US Patent 7,197,459, filed March 19, 2001, and issued March 27, 2007.

121 "The New Secret Sauce": Tim O'Reilly, "Operations: The New Secret Sauce," *O'Reilly Radar*, July 10, 2006, http://radar.oreilly.com/2006/07/operations-the-new-secret-sauc.html.

122 advantages that DevOps brings to an organization: Gene Kim, Kevin Behr, and George Spafford, *The Phoenix Project*, rev. ed. (Portland, OR: IT Revolution Press, 2014), 348–50.

122 "computer kaizen": Hal Varian, "Beyond Big Data," presented at the National Association of Business Economists Annual Meeting, September 10, 2013, San Francisco, http://people.ischool.berkeley.edu/~hal/Papers/2013/BeyondBigDataPaperFINAL.pdf.

123 "enable organizational learning and improvement": Kim, Behr, and Spafford, *The Phoenix Project*, 350.

123 "software to replace human labor": Benjamin Treynor Sloss, "Google's Approach to Service Management: Site Reliability Engineering," in *Site Reliability Engineering*, ed. Betsy Beyer, Chris Jones, Jennifer Petoff, and Niall Richard Murphy (Sebastopol, CA: O'Reilly, 2016), online at https://www.safaribooksonline.com/library/view/site-reliability-engineering/9781491929117/ch01.html.

CHAPTER 7: GOVERNMENT AS A PLATFORM

125 "subsidized access to data they were willing to pay for": Carl Malamud, "How EDGAR Met the Internet," *media.org*, retrieved March 30, 2017, http://museum.media.org/edgar/.

126 freely available on the Internet: Steven Levy, "The Internet's Own Instigator," *Backchannel*, September 12, 2016, https://backchannel.com/the-internets-own-instigator-cb6347e693b.

128 "the first Internet president": Omar Wasow, "The First Internet President," *The Root*, November 5, 2008, http://www.theroot.com/the-first-internet-president-1790900348.

129 "vending machine government": "The Next Government: Donald Kettl," IBM Center for the Business of Government, retrieved March 30, 2017, http://www.businessofgovernment.org/blog

/presidential-transition/next-government
-donald-kettl.

130 "not merely at an election one day in the year, but every day": "Thomas Jefferson to Joseph C. Cabell, February 2, 1816," in *Republican Government*, http://press -pubs.uchicago.edu/founders/documents /v1ch4s34.html. Reprinted from *The Writings of Thomas Jefferson*, ed. Andrew A. Lipscomb and Albert Ellery Bergh, 20 vols. (Washington, DC: Thomas Jefferson Memorial Association, 1905), vol. 14, 421–23.

130 supported by a variety of business models: David Robinson, Harlan Yu, William Zeller, and Ed Felten, "Government Data and the Invisible Hand," *Yale Journal of Law & Technology* 11, no. 1 (2009), art. 4, available at http://digital commons.law.yale.edu/yjolt/vol11 /iss1/4.

130 a working group meeting in December 2007: "Eight Principles of Open Government Data," *public.resource.org*, December 8, 2007, retrieved March 30, 2017, https://public.resource.org/8 _principles.html.

131 more useful to citizens and society: Andrew Young and Stefan Verhulst, *The Global Impact of Open Data* (Sebastopol, CA: O'Reilly, 2016), available for free download at http://www.oreilly.com/data /free/the-global-impact-of-open-data.csp.

131 a market now worth more than $26 billion: "Global GPS Market 2016–2022: Market Has Generated Revenue of $26.36 Billion in 2016 and Is Anticipated to Reach Up to $94.44 Billion by 2022," *Business Wire*, October 18, 2017, http://www.businesswire.com/news /home/20161018006653/en/Global -GPS-Market-2016-2022-Market -Generated-Revenue.

132 since it was first founded in 1952: Sean Pool and Jennifer Erickson, "The High Return on Investment for Publicly Funded Research," Center for American Progress, December 10, 2012, https:// www.americanprogress.org/issues /economy/reports/2012/12/10/47481 /the-high-return-on-investment-for -publicly-funded-research/.

132 and commercial space travel: It is worth noting, though, that all of Musk's big

bets have been subsidized by forward thinkers in government, as has so often been the case before. See "Elon Musk's Growing Empire Is Fueled by $4.9 Billion in Government Subsidies," *Los Angeles Times*, May 30, 2015, http://www .latimes.com/business/la-fi-hy-musk -subsidies-20150531-story.html.

132 Forty-two million: "About Us," Central Park Conservancy, http://www.central parknyc.org/about/. Seventy-five percent of the $65 million/year upkeep of the park comes from a nonprofit called the Central Park Conservancy, which was founded in 1980 as a result of declines in the quality of the park due to lack of city funding. The Central Park Conservancy could be read as a failure of government to provide the services that we pay for. It's rather a testament to the fact that sometimes, concerned citizens are willing to step forward and effectively tax themselves to pay for something that matters. In many ways, you can look at the Central Park Conservancy as a special kind of "local government" funded by concerned citizens.

134 "born when George III was on the throne": Ha-Joon Chang, *Bad Samaritans* (New York: Bloomsbury Press, 2008), 3–4.

134 great step forward in the American economy: Stephen S. Cohen and J. Bradford DeLong, *Concrete Economics* (Boston: Harvard Business Review Press, 2016).

139 families trying to navigate the maze: Stephanie Ebbert and Jenna Russell, "A Daily Diaspora, a Scattered Street," *Boston Globe*, June 12, 2011, http://archive .boston.com/news/education/k_12/articles /2011/06/12/on_one_city_street_school _choice_creates_a_gap/?page=full.

143 in the shoes of those they mean to serve: Jake Solomon, "People, Not Data," *Medium*, January 5, 2014, https://medium .com/@lippytak/people-not-data-47434 acb50a8.

143 "the poor struggle with daily": Ezra Klein, "Sorry Liberals, Obamacare's Problems Go Much Deeper than the Web Site," *Washington Post*, October 25, 2013, https://www.washingtonpost .com/news/wonk/wp/2013/10/25/oba

macares-problems-go-much-deeper
-than-the-web-site/.

144 "the best startup in Europe we can't invest in": Saul Klein, "Government Digital Service: The Best Startup in Europe
We Can't Invest In," *Guardian*, November 25, 2013, https://www.theguardian
.com/technology/2013/nov/15/govern
ment-digital-service-best-startup-europe
-invest.

145 GDS Design Principles: "GDS Design
Principles," UK Government Digital Service, retrieved March 31, 2017,
http://www.gov.uk/design-principles.

145 "Start with needs": After Mike Bracken
left the GDS, the first principle was
rewritten to leave out the revolutionary idea that existing government processes might be getting in the way of
user needs. For the original, whose first
principle is reproduced here, see "UK
Government Service Design Principles," Internet Archive, retrieved July 3,
2014, https://web-beta.archive.org/web
/20140703190229/https://www.gov.uk
/design-principles#first. The second principle reproduced here is from the current
version, footnoted above. It is actually
stronger and clearer than the original.

146 Casey Burns, and others: "The Digital
Services Playbook," United States Digital Service, retrieved March 31, 2017,
https://playbook.cio.gov. Authorship is
as recalled by Jen Pahlka.

146 "inextricably linked to an understanding of the digital": Tom Steinberg,
"5 Years On: Why Understanding
Chris Lightfoot Matters Now More
Than Ever," My Society, February
11, 2012, https://www.mysociety.org
/2012/02/11/5-years-on-why-under
standing-chris-lightfoot-matters-now
-more-than-ever/.

147 vulnerabilities in Department of Defense
websites: "2016 Report to Congress: High
Priority Projects," United States Digital Service, December 2016, retrieved
March 31, 2017, https://www.usds.gov
/report-to-congress/2016/projects/.

147 within and for the agencies: "2016 Report to Congress," United States Digital Service, December 2016, retrieved
March 31, 2017, https://www.usds.gov
/report-to-congress/2016/.

148 "to invite the tech people to the table":
As recalled by Jen Pahlka, who attended
the event.

149 "things you can fix if you choose to":
Mikey Dickerson, "Mikey Dickerson to
SXSW: Why We Need You in Government," *Medium*, retrieved March 31, 2017,
https://medium.com/the-u-s-digital
-service/mikey-dickerson-to-sxsw-why
-we-need-you-in-government-f31dab
3263a0. These were Mikey's opening re-
marks from a joint talk with Jen Pahlka
at the 2015 SXSW Interactive Festival
titled "How Government Fails and How
You Can Fix It."

150 "do at all, or do so well, for themselves":
Abraham Lincoln, "Fragment on Government," in *Collected Works of Abraham
Lincoln*, vol. 2 (Springfield, IL: Abraham Lincoln Association, 1953), 222,
reproduced at http://quod.lib.umich.edu
/l/lincoln/lincoln2/1:262?rgn=div1
;view=fulltext. There is a related fragment on page 221, reproduced at http://
quod.lib.umich.edu/l/lincoln/lincoln2
/1:261?rgn=div1;view=fulltext.

CHAPTER 8: MANAGING A WORKFORCE OF DJINNS

155 breakthroughs and business processes:
Steve Lohr, "The Origins of 'Big Data':
An Etymological Detective Story," *New
York Times*, February 1, 2013, https://
bits.blogs.nytimes.com/2013/02/01
/the-origins-of-big-data-an-etymological
-detective-story/.

155 speech recognition and machine translation: Alon Halevy, Peter Norvig, and
Fernando Pereira, "The Unreasonable
Effectiveness of Data," *IEEE Intelligent
Systems*, 1541–1672/09, retrieved March
31, 2017, https://static.googleusercontent
.com/media/research.google.com/en//
pubs/archive/35179.pdf.

156 "the sexiest job of the 21st century":
Thomas Davenport and D. J. Patil,
"Data Scientist: The Sexiest Job of the
21st Century," *Harvard Business Review*,
October 2012, https://hbr.org/2012/10
/data-scientist-the-sexiest-job-of-the
-21st-century. Hal Varian had used this
same phrase about statistics in 2009.
See "Hal Varian on How the Web
Challenges Managers," McKinsey &

Company, January 2009, http://www
.mckinsey.com/industries/high-tech
/our-insights/hal-varian-on-how-the
-web-challenges-managers.

157 "the right values for these parameters is
something of a black art": Sergey Brin
and Larry Page, "The Anatomy of a
Large-Scale Hypertextual Web Search
Engine," Stanford University, retrieved
March 31, 2017, http://infolab.stanford
.edu/~backrub/google.html.

158 as many as 50,000 subsignals: Danny
Sullivan, "FAQ: All About the Google
RankBrain Algorithm," *Search Engine
Land*, June 23, 2016, http://searchengine
land.com/faq-all-about-the-new-google
-rankbrain-algorithm-234440.

158 "new synapses for the global brain": Tim
O'Reilly, "Freebase Will Prove Addictive," *O'Reilly Radar*, March 8, 2007,
http://radar.oreilly.com/2007/03/free
base-will-prove-addictive.html.

158 "10 experiments for every successful launch": Matt McGee, "*Business-
Week* Dives Deep into Google's Search
Quality," *Search Engine Land*, October
6, 2009, http://searchengineland.com
/businessweek-dives-deep-into-googles
-search-quality-27317.

159 the manual that they provide: *Search
Quality Evaluator Guide*, Google, March
14, 2017, http://static.googleusercontent
.com/media/www.google.com/en//inside
search/howsearchworks/assets/search
qualityevaluatorguidelines.pdf.

160 "Another big difference": Brin and Page,
"The Anatomy of a Large-Scale Hypertextual Web Search Engine," Section
3.2. They expand on the problem in Appendix A.

161 "the database of intentions": John Battelle, "The Database of Intentions," *John
Batelle's Searchblog*, November 13, 2003,
http://battellemedia.com/archives
/2003/11/the_database_of_intentions
.php.

162 how Google's ad auction actually works:
Hal Varian, "Online Ad Auctions,"
draft, February 16, 2009, http://people
.ischool.berkeley.edu/~hal/Papers
/2009/online-ad-auctions.pdf.

162 users find "meaningful": Farhad Man-
joo, "Social Insecurity," *The New York
Times Magazine*, April 30, 2017, https://

www.nytimes.com/2017/04/25/mag
azine/can-facebook-fix-its-own-worst
-bug.html.

163 "We shape our tools": This quote is often mistakenly attributed to McLuhan
himself. See "We shape our tools and
thereafter our tools shape us," *McLuhan
Galaxy*, April 1, 2013, https://mcluhan
galaxy.wordpress.com/2013/04/01/we
-shape-our-tools-and-thereafter-our
-tools-shape-us/.

165 "what a typical Deep Learning system is":
Lee Gomes, "Facebook AI Director Yann
LeCun on His Quest to Unleash Deep
Learning and Make Machines Smarter,"
IEEE Spectrum, February 28, 2015,
http://spectrum.ieee.org/automaton
/robotics/artificial-intelligence/face
book-ai-director-yann-lecun-on-deep
-learning.

165 "can't run it faster than real time": Yann
LeCun, Facebook post, December 5,
2016, retrieved March 31, 2017, https://m
.facebook.com/story.php?story_fbid
=10154017359117143&id=722677142.

165 third most important: Sullivan, "FAQ:
All About the Google RankBrain Algorithm."

165 stopped all work on the old Google
Translate system: Gideon Lewis-Kraus,
"The Great A.I. Awakening," *New York
Times Magazine*, December 14, 2016,
https://www.nytimes.com/2016/12/14
/magazine/the-great-ai-awakening.html.

167 algorithmic detection of fake news:
Jennifer Slegg, "Google Tackles Fake
News, Inaccurate Content & Hate Sites
in Rater Guidelines Update," *SEM
Post*, March 14, 2017, http://www
.thesempost.com/google-tackles-fake
-news-inaccurate-content-hate-sites-rater
-guidelines-update/.

167 "directly from raw experience or data":
This claim has been removed from the
deepmind.com website, but it can still
be found via the Internet Archive. Retrieved March 28, 2016, https://web
-beta.archive.org/web/20160328210752
/https://deepmind.com/.

167 "the hallmark of true artificial general
intelligence": Demis Hassabis, "What
We Learned in Seoul with AlphaGo,"
Google Blog, March 16, 2016, https://
blog.google/topics/machine-learning

/what-we-learned-in-seoul-with -alphago/.

167 "getting to true AI": Ben Rossi, "Google DeepMind's AlphaGo Victory Not 'True AI,' Says Facebook's AI Chief," *Information Age*, March 14, 2016, http://www .information-age.com/google-deepminds -alphago-victory-not-true-ai-says-face books-ai-chief-123461099/.

169 "thinking about how to make people click ads": Ashlee Vance, "This Tech Bubble Is Different," *Bloomberg Businessweek*, April 14, 2011, https://www.bloomberg.com /news/articles/2011-04-14/this-tech -bubble-is-different.

CHAPTER 9: "A HOT TEMPER LEAPS O'ER A COLD DECREE"

172 as much as 30,000 pages of regulations: Andrew Haldane, "The Dog and the Frisbee," speech at the Federal Reserve Bank of Kansas City's 366th economic policy symposium, Jackson Hole, Wyoming, August 31, 2012, http://www.bis .org/review/r120905a.pdf.

177 Brin answers, "All of us": David Brin, *The Transparent Society* (New York: Perseus, 1998). See also http://www .davidbrin.com/transparentsociety.html.

177 "is generally bad": Bruce Schneier, "The Myth of the 'Transparent Society,'" *Bruce Schneier on Security*, March 6, 2008, https://www.schneier.com/essays /archives/2008/03/the_myth_of_the _tran.html.

178 "we all know where the creep factor comes in": Alexis Madrigal, "Get Ready to Roboshop," *Atlantic*, March 2014, https://www.theatlantic.com/magazine /archive/2014/03/get-ready-to-roboshop /357569/.

178 steering Mac users to higher-priced hotels: Dana Mattioli, "On Orbitz, Mac Users Steered to Pricier Hotels," *Wall Street Journal*, August 23, 2012, https:// www.wsj.com/articles/SB10001424052 7023044586045774888226667325882.

180 express their intent clearly and simply: "Share Your Work," Creative Commons, retrieved March 31, 2017, https://crea tivecommons.org/share-your-work/.

180 "Smart Disclosure": "Smart Disclosure Policy Resources," *data.gov*, retrieved March 31, 2017, https://www.data.gov

/consumer/smart-disclosure-policy -resources.

180 "smart contracts": Josh Stark, "Making Sense of Blockchain Smart Contracts," June 4, 2016, http://www.coindesk.com /making-sense-smart-contracts/.

181 a requirement for interpretability: Tal Zarsky, "Transparency in Data Mining: From Theory to Practice," in *Discrimination and Privacy in the Information Society*, ed. Bart Custers, Toon Calders, Bart Schermer, and Tal Zarsky (New York: Springer, 2012), 306.

182 a perfect marketplace: Adam Cohen, "'The Perfect Store," *New York Times*, June 16, 2002, http://www.nytimes.com /2002/06/16/books/chapters/the-perfect -store.html.

182 nothing was known about the sellers: Paul Resnick and Richard Zeckhauser, "Trust Among Strangers in Internet Transactions: Empirical Analysis of eBay's Reputation System," draft of February 5, 2001, version for review by NBER workshop participants, http://www.presnick .people.si.umich.edu/papers/ebay NBER/RZNBERBodegaBay.pdf.

183 "the apps and algorithms provide a filter": David Lang, "The Life-Changing Magic of Small Amounts of Money," *Medium*, unpublished post retrieved April 5, 2017, https://medium.com/@davidtlang /cacb7277ee9f.

184 "the multitude and promiscuous use of coaches": Steven Hill, "Our Streets as a Public Utility: How UBER Could Be Part of the Solution," *Medium*, September 2, 2015, https://medium.com /the-wtf-economy/our-streets-as-a -public-utility-how-uber-could-be-part -of-the-solution-65772bdf5dcf.

184 "cried out for public control over the taxi industry": Steven Hill, "Rethinking the Uber vs. Taxi Battle," *Globalist*, September 27, 2015, https://www.theglobalist .com/uber-taxi-battle-commercial -transport/.

185 "The entire transaction": Varian, "Beyond Big Data," 9.

186 "maximum amount of validated learning about customers": Eric Ries, "Minimum Viable Product: A Guide," *Startup Lessons Learned*, August 3, 2009, http:// www.startuplessonslearned.com/2009

/08/minimum-viable-product-guide .html.

186 Feedback loops are tight: For an excellent account of this process, see Chris Anderson, "Closing the Loop," *Edge*, retrieved March 31, 2017, https://www .edge.org/conversation/chris_anderson -closing-the-loop.

187 "500 pages of untested assumptions": Tom Loosemore, "Government as a Platform: How New Foundations Can Support Natively Digital Public Services," presented at the Code for America Summit in San Francisco, September 30– October 2, 2015, https://www.youtube .com/watch?v=VjE_zj-7A7A&feature =youtu.be.

189 "fall back to a minimal risk condition": NHTSA Federal Automated Vehicles Policy, National Highway Traffic Safety Administration, September 2016, https://www.nhtsa.gov/sites/nhtsa.dot .gov/files/federal_automated_vehicles _policy.pdf, 14.

189 "regulators need to accept a new model": Nick Grossman, "Here's the Solution to the Uber and Airbnb Problems— and No One Will Like It," *The Slow Hunch*, July 23, 2015, http://www .nickgrossman.is/2015/heres-the-solu tion-to-the-uber-and-airbnb-problems -and-no-one-will-like-it/.

190 Hospitality Staffing Solutions: Dave Jamieson, "As Hotels Outsource Jobs, Workers Lose Hold on Living Wage," *Huffington Post*, October 24, 2011, http://www.huffingtonpost.com/2011 /08/24/-hotel-labor-living-wage -outsourcing-indianapolis_n_934667 .html.

191 Integrity Staffing Solutions: Dave Jamieson, "The Life and Death of an Amazon Warehouse Temp," *Medium*, October 23, 2015, https://medium.com /the-wtf-economy/the-life-and-death-of -an-amazon-warehouse-temp-8168c 4702049.

191 the Gap: R. L. Stephens II, "I Often Can't Afford Groceries Because of Volatile Work Schedules at Gap," *Guardian*, August 17, 2015, https://www .theguardian.com/commentisfree/2015 /aug/17/cant-afford-groceries-volatile -work-schedules-gap.

191 Starbucks: Jodi Cantor, "Working Anything but 9 to 5," *New York Times*, August 13, 2014, https://www.nytimes .com/interactive/2014/08/13/us/star bucks-workers-scheduling-hours.html.

192 Starbucks only banned in mid-2014: Jodi Cantor, "Starbucks to Revise Policies to End Irregular Schedules for Its 130,000 Baristas," *New York Times*, August 15, 2014, https://www.nytimes .com/2014/08/15/us/starbucks-to-revise -work-scheduling-policies.html.

192 "not enough hours": Jodi Lambert, "The Real Low-Wage Issue: Not Enough Hours," *CNN*, January 13, 2014, http:// money.cnn.com/2014/01/13/news /economy/minimum-wage-hours/.

192 a host of other labor woes: Carrie Gleason and Susan Lambert, "Uncertainty by the Hour," Future of Work Project, retrieved March 31, 2017, http://static.opensocietyfoundations .org/misc/future-of-work/just-in-time -workforce-technologies-and-low-wage -workers.pdf.

192 a study of Uber drivers: Jonathan Hall and Alan Krueger, "An Analysis of the Labor Market for Uber's Driver-Partners in the United States," Uber, January 22, 2015, https://s3.amazonaws.com/uber -static/comms/PDF/Uber_Driver-Partners _Hall_Kreuger_2015.pdf.

193 rather than to increase hours for individual workers: Susan Lambert, "Work Scheduling Study," University of Chicago School of Social Service Administration, May 2010, retrieved March 31, 2017, https://ssascholars.uchicago .edu/sites/default/files/work-scheduling -study/files/univ_of_chicago_work _scheduling_manager_report_6_25_0 .pdf.

193 "In August 2013": Esther Kaplan, "The Spy Who Fired Me," *Harper's*, March 2015, 36, available at http://populardemo cracy.org/sites/default/files/Harpers Magazine-2015-03-0085373.pdf.

194 "new jobs fall on that spectrum": Lauren Smiley, "Grilling the Government About the On-Demand Economy," *Backchannel*, August 23, 2015, https://backchannel.com/why-the-us -secretary-or-labor-doesn-t-uber -272f18799f1a.

194 They became part-time employees: Brad Stone, "Instacart Reclassifies Part of Its Workforce Amid Regulatory Pressure on Uber," *Bloomberg Technology*, June 22, 2015, https://www.bloomberg.com/news/articles/2015-06-22/instacart-reclassifies-part-of-its-workforce-amid-regulatory-pressure-on-uber.

195 present for their children's birthdays: Noam Scheiber, "The Perils of Ever-Changing Work Schedules Extend to Children's Well-Being," *New York Times*, August 12, 2015, https://www.nytimes.com/2015/08/13/business/economy/the-perils-of-ever-changin-work-schedules-extend-to-childrens-well-being.html.

196 writing in *Harvard Business Review*: Andrei Hagiu and Rob Biederman, "Companies Need an Option Between Contractor and Employee," *Harvard Business Review*, August 21, 2015, https://hbr.org/2015/08/companies-need-an-option-between-contractor-and-employee.

196 writing on *Medium*: Simon Rothman, "The Rise of the Uncollared Worker and the Future of the Middle Class," *Medium*, July 7, 2015, https://news.greylock.com/the-rise-of-the-uncollared-worker-and-the-future-of-the-middle-class-860a928357b7.

196 "Shared Security Account": Nick Hanauer and David Rolf, "Shared Security, Shared Growth," *Democracy*, no. 37 (Summer 2015), http://democracyjournal.org/magazine/37/shared-security-shared-growth/?page=all.

196 policy proposal for portable benefits: Steven Hill, "New Economy, New Social Contract," New America, August 4, 2015, https://www.newamerica.org/economic-growth/policy-papers/new-economy-new-social-contract/.

197 "the only game being played": Zeynep Ton, *The Good Jobs Strategy* (Boston: New Harvest, 2014). This quote appears at http://zeynepton.com/book/.

CHAPTER 10: MEDIA IN THE AGE OF ALGORITHMS

199 Macedonian teens out to make a buck: Craig Silverman and Lawrence Alexander, "How Teens in the Balkans Are Duping Trump Supporters with Fake News," *BuzzFeed*, November 3, 2016, https://www.buzzfeed.com/craigsilverman/how-macedonia-became-a-global-hub-for-pro-trump-misinfo.

199 to churn out the stuff: Laura Sydell, "We Tracked Down a Fake-News Creator in the Suburbs. Here's What We Learned," *NPR All Tech Considered*, November 23, 2016, http://www.npr.org/sections/alltechconsidered/2016/11/23/503146770/npr-finds-the-head-of-a-covert-fake-news-operation-in-the-suburbs.

199 "a pretty crazy idea": Aarti Shahani, "Zuckerberg Denies Fake News on Facebook Had Impact on the Election," *NPR All Tech Considered*, November 11, 2016, http://www.npr.org/sections/alltechconsidered/2016/11/11/501743684/zuckerberg-denies-fake-news-on-facebook-had-impact-on-the-election.

200 hyperpartisan stories shown to each group: "Blue Feed/Red Feed," *Wall Street Journal*, May 18, 2016, updated hourly, retrieved March 31, 2007, http://graphics.wsj.com/blue-feed-red-feed/.

200 video claiming that her aide Huma Abedin: "Huma Kidding?," *Snopes.com*, November 2, 2016, http://www.snopes.com/huma-abedin-ties-to-terrorists/.

201 planted or amplified by Russia: Joseph Menn, "U.S. Government Loses to Russia's Disinformation Campaign," Reuters, December 21, 2016, http://www.reuters.com/article/us-usa-russia-disinformation-analysis-idUSKBN1492PA.

201 the week after his dismissive comments: Mark Zuckerberg, Facebook post, November 12, 2016, https://www.facebook.com/zuck/posts/10103253901916271.

202 autocomplete for "Jews are . . .": Carole Cadwalladr, "Google, Democracy and the Truth About Internet Search," *Guardian*, December 4, 2016, https://www.theguardian.com/technology/2016/dec/04/google-democracy-truth-internet-search-facebook.

203 again topped by a page from Stormfront: Carole Cadwalladr, "How to Bump Holocaust Deniers off Google's Top Spot? Pay Google," *Guardian*, December 17, 2016, https://www.theguardian

.com/technology/2016/dec/17/holocaust-deniers-google-search-top-spot.

203 "this is exactly what it is": Carole Cadwalladr, "Google Is Not 'Just' a Platform. It Frames, Shapes and Distorts How We See the World," *Guardian*, December 11, 2016, https://www.theguardian.com/commentisfree/2016/dec/11/google-frames-shapes-and-distorts-how-we-see-world.

203 from 250 billion unique web domain names: "A Look at the Future of Search with Google's Amit Singhal at SXSW," *PR Newswire*, March 10, 2013, http://www.prnewswire.com/blog/a-look-at-the-future-of-search-with-googles-amit-singhal-at-sxsw-6602.html.

203 more than 5 billion searches a day: Danny Sullivan, "Google Now Handles At Least 2 Trillion Searches per Year," *Search Engine Land*, May 24, 2016, http://searchengineland.com/google-now-handles-2-999-trillion-searches-per-year-250247.

204 made only about 300 times a day: Danny Sullivan, "Official: Google Makes Change, Results Are No Longer in Denial over 'Did the Holocaust Happen?,'" *Search Engine Land*, December 20, 2016, http://searchengineland.com/googles-results-no-longer-in-denial-over-holocaust-265832.

204 ends up with all the monkeys: William Oncken Jr. and Donald L. Wass, "Who's Got the Monkey?," *Harvard Business Review*, November–December 1999, https://hbr.org/1999/11/management-time-whos-got-the-monkey#comment-section.

205 Holocaust denial had been improved: Danny Sullivan, "Google's Top Results for 'Did the Holocaust Happen' Now Expunged of Denial Sites," *Search Engine Land*, December 24, 2016, http://searchengineland.com/google-holocaust-denial-site-gone-266353.

205 "Meme magic is real": Milo Yiannopoulos, "Meme Magic: Donald Trump Is the Internet's Revenge on Lazy Elites," *Breitbart*, May 4, 2016, http://www.breitbart.com/milo/2016/05/04/meme-magic-donald-trump-internets-revenge-lazy-entitled-elites/.

205 patent filed in June 2015: Erez Laks, Adam Stopek, Adi Masad, Israel Nir, Systems and Methods to Identify Objectionable Content, US Patent Application 20160350675, filed June 1, 2016, published December 1, 2016, http://pdfaiw.uspto.gov/.aiw?PageNum=0&docid=20160350675&IDKey=B0738725A3CA.

206 flagged by fact checkers or the community: Mark Zuckerberg, Facebook post, November 18, 2016, https://www.facebook.com/zuck/posts/10103269806149061.

206 350,000 times on Facebook: Sapna Maheshwari, "How Fake News Goes Viral: A Case Study," *New York Times*, November 20, 2016, https://www.nytimes.com/2016/11/20/business/media/how-fake-news-spreads.html.

206 "social listening tools": Alexis Sobel Fitts, "The New Importance of 'Social Listening' Tools," *Columbia Journalism Review*, July/August 2015, http://www.cjr.org/analysis/the_new_importance_of_social_listening_tools.php.

207 Tucker had deleted the original tweet: Eric Tucker, "Why I'm Removing the 'Fake Protests' Twitter Post," *Eric Tucker* (blog), November 11, 2016, https://blog.erictucker.com/2016/11/11/why-im-considering-to-remove-the-fake-protests-twitter-post/.

207 recognized the significance of the blue check mark: Brooke Donald, "Stanford Researchers Find Students Have Trouble Judging the Credibility of Information Online," Stanford Graduate School of Education, November 22, 2016, https://ed.stanford.edu/news/stanford-researchers-find-students-have-trouble-judging-credibility-information-online.

207 "only the people we want to see it, see it": Joshua Green and Sissa Isenberg, "Inside the Trump Bunker, with Days to Go," *Bloomberg Businessweek*, October 27, 2016, https://www.bloomberg.com/news/articles/2016-10-27/inside-the-trump-bunker-with-12-days-to-go.

208 "never letting them go": Cadwalladr, "Google, Democracy and the Truth About Internet Search."

208 programs masquerading as users: Vindu Goel, "Russian Cyberforgers Steal Millions a Day with Fake Sites," *New York Times*, December 20, 2016, http://www

.nytimes.com/2016/12/20/technology
/forgers-use-fake-web-users-to-steal
-real-ad-revenue.html.

209 faster than humans can patch them: Cyber Grand Challenge Rules, Version 3, November 18, 2014, DARPA, http://archive.darpa.mil/CyberGrandChallenge_CompetitorSite/Files/CGC_Rules_18_Nov_14_Version_3.pdf.

209 "uncertain, ambiguous, or incomprehensible": Harry Hillaker, "Tribute to John R. Boyd," *Code One*, July 1997, retrieved April 1, 2017, https://web.archive.org/web/20070917232626/http://www.codeonemagazine.com/archives/1997/articles/jul_97/july2a_97.html.

210 decry the result as biased: Raj Shah, "PolitiFact's So-Called Fact-Checks Show Bias, Incompetence, or Both," Republican National Committee, August 30, 2016, https://gop.com/politifacts-so-called-fact-checks-show-bias-incompetence-or-both/.

210 "reflexive knowledge": George Soros, *The Crisis of Global Capitalism* (New York: PublicAffairs, 1998), 6–18.

210 "shaping the events in which we participate": George Soros, *Open Society* (New York: PublicAffairs, 2000), xii.

212 painted a very different picture: Gus Lubin, Mike Nudelman, and Erin Fuchs, "9 Maps That Show How Americans Commit Crime," *Business Insider*, September 25, 2013, http://www.businessinsider.com/maps-on-fbis-uniform-crime-report-2013-9.

212 "clickbait" headlines: Alex Peysakhovich and Kristin Hendrix, "News Feed FYI: Further Reducing Clickbait in Feed," Facebook newsroom, August 24, 2016, http://newsroom.fb.com/news/2016/08/news-feed-fyi-further-reducing-clickbait-in-feed/.

213 "legalizing child prostitution": Travis Allen, "California Democrats Legalize Child Prostitution," December 29, 2016, http://www.washingtonexaminer.com/california-democrats-legalize-child-prostitution/article/2610540.

214 "maybe you recognize the source, maybe you don't": John Borthwick, "Media Hacking," *Render*, March 7, 2015, https://render.betaworks.com/media-hacking-3b1e350d619c.

214 "For the reason that it makes Google more money": Cadwalladr, "How to Bump Holocaust Deniers off Google's Top Spot? Pay Google."

215 Facebook's desire not to be the arbiter of truth: Peter Kafka, "Facebook Has Started to Flag Fake News Stories," *Recode*, March 4, 2017, https://www.recode.net/2017/3/4/14816254/facebook-fake-news-disputed-trump-snopes-politifact-seattle-tribune.

215 "before it turns into a tsunami": Krishna Bharat, "How to Detect Fake News in Real-Time," *NewCo Shift*, April 27, 2017, https://shift.newco.co/how-to-detect-fake-news-in-real-time-9fdae0197bfd. Bharat's article has many additional practical suggestions for how to algorithmically detect fake news in addition to some of those that I outline in this chapter.

215 "the overall logic is the same": Bharat, "How to Detect Fake News in Real-Time."

217 "to make it tolerate them": Michael Marder, "Failure of U.S. Public Secondary Schools in Mathematics," University of Texas UTeach, retrieved April 1, 2017, https://uteach.utexas.edu/sites/default/files/BrokenEducation2011.pdf, 3.

218 work together for the common good: Mark Zuckerberg, "Building Global Community," *Facebook*, February 16, 2017, https://www.facebook.com/notes/mark-zuckerberg/building-global-community/10154544292806634/.

219 "They have become rich because they were civic": Robert Putnam, "The Prosperous Community: Social Capital and Public Life," *American Prospect*, Spring 1993, retrieved April 1, 2017, http://prospect.org/article/prosperous-community-social-capital-and-public-life.

219 "for inclusion of all": Zuckerberg, "Building Global Community."

220 his experience with the Egyptian revolution: Wael Ghonim, *Revolution 2.0* (New York: Houghton Mifflin, 2012).

221 "engage in rational discussion": Colin Megill, "pol.is in Taiwan," *pol.is blog*, May 25, 2016, https://blog.pol.is/pol-is-in-taiwan-da7570d372b5.

221 "love comedies and horror but hate documentaries": Ibid.

222 you move accordingly: "Human Spectrogram," *Knowledge Sharing Tools and Method Toolkit*, wiki retrieved April 1, 2017, http://www.kstoolkit.org/Human +Spectrogram.

222 "mandatory for riders on uberX private vehicles": Audrey Tang, "Uber Responds to vTaiwan's Coherent Blended Volition," *pol.is blog*, May 23, 2016, https://blog .pol.is/uber-responds-to-vtaiwans-coherent -blended-volition-3e9b75102b9b.

223 points of agreement and disagreement: Ray Dalio, TED, April 24, 2017, https:// ted2017.ted.com/program.

224 "figure out how to get around it": Josh Constine, "Facebook's New Anti-Clickbait Algorithm Buries Bogus Headlines," *TechCrunch*, August 4, 2016, https:// techcrunch.com/2016/08/04/facebook -clickbait/.

226 half of digital ad spending: Greg Sterling, "Search Ads Generated 50 Percent of Digital Revenue in First Half of 2016," *Search Engine Land*, November 1, 2016, http://searchengineland.com/search -ads-1h-generated-16-3-billion-50 -percent-total-digital-revenue-262217.

226 "We need a new model": Evan Williams, "Renewing Medium's Focus," *Medium*, January 4, 2017, https://blog .medium.com/renewing-mediums-focus -98f374a960be.

227 "To continue on this trajectory": Ibid.

227 "In an experiment": Adam D. I. Kramer, Jamie E. Guillory, and Jeffrey T. Hancock, "Experimental Evidence of Massive-Scale Emotional Contagion Through Social Networks," *Proceedings of the National Academy of Sciences*, June 17, 2014, updated with *PNAS* "Editorial Expression of Concern and Correction," July 22, 2014, http://www.pnas.org/content/111/24 /8788.full.pdf.

227 "we are all lab rats": Vindu Goel, "Facebook Tinkers with Users' Emotions in News Feed Experiment, Stirring Outcry," *New York Times*, June 29, 2014, https://www .nytimes.com/2014/06/30/technology/face book-tinkers-with-users-emotions-in-news -feed-experiment-stirring-outcry.html.

228 with apologies to Pedro Domingos: This is a reference to the title of Domingos's book, *The Master Algorithm* (New York: Basic Books, 2015).

228 "damn good for CBS": Eliza Collins, "Les Moonves: Trump's Run Is 'Damn Good for CBS,'" *Politico*, June 29, 2016, http:// www.politico.com/blogs/on-media/2016 /02/les-moonves-trump-cbs-220001.

CHAPTER 11: OUR SKYNET MOMENT

230 The messages were powerful and personal: "We Are the 99 Percent," *tumblr .com*, September 14, 2011, http://weare the99percent.tumblr.com/page/231.

231 "AI systems must do what we want them to do": "An Open Letter: Research Priorities for Robust and Beneficial Artificial Intelligence," Future of Life Institute, retrieved April 1, 2017, https://futureoflife.org/ai-open-letter/.

231 "unconstrained by a need to generate financial return": Greg Brockman, Ilya Sutskever, and OpenAI, "Introducing OpenAI," *OpenAI Blog*, December 11, 2015, https://blog.openai.com/introduc ing-openai/.

232 best friend of one autistic boy: Judith Newman, "To Siri, with Love," *New York Times*, October 17, 2014, https:// www.nytimes.com/2014/10/19/fashion /how-apples-siri-became-one-autistic -boys-bff.html.

234 overpopulation on Mars: "Andrew Ng: Why 'Deep Learning' Is a Mandate for Humans, Not Just Machines," *Wired*, May 2015, retrieved April 1, 2017, https:// www.wired.com/brandlab/2015/05 /andrew-ng-deep-learning-mandate -humans-not-just-machines/.

235 change how we think and how we feel: Emeran A. Mayer, Rob Knight, Sarkis K. Mazmanian, John F. Cryan, and Kirsten Tillisch, "Gut Microbes and the Brain: Paradigm Shift in Neuroscience," *Journal of Neuroscience*, 34, no. 46 (2014): 15490–96, doi:10.1523/JNEU ROSCI.3299-14.2014.

235 "games of self-play": David Silver et al., "Mastering the Game of Go with Deep Neural Networks and Tree Search," *Nature* 529 (2016): 484–89, doi:10.1038 /nature16961.

236 "previously encountered examples": Beau Cronin, "Untapped Opportunities in AI," *O'Reilly Ideas*, June 4, 2014, https:// www.oreilly.com/ideas/untapped -opportunities-in-ai.

237 "that for a computer is plenty of time": Michael Lewis interviewed by Terry Gross, "On a 'Rigged' Wall Street, Milliseconds Make All the Difference," *NPR Fresh Air*, April 1, 2014, http://www.npr.org/2014/04/01/297686724/on-a-rigged-wall-street-milliseconds-make-all-the-difference.

238 "Creating things that you don't understand is really not a good idea": Felix Salmon, "John Thain Comes Clean," Reuters, October 7, 2009, http://blogs.reuters.com/felix-salmon/2009/10/07/john-thain-comes-clean/.

238 credit far in excess of the underlying real assets: Gary Gorton, "Shadow Banking," *The Region* (Federal Reserve Bank of Minneapolis), December 2010, retrieved April 2, 2017, http://faculty.som.yale.edu/garygorton/documents/InterviewwithTheRegionFRBofMinneapolis.pdf.

239 "an existential threat to capitalism": Mark Blyth, "Global Trumpism," *Foreign Affairs*, November 15, 2016, https://www.foreignaffairs.com/articles/2016-11-15/global-trumpism.

240 "pure and unadulterated socialism": Milton Friedman, "The Social Responsibility of Business Is to Increase Its Profits," *New York Times Magazine*, September 13, 1970, retrieved April 2, 2017, http://www.colorado.edu/studentgroups/libertarians/issues/friedman-soc-resp-business.html.

241 benefit the business and its actual owners: Michael C. Jensen, and William H. Meckling, "Theory of the Firm: Managerial Behavior, Agency Costs and Ownership Structure," *Journal of Financial Economics* 3, no. 4 (1976), http://dx.doi.org/10.2139/ssrn.94043.

241 sold or shuttered: Jack Welch, "Growing Fast in a Slow-Growth Economy," Appendix A in Jack Welch and John Byrne, *Jack: Straight from the Gut* (New York: Warner Books, 2001).

242 "above which repurchases will be eschewed": Warren Buffett, "Berkshire Hathaway Shareholder Letters: 2016," Berkshire Hathaway, February 25, 2017, http://berkshirehathaway.com/letters/2016ltr.pdf.

243 "fulfill their responsibilities to their employees": Larry Fink, "I write on behalf of our clients . . . ," BlackRock, January 24, 2017, https://www.blackrock.com/corporate/en-us/investor-relations/larry-fink-ceo-letter.

243 growth of productivity in the US economy slowed substantially after 1970: Robert J. Gordon, *The Rise and Fall of American Growth* (Princeton, NJ: Princeton University Press, 2016).

244 half of all Americans are shareholders in any form: Justin McCarthy, "Little Change in Percentage of Americans Who Own Stocks," Gallup, April 22, 2015, http://www.gallup.com/poll/182816/little-change-percentage-americans-invested-market.aspx.

244 outperforms both its publicly traded competitors and the entire S&P 500 retail index: Kyle Stock, "REI's Crunchy Business Model Is Crushing Retail Competitors," *Bloomberg*, March 27, 2015, https://www.bloomberg.com/news/articles/2015-03-27/rei-s-crunchy-business-model-is-crushing-retail-competitors.

244 from money managers to its customers: "Why Ownership Matters," Vanguard, retrieved April 4, 2017, https://about.vanguard.com/what-sets-vanguard-apart/why-ownership-matters/.

245 an astonishing $3.4 trillion on stock buybacks: William Lazonick, "Stock Buybacks: From Retain-and-Reinvest to Downsize-and-Distribute," Brookings Center for Effective Public Management, April 2015, https://www.brookings.edu/wp-content/uploads/2016/06/lazonick.pdf.

245 "investment in productive assets": Ibid., 4.

245 "distributes corporate cash to shareholders": Ibid., 2.

246 "social rate of return" from innovation: Charles Jones and John Williams, "Measuring the Social Return to R&D," Federal Reserve Board of Governors, February 1997, https://www.federalreserve.gov/pubs/feds/1997/199712/199712pap.pdf.

246 "not the golden goose itself": Ashish Arora, Sharon Belenzon, and Andrea Patacconi, "Killing the Golden Goose? The Decline of Science in Corporate R&D," National Bureau of Economic Research, January 2015, doi:10.3386/w20902.

246 from about 4% to nearly 11%: Derek Thompson, "Corporate Profits Are Eating the Economy," *Atlantic*, March 4, 2013, https://www.theatlantic.com /business/archive/2013/03/corporate -profits-are-eating-the-economy /273687/. Updated figures on which this graph is based are available from US Bureau of Economic Analysis, Compensation of Employees: Wages and Salary Accruals (WASCUR), https://fred.st-louisfed.org/series/WASCUR, retrieved from FRED, Federal Reserve Bank of St. Louis, April 2, 2017; Corporate Profits After Tax (without IVA and CCAdj) (CP), retrieved from FRED, Federal Reserve Bank of St. Louis; https://fred .stlouisfed.org/series/CP, April 2, 2017; Gross Domestic Product (GDP), retrieved from FRED, Federal Reserve Bank of St. Louis; https://fred.stlouisfed .org/series/GDP, April 2, 2017.

246 "something approaching a zero-sum game": Rana Foroohar, *Makers and Takers* (New York: Crown, 2016), 18.

246 "the one percent in pre-revolutionary France": Rana Foroohar, "Thomas Piketty: Marx 2.0," *Time*, May 9, 2014, http://time .com/92087/thomas-piketty-marx-2-0/. Retrieved April 2, 2017, http://piketty .pse.ens.fr/files/capital21c/en/media /Time%20-%20Capital%20in%20the %20Twenty-First%20Century.pdf.

247 "'sustainable prosperity'": Lazonick, "Stock Buybacks," 2.

247 more of the compensation moved to stock: Foroohar, *Makers and Takers*, 280.

247 options had to be disclosed, but not valued: Hal Varian, "Economic Scene," *New York Times*, April 8, 2004, retrieved April 2, 2017, http://people.ischool.berkeley .edu/~hal/people/hal/NYTimes/2004 -04-08.html.

248 "profit extracted through harm to others": Umair Haque, "The Value Every Business Needs to Create Now," *Harvard Business Review*, July 31, 2009, https://hbr.org /2009/07/the-value-every-business-needs.

248 disinformation firms used by the tobacco industry: Naomi Oreskes and Erik Conway, *Merchants of Doubt* (New York: Bloomsbury Press, 2011).

249 "left holding the bag": George Akerlof and Paul Romer, "Looting: The Economic Underworld of Bankruptcy for Profit," *Brookings Papers on Economic Activity* 2 (1993), http://pages.stern.nyu .edu/~promer/Looting.pdf.

250 "The customer is the foundation of a business": Peter F. Drucker, *The Practice of Management* (New York: Routledge, 2007), 31–32.

251 "a dumb idea": Francesco Guerrera, "Welch Condemns Share Price Focus," *Financial Times*, March 12, 2009, https:// www.ft.com/content/294ff1f2-0f27 -11de-ba10-0000779fd2ac.

252 "now serves mainly itself": Rana Foroohar, "American Capitalisms's Great Crisis," *Time*, May 11, 2016, http:// time.com/4327419/american-capitalisms -great-crisis/.

CHAPTER 12: REWRITING THE RULES

255 "improve life for people in general": Joseph E. Stiglitz, "Of the 1%, by the 1%, for the 1%," *Vanity Fair*, May 2011, http://www.vanityfair.com/news /2011/05/top-one-percent-201105.

256 "and shareholder value creation": Nelson D. Schwartz, "Carrier Workers See Costs, Not Benefits, of Global Trade," *New York Times*, March 19, 2016, https:// www.nytimes.com/2016/03/20/business /economy/carrier-workers-see-costs-not -benefits-of-global-trade.html.

256 $12 billion to buy back their stock: Tedd Mann and Ezekiel Minaya, "United Technologies Unveils $12 Billion Buyback," *Wall Street Journal*, October 20, 2015, https://www.wsj.com/articles/uni ted-technologies-unveils-12-billion -buyback-1445343580.

258 "economism": James Kwak, *Economism* (New York: Random House, 2016). This is often referred to as "market fundamentalism."

260 scheduled to leave by calling a dispatcher: Stone, *The Upstarts*, 43.

261 "extract all of his surplus": Hal Varian, "Economic Mechanism Design for Computerized Agents," *Proceedings of the First USENIX Workshop on Electronic Commerce* (New York: Usenix, 1995), retrieved April 2, 2015, http://people.ischool.berkeley.edu /~hal/Papers/mechanism-design.pdf.

262 "It is not from the benevolence of the butcher": Russ Roberts, *How Adam*

Smith Can Change Your Life (New York: Penguin, 2014), 21.

262 who would set the rules of the game: There is a growing nostalgia for unions, very different from the disdain that they were held in twenty years ago. See for example Ben Casselman, "Americans Don't Miss Manufacturing—They Miss Unions," *FiveThirtyEight*, May 13, 2016, https://fivethirtyeight.com/features /americans-dont-miss-manufacturing -they-miss-unions/.

263 rethink the labor movement as well: Harold Meyerson, "The Seeds of a New Labor Movement," *American Prospect*, October 30, 2014, http://prospect.org/article/labor -crossroads-seeds-new-movement.

263 "redistribute so that overall people are better off": Pia Malaney, "The Economic Origins of the Populist Backlash," *BigThink*, March 5, 2017, http:// bigthink.com/videos/pia-malaney-on -the-economics-of-rust-belt-populism.

264 "the federal minimum would be $15.34": John Schmitt: "The Minimum Wage Is Too Damn Low," Center for Economic Policy Research, March 2012, http:// cepr.net/documents/publications/min -wage1-2012-03.pdf.

265 optimized for those with more to spend: Xavier Jaravel, "The Unequal Gains from Product Innovations: Evidence from the US Retail Sector," 2016, http://scholar .harvard.edu/xavier/publications/unequal -gains-product-innovations-evidence -us-retail-sector.

265 "Even the richest people only sleep on one or two pillows": Nick Hanauer, in the film *Inequality for All*, http://inequality forall.com. The clip containing Nick's comments can be found at http://www .upworthy.com/when-they-say-cutting -taxes-on-the-rich-means-job-creation -theyre-lying-just-ask-this-rich-guy.

265 "go to the casino than the restaurant": Foroohar, *Makers and Takers*, 14.

266 approximately $5 billion per year: "Study Shows Walmart Can 'Easily Afford' $15 Minimum Wage," *Fortune*, June 11, 2016, http://fortune.com/2016/06/11/wal mart-minimum-wage-study/.

266 Supplemental Nutrition Assistance: "WAL-MART ON TAX DAY," Americans for Tax Fairness, retrieved April 2, 2017, https://

americansfortaxfairness.org/files/Wal mart-on-Tax-Day-Americans-for-Tax -Fairness-1.pdf.

266 $153 billion per year: Ken Jacobs, "Americans Are Spending $153 Billion a Year to Subsidize McDonald's and Wal-Mart's Low Wage Workers," *Washington Post*, April 15, 2015, https://www .washingtonpost.com/posteverything/wp /2015/04/15/we-are-spending-153 -billion-a-year-to-subsidize-mcdonalds -and-walmarts-low-wage-workers/.

266 costing the company $2.6 billion: Neil Irwin, "How Did Walmart Get Cleaner Stores and Higher Sales? It Paid Its People More," *New York Times*, October 25, 2016, https://www.nytimes.com /2016/10/16/upshot/how-did-walmart -get-cleaner-stores-and-higher-sales-it -paid-its-people-more.html.

266 rewrite the rules: Joseph Stiglitz, *Rewriting the Rules of the American Economy* (New York: Roosevelt Institute, 2015), http://rooseveltinstitute.org/rewrite -rules/.

267 a national $15 minimum wage: David Rolf, *The Fight for $15* (New York: New Press, 2016).

267 "an intimidation tactic masquerading as an economic theory": Nick Hanauer in conversation with Tim O'Reilly, Next:Economy Summit, San Francisco, November 12–13, 2015, video at https://www.safari booksonline.com/library/view/nexteconomy -2015-/9781491944547/video231634 .html.

267 not have much impact in major cities: Paul K. Sonn and Yannet Lathrop, "Raise Wages, Kill Jobs? Seven Decades of Historical Data Find No Correlation Between Minimum Wage Increases and Employment Levels," National Employment Law Project, May 5, 2016, http:// www.nelp.org/publication/raise-wages -kill-jobs-no-correlation-minimum -wage-increases-employment-levels/.

269 Summit on Technology and Opportunity: Summit on Technology and Opportunity, Stanford University, November 29, 30, 2016, http://inequality .stanford.edu/sites/default/files/Agenda _Summit-Tech-Opportunity_2.pdf.

269 a lunchtime debate with Martin Ford: Martin Ford and Tim O'Reilly, "Two

(Contrasting) Views of the Future," Stanford Center on Poverty and Inequality, a conversation at the Summit on Technology and Opportunity, Stanford University, November 29, 30, 2016, video published December 16, 2016, https://www.youtube.com/watch?v=F7vJDtwidWU.

269 including knowledge work: Martin Ford, *The Rise of the Robots* (New York: Basic Books, 2015).

271 lack of aggregate consumer demand: Bill Gross, "America's Debt Is Not Its Biggest Problem," *Washington Post*, August 10, 2011, https://www.washingtonpost.com/opinions/americas-debt-is-not-its-biggest-problem/2011/08/10/gIQAgYvE7I_story.html.

271 "on the demand, rather than the supply, side": Robert Summers, "The Age of Secular Stagnation: What It Is and What to Do About It," *Foreign Affairs*, February 15, 2016, retrieved from http://larrysummers.com/2016/02/17/the-age-of-secular-stagnation/.

271 "Only around 15%": Rana Foroohar, "The Economy's Hidden Illness—One Even Trump Failed to Address," *LinkedIn Pulse*, November 12, 2016, https://www.linkedin.com/pulse/economys-hidden-illness-one-even-trump-failed-address-rana-foroohar.

272 "the job is likely to be ill-done": John Maynard Keynes, *The General Theory of Employment, Interest, and Money* (New York: Harcourt Brace, 1964), 159.

272 "long-term prospects and to those only": Ibid., 160.

272 each additional year that an asset is held: Fink, "I write on behalf of our clients . . ."

272 such as that proposed by Thomas Piketty: Michelle Fox, "Why We Need a Global Wealth Tax: Piketty," *CNBC*, March 10, 2015, http://www.cnbc.com/2015/03/10/why-we-need-a-global-wealth-tax-piketty.html.

273 "it's good for business": Stiglitz, "Of the 1%, by the 1%, for the 1%."

CHAPTER 13: SUPERMONEY

274 "games played between the state, the market economy, and financial capitalism": William H. Janeway, *Doing Capitalism in the Innovation Economy* (Cambridge: Cambridge University Press, 2012), 3.

277 pays off all the failed bets: Carlota Perez, *Technological Revolutions and Financial Capital* (Cheltenham, England: Edward Elgar, 2002).

277 "Occasionally, decisively": Bill Janeway, "What I Learned by Doing Capitalism," LSE Public Lecture, London School of Economics and Political Science, October 11, 2012, transcript retrieved April 4, 2017.

278 Goodman's 1972 book of that name: Adam Smith, *Supermoney* (Hoboken, NJ: Wiley, 2006).

281 the rise of superstar firms: Bouree Lam, "One Reason Workers Are Struggling Even When Companies Are Doing Well," *Atlantic*, February 1, 2017, https://www.theatlantic.com/business/archive/2017/02/labors-share/515211/.

281 when stock-based compensation is taken into account: Bloomberg News, "Amazon, Facebook Admit Stock Compensation Is a Normal Cost," *Investor's Business Daily*, May 3, 2016, http://www.investors.com/news/technology/amazon-stops-pretending-that-stock-compensation-isnt-a-normal-cost/.

282 no longer afford to live in a city like San Francisco: Hal Varian, "Is Affordable Housing Becoming an Oxymoron?," *New York Times*, October 20, 2005, http://people.ischool.berkeley.edu/~hal/people/hal/NYTimes/2005-10-21.html. Hal called this in 2005!

282 venture capital investment in 2015: Press release, "$58.8 Billion in Venture Capital Invested Across U.S. in 2015," National Venture Capital Association, January 15, 2016, http://nvca.org/pressreleases/58-8-billion-in-venture-capital-invested-across-u-s-in-2015-according-to-the-moneytree-report-2/.

282 smaller funds typically deliver better results: Kauffman Foundation, "WE HAVE MET THE ENEMY . . . AND HE IS US: Lessons from Twenty Years of the Kauffman Foundation's Investments in Venture Capital Funds and the Triumph of Hope over Experience," Ewing Marion Kauffman Foundation, May 2012, http://www.kauffman.org/~/media/kauffman_org/research%20reports%20and

%20covers/2012/05/we_have_met_the
_enemy_and_he_is_us.pdf.

283 "it's one of a hundred games in town":
Jon Oringer in conversation with
Charlie Herman, "Failure Is Not an
Option. . . . But it Should Be," *Money
Talking*, WNYC, January 16, 2015,
http://www.wnyc.org/story/failure-not
-an-option-but-it-should-be/.

286 Three had no investment at all from
VCs: Bryce Roberts, "Helluva Lifestyle
Business You Got There," *Medium*,
January 31, 2017, https://medium.com
/strong-words/helluva-lifestyle-business
-you-got-there-e1ebd3104a95.

286 which he called indie.vc: Bryce Roberts,
"We Invest in Real Businesses," *indie.vc*,
retrieved April 3, 2017, http://www
.indie.vc.

287 tens of millions in distribution: Jason
Fried, "Jason Fried on Valuations, Basecamp, and Why He's No Longer Poking
the World in the Eye," interview with
Mixergy, April 4, 2016, https://mixergy
.com/interviews/basecamp-with-jason
-fried/.

287 "if growth is not immediate and meteoric": Marc Hedlund, "Indie.vc, and focus," *Skyliner* (blog), December 14, 2016,
https://blog.skyliner.io/indie-vc-and
-focus-8e833d8680d4.

289 "faster than any company in Silicon
Valley": Hank Green, "Introducing the
Internet Creators Guild," June 15, 2016,
https://medium.com/internet-creators
-guild/introducing-the-internet-creators
-guild-e0db6867e0c3.

290 at the Vatican in November 2016: Fortune
+Time Global Forum 2016, "The 21st
Century Challenge: Forging a New
Social Compact," Rome and Vatican
City, December 2–3, 2016, http://www
.fortuneconferences.com/wp-content
/uploads/2016/12/Fortune-Time-Global
-Forum-2016-Working-Group-Solutions
.pdf.

290 by \$165 billion: Google, *Economic Im-
pact, United States 2015*, retrieved December 12, 2016, https://economicimpact
.google.com/#/.

290 more than 60% of their traffic came from
search: Nathan Safran, "Organic Search
Is Actually Responsible for 64% of Your
Web Traffic (Thought Experiment),"

July 10, 2014, https://www.conductor
.com/blog/2014/07/organic-search
-actually-responsible-64-web-traffic/.

291 commissioned a report: Yancey Strickler, "Kickstarter's Impact on the Creative Economy," *The Kickstarter Blog*, July
28, 2016, https://www.kickstarter.com
/blog/kickstarters-impact-on-the-creative
-economy.

291 have gone on to great success: Amy Feld-
man, "Ten of the Most Successful Companies Built on Kickstarter," *Forbes*,
April 14, 2016, https://www.forbes.com
/sites/amyfeldman/2016/04/14/ten-of
-the-most-successful-companies-built
-on-kickstarter/#4dec455f69e8.

292 register as a public benefit corporation: Yancey Strickler, Perry Chen, and
Charles Adler, "Kickstarter Is Now a Benefit Corporation," *The Kickstarter Blog*,
September 21, 2015, https://www.kick
starter.com/blog/kickstarter-is-now
-a-benefit-corporation.

292 regular cash distributions to their shareholders: Joshua Brustein, "Kickstarter
Just Did Something Tech Startups Never
Do: It Paid a Dividend," *Bloomberg*, June
17, 2016, https://www.bloomberg.com
/news/articles/2016-06-17/kickstarter
-just-did-something-tech-startups-never
-do-it-paid-a-dividend.

292 shareholder value primacy has no legal
basis: Lynn Stout, *The Shareholder Value
Myth* (San Francisco: Berrett-Koehler,
2012).

292 argues otherwise: Leo E. Strine, "Making It Easier for Directors to 'Do the
Right Thing'?," *Harvard Business Law
Review* 4 (2014): 235, University of
Pennsylvania Institute for Law & Economics, Research Paper No. 14–41,
posted December 18, 2014, https://ssrn
.com/abstract=2539098.

293 "if not more than, the bottom line":
Etsy, "Building an Etsy Economy: The
New Face of Creative Entrepreneurship," 2015, retrieved April 4, 2017,
https://extfiles.etsy.com/Press/reports
/Etsy_NewFaceofCreativeEntrepreneur
ship_2015.pdf.

293 the ouster of Chad Dickerson, Etsy's
CEO: The Associated Press, "Etsy
Replaces CEO, Cuts Jobs Amid Share-
holder Pressure," *ABC News*, May 2,

2017, http://abcnews.go.com/Business/wireStory/etsy-replaces-ceo-cuts-jobs-amid-shareholder-pressure-47167426.

293 supported more than 10,000 jobs: "Airbnb Community Tops $1.15 Billion in Economic Activity in New York City," *Airbnb*, May 12, 2015, https://www.airbnb.com/press/news/airbnb-community-tops-1-15-billion-in-economic-activity-in-new-york-city.

293 helped them stay in their home: "Airbnb Economic Impact," *Airbnb*, retrieved April 4, 2017, http://blog.airbnb.com/economic-impact-airbnb/.

294 A third-party economic study: Peter Cohen, Robert Hahn, Jonathan Hall, Steven Levitt, and Robert Metcalfe, "Using Big Data to Estimate Consumer Surplus: The Case of Uber," National Bureau of Economic Research, Working Paper No. 22627, September 2016, doi:10.3386/w22627.

294 nine million third-party sellers: Duncan Clark, *Alibaba: The House That Jack Built* (New York: Harper, 2016), 5.

294 in order to favor more lucrative sales by big brands: Ina Steiner, "eBay Makes Big Promises to Small Sellers as SEO Penalty Still Stings," *eCommerce Bytes*, April 23, 2015, http://www.ecommercebytes.com/cab/abn/y15/m04/i23/s02.

294 half of all private sector employment: "SBA Advocacy: Frequently Asked Questions" Small Business Administration, September 2012, retrieved May 12, 2017, https://www.sba.gov/sites/default/files/FAQ_Sept_2012.pdf.

295 "now drying the clothes": Steve Baer, "The Clothesline Paradox," *CoEvolution Quarterly*, Winter 1975, retrieved April 3, 2017, http://www.wholeearth.com/issue/2008/article/358/the.clothesline.paradox.

296 value as and if created: Mariana Mazzucato, *The Entrepreneurial State* (London: Anthem, 2013), 185–87.

296 "a minuscule fraction": William D. Nordhaus, "Schumpeterian Profits in the American Economy: Theory and Measurement," National Bureau of Economic Research, NBER Working Paper No. 10433, issued April 2004, doi:10.3386/w10433.

297 had to change its policies: Sam Shead, "Apple Is Finally Going to Start Publishing Its AI Research," *Business Insider*, December 6, 2016, http://www.businessinsider.com/apple-is-finally-going-to-start-publishing-its-artificial-intelligence-research-2016-12.

CHAPTER 14: WE DON'T HAVE TO RUN OUT OF JOBS

298 "live wisely and agreeably and well": John Maynard Keynes, "Economic Possibilities for Our Grandchildren," in *Essays in Persuasion* (New York: Harcourt Brace, 1932), 358–73, available online from http://www.econ.yale.edu/smith/econ116a/keynes1.pdf.

298 the world has been getting better: Max Roser and Esteban Ortiz-Ospina, "Global Extreme Poverty," *OurWorldInData.org*, first published in 2013; substantive revision March 27, 2017, retrieved April 4, 2017, https://ourworldindata.org/extreme-poverty/.

299 once destroyed factory jobs: Carl Benedikt Frey and Michael A. Osborne, "The Future of Employment: How Susceptible Are Jobs to Computerisation," Oxford Martin Institute, September 17, 2013, http://www.oxfordmartin.ox.ac.uk/downloads/academic/The_Future_of_Employment.pdf.

299 the age of growth is over: Robert Gordon, "The Death of Innovation, the End of Growth," *TED 2013*, https://www.ted.com/talks/robert_gordon_the_death_of_innovation_the_end_of_growth.

301 something that had never been done before: Margot Lee Shetterly, *Hidden Figures* (New York: William Morrow, 2016).

302 well-paid human jobs: Already, in the US, 43% of the electric power generation workforce is employed in solar technologies, versus 22% of electric power generation via fossil fuels. US Energy and Employment Report, Department of Energy, 2016, 28, https://www.energy.gov/sites/prod/files/2017/01/f34/2017%20US%20Energy%20and%20Jobs%20Report_0.pdf.

302 cure all disease within their children's lifetimes: Mark Zuckerberg, "Can we cure all diseases in our children's lifetime?," Facebook post, September 21,

2016, https://www.facebook.com/notes/mark-zuckerberg/can-we-cure-all-diseases-in-our-childrens-lifetime/10154087783966634/.

303 "the leading edge of who is going to create that": Jeff Immelt, in conversation with Tim O'Reilly, Next:Economy Summit, San Francisco, November 12, 2015, https://www.oreilly.com/ideas/ges-digital-transformation.

304 it has roughly hovered since: Max Roser, "Working Hours," OurWorldInData.org, 2016, retrieved April 4, 2017, https://ourworldindata.org/working-hours/.

304 generosity is the robust strategy: Ryan Avent, The Wealth of Humans (New York: St. Martin's, 2016), 242.

305 making the case for UBI: Andy Stern, Raising the Floor (New York, Public Affairs, 2016).

305 a pilot program in Oakland, California: Sam Altman, "Moving Forward on Basic Income," Y Combinator (blog), May 31, 2016, https://blog.ycombinator.com/moving-forward-on-basic-income/.

305 a true randomized control trial: "Launch a basic income," GiveDirectly, retrieved April 4, 2017, https://www.givedirectly.org/basic-income.

305 proposed by Thomas Paine in 1795: "Agrarian Justice," The Writings of Thomas Paine, vol. 3, 1791–1804 (New York: G. P. Putnam's Sons, 1895), Project Gutenberg ebook edition retrieved April 4, 2017, http://www.gutenberg.org/files/31271/31271-h/31271-h.htm#link2H_4_0029.

305 Paul Ryan in 2014: Noah Gordon, "The Conservative Case for a Guaranteed Basic Income," Atlantic, August 6, 2014, https://www.theatlantic.com/politics/archive/2014/08/why-arent-reformicons-pushing-a-guaranteed-basic-income/375600/.

305 arguments against UBI: Charles Murray and Andrews Stern (For), Jared Bernstein and Jason Furman (Against), "Universal Basic Income Is the Safety Net of the Future," Intelligence Squared Debates, March 22, 2017, http://www.intelligencesquaredus.org/debates/universal-basic-income-safety-net-future. The audience was persuaded 41% to 4% against the motion.

307 Bill Gates proposed a "robot tax": Kevin

J. Delaney, "The Robot That Takes Your Job Should Pay Taxes, Says Bill Gates," Quartz, February 17, 2017, https://qz.com/911968/bill-gates-the-robot-that-takes-your-job-should-pay-taxes/.

307 only $2,400 per person: Ed Dolan, "Could We Afford a Universal Basic Income?," EconoMonitor, January 13, 2014, revised June 25, 2014, http://www.economonitor.com/dolanecon/2014/01/13/could-we-afford-a-universal-basic-income/.

307 would cost only $175 billion: Matt Bruenig and Elizabeth Stoker, "How to Cut the Poverty Rate in Half (It's Easy)," The Atlantic, October 29, 2013, https://www.theatlantic.com/business/archive/2013/10/how-to-cut-the-poverty-rate-in-half-its-easy/280971/.

307 "I am confident": "The Future of Work and the Proposal for a Universal Basic Income: A Discussion with Andy Stern, Natalie Foster, and Sam Altman," held at Bloomberg Beta in San Francisco on June 27, 2016, https://raisingthefloor.splashthat.com.

309 Anne-Marie Slaughter: Anne-Marie Slaughter, Unfinished Business (New York: Random House, 2015).

309 "patterns of consumption": Anne-Marie Slaughter, "How the Future of Work May Make Many of Us Happier," Huffington Post, retrieved April 4, 2017, http://www.huffingtonpost.com/annemarie-slaughter/future-of-work-happier_b_6453594.html.

309 "support the families they are caring for": Anne-Marie Slaughter, in conversation with Tim O'Reilly and Lauren Smiley, "Flexibility Needed: Not Just for On Demand Workers," Next:Economy Summit, San Francisco, October 10–11, 2015. Video retrieved April 4, 2017, https://www.safaribooksonline.com/library/view/next economy-2015-/9781491944547/video231631.html.

310 300,000 "fitness trainers" in the United States: "Fitness Trainers and Instructors," Occupational Outlook Handbook, US Department of Labor, Bureau of Labor Statistics, retrieved April 4, 2017, https://www.bls.gov/ooh/personal-care-and-service/fitness-trainers-and-instructors.htm.

310 By 2011, they were 12.2%: Ian Stewart, Debapratim De, and Alex Cole, "Technology and People: The Great Job-Creating Machine," Deloitte, August 2015, https://www2.deloitte.com /uk/en/pages/finance/articles/technology -and-people.html.

310 an eager consumer of caring services: Zoë Baird and Emily Parker, "A Surprising New Source of American Jobs: China," *Wall Street Journal*, May 29, 2015, https://www.wsj.com/articles/a -surprising-new-source-of-american -jobs-china-1432922899.

311 the other is Papua New Guinea: Laura Addati, Naomi Cassirer, and Katherine Gilchrist, *Maternity and Paternity at Work: Law and Practice Across the World* (Geneva: International Labor Organization, 2014), http://www.ilo.org/wcmsp5 /groups/public/---dgreports/---dcomm /---publ/documents/publication/wcms _242615.pdf.

311 prison costs later in life: "Education vs Prison Costs," *CNN Money*, retrieved April 4, 2017, http://money.cnn.com/info graphic/economy/education-vs-prison -costs/.

313 what they *mean*, not just what they *do*: Dave Hickey, "The Birth of the Big Beautiful Art Market," *Air Guitar* (Los Angeles: Art Issues Press, 1997), 66–67.

313 "he may not be soon reduced to form another wish": Samuel Johnson, *Rasselas*, in *Rasselas, Poems, and Selected Prose*, ed. Bertrand H. Bronson (New York: Holt Rinehart & Winston, 1958), 572–73.

315 price double that of a mass-produced beer: John Kell, "What You Didn't Know About the Boom in Craft Beer," *Fortune*, March 22, 2015, http://fortune .com/2016/03/22/craft-beer-sales-rise -2015/.

315 artisan goods on Etsy: Fareeha Ali, "Etsy's Sales, Sellers and Buyers Grow in Q1," *Internet Retailer*, May 4, 2016, https://www.internetretailer.com/2016 /05/04/etsys-sales-sellers-and-buyers -grow-q1.

315 "more time and less stuff": Slaughter, "How the Future of Work May Make Many of Us Happier."

316 "I hit a million views a month": Green, "Introducing the Internet Creators Guild."

316 six-figure earnings playing video games: John Egger, "How Exactly Do Twitch Streamers Make a Living? Destiny Breaks It Down," *Dot Esports*, April 21, 2015, https://dotesports.com/general/twitch -streaming-money-careers-destiny -1785.

317 get other people to approve of and support your creative projects: Cory Doctorow, *Down and Out in the Magic Kingdom* (New York: Tor Books, 2003).

318 "something you turned into money": Hickey, *Air Guitar*, 45.

318 "Eat food. Not too much. Mostly plants": Michael Pollan, *In Defense of Food* (New York: Penguin, 2008).

318 in family and social life: Dan Buettner, *The Blue Zones*, 2nd ed. (Washington, DC: National Geographic Society, 2012).

319 "something we haven't seen yet, something we have to invent": Jennifer Pahlka, "Day One," January 21, 2017, *Medium*, https://medium.com/@pahlkadot/day -one-39a0cd5bd886.

CHAPTER 15: DON'T REPLACE PEOPLE, AUGMENT THEM

320 Markle Foundation Rework America task force: For more information, see "AMERICA'S MOMENT: Creating Opportunity in the Connected Age," Markle Foundation, https://www.mar kle.org/rework-america/americas-mo ment.

320 transition from wartime to peaceful employment: Claudia Goldin and Lawrence F. Katz, "Human Capital and Social Capital: The Rise of Secondary Schooling in America, 1910 to 1940," National Bureau of Economic Research, NBER Working Paper No. 6439, March 1998, doi:10.3386/w6439.

321 free for all residents: Nanette Asimov, "SF Reaches Deal for Free Tuition at City College," *SFGate*, February 27, 2017, http://www.sfgate.com/bayarea/article /SF-reaches-deal-for-free-tuition-at -City-College-10912051.php.

321 "then we have a problem": "Former ambassador Jeffrey Bleich speaks on Trump, disruptive technology, and the role of education in a changing economy," an edited transcript of the keynote address delivered by Jeffrey Bleich at

Universities Australia's higher education conference in Canberra on March 1, 2017, *The Conversation*, updated March 6, 2017, http://theconversation.com/former-ambassador-jeffrey-bleich-speaks-on-trump-disruptive-technology-and-the-role-of-education-in-a-changing-economy-73957.

323 "he effects by Discoveries and Inventions": Abraham Lincoln, "Lecture on Discoveries and Inventions," April 6, 1858, *Abraham Lincoln Online*, retrieved April 4, 2017, https://www.abrahamlincolnonline.org/lincoln/speeches/discoveries.htm.

327 man and machine together design new forms: "Generative Design," *autodesk.com*, retrieved April 4, 2017, http://www.autodesk.com/solutions/generative-design.

328 half the size and uses half the material: "3D Makeover for Hyper-efficient Metalwork," Arup, May 11, 2015, http://www.arup.com/news/2015_05_may/11_may_3d_makeover_for_hyper-efficient_metalwork.

328 create from scratch a complete human genome: Jef D. Boeke, George Church, Andrew Hessel, Nancy J. Kelley, et al., "The Genome Project-Write," *Science*, July 8, 2016, 126–27, doi:10.1126/science.aaf6850. For a popular account, see Sharon Begley, "Audacious Project Plans to Create Human Genomes from Scratch," *Stat*, June 2, 2016, https://www.statnews.com/2016/06/02/project-human-genome-synthesis/.

328 bring extinct species back to life: "Revive & Restore: Genetic Rescue for Endangered and Extinct Species," retrieved April 4, 2017, http://reviverestore.org.

328 rewrite the DNA inside living organisms: "CRISPR/Cas9 and Targeted Genome Editing: A New Era in Molecular Biology," New England Biolabs, retrieved April 4, 2017, https://www.neb.com/tools-and-resources/feature-articles/crispr-cas9-and-targeted-genome-editing-a-new-era-in-molecular-biology.

328 prosthetic limbs that provide sensory feedback: "Neurotechnology Provides Near-Natural Sense of Touch," DARPA, September 11, 2015, http://www.darpa.mil/news-events/2015-09-11.

328 respond directly to the mind: Emily Reynolds, "This Mind-Controlled Limb Can Move Individual Fingers," *Wired*, February 11, 2016, http://www.wired.co.uk/article/mind-controlled-prosthetics.

328 a neural memory implant as a cure for Alzheimer's: Elizabeth Dwoskin, "Putting a Computer in Your Brain Is No Longer Science Fiction," *Washington Post*, August 25, 2016, https://www.washingtonpost.com/news/the-switch/wp/2016/08/15/putting-a-computer-in-your-brain-is-no-longer-science-fiction/.

329 enhancing human intelligence: Bryan Johnson, "The Combination of Human and Artificial Intelligence Will Define Humanity's Future," *TechCrunch*, October 12, 2016, https://techcrunch.com/2016/10/12/the-combination-of-human-and-artificial-intelligence-will-define-humanitys-future/.

329 "helps with certain severe brain injuries": Tim Urban, "Neuralink and the Brain's Magical Future," *Wait But Why*, April 20, 2017, http://waitbutwhy.com/2017/04/neuralink.html.

329 "get the Human Colossus working on the cause": Ibid.

329 a direct neural interface: Elon Musk, quoted in Tim Urban, "Neuralink and the Brain's Magical Future."

330 "the capacity of people to take in, process, and use information": Ibid.

330 "I want to make humans cool again": Janelle Nanos, "Is Paul English the Soul of the New Machine?," *Boston Globe*, May 12, 2016, http://www.bostonglobe.com/business/2016/05/12/drives-uber-helps-haiti-and-may-revolutionize-how-travel-paul-english-soul-new-machine/R2vThUDvRMckM5KoPIjVKK/story.html.

332 "the Robot Lawyer": Josh Browder, "Will Bots Replace Lawyers?," talk given at Next:Economy Summit, San Francisco, October 10–11, 2016, https://www.safaribooksonline.com/library/view/next-economy-summit-2016/9781491976067/video282513.html.

332 automate the application for asylum: Elena Cresci, "Chatbot That Overturned 160,000 Parking Fines Now Helping Refugees Claim Asylum," *Guardian*, March 6, 2017, https://www.theguardian.com/tech

nology/2017/mar/06/chatbot-donotpay-refugees-claim-asylum-legal-aid.

335 with an apprenticeship: Steven Levy, "How Google Is Remaking Itself as a 'Machine Learning First' Company," *Backchannel*, June 22, 2016, https://backchannel.com/how-google-is-remaking-itself-as-a-machine-learning-first-company-ada63defcb70.

336 "The Internet Was Built on O'Reilly Books": *Publishers Weekly*, February 21, 2000. That cover was reproduced in a blog post by brian d. foy, "The Internet Was Built on O'Reilly Books," *programmingperl.com*, October 28, 2015, https://www.programmingperl.org/2015/10/the-internet-was-built-on-oreilly-books/.

337 the cover story featured Charles Benton: *Make*, January 2005, https://www.scribd.com/doc/33542837/MAKE-Magazine-Volume-1.

337 "If you can't open it, you don't own it": Phil Torrone, "Owner's Manifesto," *Make*, November 26, 2006, http://makezine.com/2006/11/26/owners-manifesto/.

337 denying them the right to repair: Cory Doctorow, *Information Doesn't Want to Be Free: Laws for the Internet Age* (San Francisco: McSweeney's, 2014).

338 who controls products that the consumers nominally own: Jason Koebler, "Why American Farmers Are Hacking Their Tractors with Ukrainian Firmware," *Vice*, March 21, 2017, https://motherboard.vice.com/en_us/article/why-american-farmers-are-hacking-their-tractors-with-ukrainian-firmware.

338 we wrote a book together: Dale Dougherty and Tim O'Reilly, *Unix Text Processing* (Indianapolis: Hayden, 1987).

339 study of motivations of people working on open source software projects: Karim Lakhani and Robert Wolf, "Why Hackers Do What They Do: Understanding Motivation and Effort in Free/Open Source Software Projects," in *Perspectives on Free and Open Source Software*, ed. J. Feller, B. Fitzgerald, S. Hissam, and K. R. Lakhani (Cambridge, MA: MIT Press, 2005), retrieved April 4, 2017, https://ocw.mit.edu/courses/sloan-school-of-management/15-352-managing-innovation-emerging-trends-spring-2005/readings/lakhaniwolf.pdf.

340 Ignorance, not knowledge, drives science: Stuart Firestein, *Ignorance* (New York: Oxford University Press, 2012).

340 "play with things for my own entertainment": Feynman, *Surely You're Joking, Mr. Feynman*, 157–58.

341 discovered by sponsors and invited to competitions: John Hagel III, John Seely Brown, and Lang Davison, *The Power of Pull* (New York: Basic Books, 2010), 1–5.

342 "far beyond teachers or textbooks": Brit Morin, "Gen Z Rising," *The Information*, February 5, 2017, https://www.theinformation.com/gen-z-rising.

342 100 million hours of how-to videos: Google, "I Want-to-Do Moments: From Home to Beauty," *Think with Google*, retrieved April 4, 2017, https://www.thinkwithgoogle.com/articles/i-want-to-do-micro-moments.html.

343 reflect the needs of the digital economy: "Skillful: Building a Skills-Based Labor Market," Markle, retrieved April 4, 2017, https://www.markle.org/rework-america/skillful.

345 augmented reality display for infantry soldiers: Adam Clark Estes, "DARPA Hacked Together a Super Cheap Google Glass-Like Display," *Gizmodo*, April 7, 2015, http://gizmodo.com/darpa-hacked-together-a-super-cheap-google-glass-like-d-1695961692.

345 deep commitment Microsoft has made to human augmentation: Satya Nadella, interviewed by Gerard Baker, "Microsoft CEO Envisions a Whole New Reality," *Wall Street Journal*, October 30, 2016, https://www.wsj.com/articles/microsoft-ceo-envisions-a-whole-new-reality-1477880580.

346 equally but differently skilled: James Bessen, *Learning by Doing* (New Haven, CT: Yale University Press, 2015), 28–29.

346 "little to do with the knowledge acquired in college": Ibid., 25.

346 "they published quality periodicals": Ibid., 24.

347 "what it takes to create a stable, trained labor force": Ibid., 36.

347 "sling JavaScript for their local bank": Clive Thompson, "The Next Big Blue-Collar Job Is Coding," *Wired*, February 2, 2017,

348 https://www.wired.com/2017/02/program ming-is-the-new-blue-collar-job/.

348 "when shared by a critical mass of people": Ryan Avent, *The Wealth of Humans* (New York: St. Martin's, 2016), 119.

348 popularized by Robert Putnam: Robert Putnam, *Bowling Alone* (New York: Simon & Schuster, 2001).

349 "you get a weekly magazine at the end of it": Avent, *The Wealth of Humans*, 105.

349 "hinder our digital efforts": Ibid., 110–11.

349 transform IBM's internal software development culture: Jeff Smith in conversation with Tim O'Reilly, "How Jeff Smith Built an Agile Culture at IBM," Next:Economy Summit, San Francisco, October 10, 2016, https://www.oreilly .com/ideas/how-jeff-smith-built-an -agile-culture-at-ibm.

350 more than two million manufacturing jobs will go unfilled: "The Skills Gap in U.S. Manufacturing: 2015 and Beyond," Deloitte Manufacturing Institute, retrieved April 4, 2017, http:// www.themanufacturinginstitute.org /~/media/827DBC76533942679A15E F7067A704CD.ashx.

CHAPTER 16: WORK ON STUFF THAT MATTERS

353 "coming away stronger from the fight": Rainer Maria Rilke, "The Man Watching," *Selected Poems of Rainer Maria Rilke*, translation and commentary by Robert Bly (New York: Harper, 1981).

353 "incremental improvements to current technology": Satya Nadella, *Hit Refresh* (New York: Harper Business, 2017), unpublished manuscript, 195.

355 "Father Madeleine made his fortune": Victor Hugo, *Les Misérables*, translated by Charles E. Wilbour, revised and edited by Frederick Mynon Cooper (New York: A. L. Burt, 1929), 156.

356 "'Well . . . that's outside!' she laughed": Brian Eno, "The Big Here and Long Now," Long Now Foundation, retrieved April 4, 2017, http://longnow.org/essays /big-here-long-now/.

356 We borrow from other countries to finance our consumption: James Fallows, "Be Nice to the Countries That Lend You Money," *Atlantic*, December 2008,

https://www.theatlantic.com/magazine /archive/2008/12/be-nice-to-the-countries -that-lend-you-money/307148/.

357 "I'm troubled by the problem of how to sell automobiles to them": "'How Will You Get Robots to Pay Union Dues?'" "'How Will You Get Robots to Buy Cars?'," *Quote Investigator*, retrieved April 4, 2017, http:// quoteinvestigator.com/2011/11/16/robots -buy-cars/.

357 "be careful about what we pretend to be": Kurt Vonnegut, *Mother Night* (New York: Avon, 1967), v.

359 "an imaginative leap into the future": Peter Schwartz, *The Art of the Long View* (New York: Crown, 1996), xiv.

360 "underestimate the change that will occur in the next ten": Bill Gates, *The Road Ahead: Completely Revised and Up-to-Date* (New York, Penguin, 1996).

362 losses to its economy as 10% of GDP: "Economic Losses from Pollution Account for 10% of GDP," *China.org.cn*, June 6, 2006, http://www.china.org.cn /english/environment/170527.htm.

363 a carbon tax whose proceeds would be rebated directly to all Americans: James A. Baker III, Martin Feldstein, Ted Halstead, N. Gregory Mankiw, Henry M. Paulson Jr, George P. Shultz, and Thomas P. Stephenson, "The Conservative Case for Carbon Dividends," Climate Leadership Council, February 2017, https://www.clcouncil.org/wp-content /uploads/2017/02/TheConservative CaseforCarbonDividends.pdf.

363 a map of all the energy sources and uses in the US economy: Adele Peters, "This Very, Very Detailed Chart Shows How All the Energy in the U.S. Is Used," *Fast Company*, August 9, 2016, https:// www.fastcompany.com/3062630 /visualizing/this-very-very-detailed -chart-shows-how-all-the-energy-in-the -us-is-used.

366 his annual shareholder letter: Jeff Bezos, "2016 Letter to Shareholders," Amazon, April 12, 2017, https://www.amazon .com/p/feature/z6o9g6sysxur57t.

369 a virtual tour of her factory: Limor Fried, "The Small Scale Factory of the Future," presentation at Next:Economy Summit, San Francisco, November 12, 2015, https:// www.safaribooksonline.com/library

/view/nexteconomy-2015-/97814
91944547/video231262.html.

370 an open field in Rwanda: Keller Rinaudo, "On-Demand Drone Delivery for
Blood and Medicine," presentation at
Next:Economy Summit, San Francisco,
October 10, 2016, https://www.safari
booksonline.com/library/view/next
economy-summit-2016/9781491976067
/video282448.html.

371 "I didn't know that anybody cared what
I was doing": Rehema Ellis, "'Humans
of New York' Raises $1 Million for
Brooklyn School," NBC News, February 4, 2015, http://www.nbcnews.com
/nightly-news/humans-new-york-raises
-1-million-brooklyn-school-n300296.

371 He ended up raising $3.8 million for
research: Eun Kyung Kim, "'Humans
of New York' Project Raises $3.8 Mil-
lion to Fight Pediatric Cancer in Just 3
Weeks," Today, May 24, 2016, http://
www.today.com/health/humans-new
-york-project-raises-3-8-million-fight
-pediatric-t94501.

INDEX